Advances in

Plant Pathology

Volume 4

Advances in Plant Pathology

series edited by

D. S. Ingram

The Botany School
The University of Cambridge
Cambridge
England

and

P. H. Williams

Department of Plant Pathology
The University of Wisconsin
Madison, Wisconsin
USA

Volume 4
Genetics of Pathogenicity Factors: Application to Phytopathogenic Bacteria

Arun K. Chatterjee

Department of Plant Pathology
Kansas State University
Manhattan, Kansas 66506
USA

and

Anne K. Vidaver

Department of Plant Pathology
University of Nebraska
Lincoln, Nebraska 68583
USA

1986

ACADEMIC PRESS

Harcourt Brace Jovanovich, Publishers

London Orlando San Diego New York
Austin Montreal Sydney Tokyo Toronto

ACADEMIC PRESS INC. (LONDON) LTD.
24/28 Oval Road,
London NW1 7DX

U.S. Edition published by
ACADEMIC PRESS INC.
Orlando, Florida 32887

ISBN 0-12-033704-5

Typeset by Latimer Trend & Company Ltd, Plymouth
Printed in Great Britain by St Edmundsbury Press,
Bury St Edmunds, Suffolk

Series Preface

Advances in Plant Pathology is a series which publishes extended review articles on topics of current or future interest and importance to plant pathologists.

Some volumes in the series, like Volumes 1 and 2, will have no particular theme, but will contain a mixture of articles drawn from a range of subject areas within or related to the general field of plant pathology. Others will deal with a single important topic in depth, either as a series of short articles assembled by a specialist editor, like volume 3, or as a complete work by only one or two authors, like the present volume. Also, occasionally, a volume may be devoted to the proceedings of a symposium or conference. The contributors to all volumes are asked to concentrate on ideas and interpretations, to be controversial where appropriate, and to write for a wide audience which includes undergraduates and postgraduate students, research scientists and college teachers.

It is through its flexibility of style, format and content, its emphasis on ideas and interpretations, and the breadth of its target audience that *Advances in Plant Pathology* is intended to differ from other review series. As the editors we are anxious to serve the needs and interests of all plant pathologists. We would therefore *welcome* comments, suggestions and ideas for future volumes. These should be sent to Dr D. S. Ingram, The Botany School, Downing Street, Cambridge CB2 3EA, England.

Cambridge and Madison, 1985

D. S. Ingram
P. H. Williams

Preface

Genetics is a powerful tool of biology, and in microbiology particularly has contributed enormously to our understanding of gene structure, function and regulation. Genes affecting pathogenicity and virulence are of particular interest to both medical microbiologists and plant pathologists. The bacterial pathogens of animals and man have been far more extensively studied than plant pathogens, with the possible exception of the crown gall bacterium, *Agrobacterium tumefaciens*. In recent years, we have seen exciting confirmation and controversy in genetic research dealing with other plant pathogenic bacteria. This treatise nevertheless draws heavily on findings with non-plant pathogenic bacteria, both as illustrative examples and to stimulate the thinking and research efforts of plant pathologists and allied professionals.

Our efforts in this treatise are at least two-fold: to analyze and critique recent work on genetics of pathogenic bacteria, with application to plant pathogens, and to bring together information that is scattered throughout the literature for plant pathologists and other plant-related professionals. The latter effort is to minimize the time and effort needed for readers to become aware of recent developments in this area. The lack of extensive discussion on crown gall may seem ironic to some readers, but it is the only area, in our opinion, that *has* been adequately reviewed and is in such a dynamic flux that other reviews will be forthcoming. Hence, the scope of our review is not inclusive of all plant pathogens.

We are particularly intrigued by the variety of comments we have received from *ad hoc* reviewers of this treatise. Plant pathologists and botanists without knowledge of plant pathogenic bacteria and bacterial genetics were particularly appreciative of our presentation; those actively working in the area of genetics and molecular biology were interested in the scope and interrelationships we discussed. It is for both groups that our review was intended, encompassing both students and research scientists in different disciplines. There are undoubtedly deficiencies, based on our collective biases and inadequacies. We hope that these are minimized through the helpful comments from our reviewers. The treatise is an assessment of current knowledge of genetics of bacterial pathogens and we will have done our work well if it is both criticized constructively and used as a basis for stimulating further research.

Research in our laboratories is supported by the National Science Foundation (grant PCM-8022003 to AKC), and the Science and Education Administration of the United States Department of Agriculture (grant 59-2201-0-1-404-0 to AKC and 87-221-402-01 to AKV, from the Competitive Research Grants Office).

We are grateful to J. Fletcher and S. Pueppke for their critical comments on the manuscript. We thank P. Meadow, G. R. K. Sastry and I. W. Sutherland for sharing with us the results of their research prior to publication. We gratefully acknowledge the skilful assistance of L. M. Ross, W. Haskins and D. Kuczmarski with the bibliography and T. Arnold, A. Bockus and M. Ward for typing the manuscript.

This review is Contribution No. 85-45-B, Department of Plant Pathology, Kansas Agricultural Experiment Station, Kansas State University, Manhattan, Kansas 66506.

Manhattan and Lincoln, 1985

Arun K. Chatterjee
Anne K. Vidaver

Contents

Abbreviations and Symbols

A	adenine
Abe	abequose
Ac	acetic acid
Ade^+	adenine synthesis phenotype
ade$^+$	adenine synthesis genotype
Agc	agropine catabolism
Agr^s	agrocin sensitivity phenotype
ant^r	antimony resistance genotype
Ap^r	ampicillin resistance phenotype
Ape	exclusion of phage API phenotype
araC	regulatory gene for arabinose catabolic enzymes
Arc	arginine catabolism phenotype
Arg^+	arginine synthesis phenotype
arg$^+$	arginine synthesis genotype
aro$^+$	aromatic synthesis genotype
asa^r	arsenate resistance genotype
asi^r	arsenite resistance genotype
asp$^+$	aspartate synthesis genotype
ATP	adenosine triphosphate
B	bacterium
C	cytosine
c	immunity locus for the *Escherichia coli* phage Mu
Cb^r	carbenicillin resistance phenotype
*c*AMP	adenosine 3′,5′-cyclic monophosphate
cap$^+$	capsule formation genotype
ccc-DNA	covalently closed circular DNA
cel$^+$	cellulase production genotype
chl^r	chloramphenicol resistance genotype according to Pischl and Farrand (1983). In *E. coli, chl* stands for chlorate (Bachmann, 1983).
CL	cellulase production
clb	cellobiose utilization genotype
clr	mutation leading to clear plaque (non-lysogenizing) genotype
Cm^r	chloramphenicol resistance phenotype
CM	cytoplasmic membrane
Cma	chromosome mobilization ability phenotype
Col	colicin

cos	DNA fragments that carry the cohesive end site of bacteriophage lambda genotype
CP	capsular polymer
cps$^+$	$=$*cap*$^+$; capsule synthesis genotype
CRP	*c*AMP receptor protein
crp$^+$	*c*AMP receptor protein genotype
CTV	carotovoricins
ctv	carotovoricin production genotype
cup$^+$	copper resistance genotype
cys$^+$	cysteine synthesis genotype
d	defective genome (generally used to designate defective bacterial viruses)
dA	deoxyribonucleotide, adenine
dT	deoxyribonucleotide, thymine
DEAE	diethylaminoethyl
DGH	3-deoxy-D-glycero-2,5-hexodiulosonate
DNA	deoxyribonucleic acid
dOclA	deoxyoctulsonic acid
DTH	4-deoxy-L-*threo*-5-hexoseulose uronate
EDTA	ethylenediaminetetraacetic acid
EGTA	ethyleneglycol-bis-(β-amino-ether)*N*,*N*′-tetraacetic acid
EMS	ethyl methanesulfonate
EPS	extracellular polysaccharide
EtN	ethanolamine
F	fertility factor
F′	F-prime; refers to the F plasmid containing genomic DNA
FA	fatty acid
Fim$^+$	fimbriae phenotype
fim$^+$	fimbriae genotype
flu$^+$	fluorescent pigment production genotype
Fru	fructose
G	guanine
Gal	galactose
gal$^+$	D-galactose utilization genotype
GalE	UDP galactose 4-epimerase
GalN	galactosamine
GalU	glucose-1-phosphate uridylyltransferase
gau	galacturonate utilization genotype
GDP	guanosine diphosphate
Glc	glucose
GlcA	glucuronic acid
GlcN	glucosamine
gliv	glutamate, leucine, isoleucine, valine synthesis genotype
glu$^+$	glutamate or glutamine synthesis genotype
gtu$^+$	galacturonate utilization genotype
GTU	D-galacturonate
gua$^+$	guanosine synthesis genotype

hep	l-glycero-D-mannoheptose (heptose)
HIC	hydrophobic interaction chromatography
His^+	histidine synthesis phenotype
his$^+$	histidine synthesis genotype
Hfr	high frequency of recombination
Hm	heat modifiable
HPRG	hydroxyproline-rich glycoprotein
HR	hypersensitivity reaction
IAA	indolacetic acid
IAA^+	indolacetic acid production phenotype
iaa$^+$	indolacetic acid production genotype
iaaH	indoleacetamide hydrolase
iaaL	lysine synthetase
iaaM	tryptophan monooxygenase
iaaP	indole-3-pyruvic acid decarboxylase
idg$^+$	indigoidine pigment production genotype
ile$^+$	isoleucine synthesis genotype
Ilv^+	isoleucine-valine synthesis phenotype
ilv$^+$	isoleucine-valine synthesis genotype
ina$^+$	ice nucleating ability genotype = *ice*$^+$
Inc	incompatibility
IP	C_{55}-isoprenyl phosphate
IPA	C_{55}-isoprenoid alcohol
IPP	C_{55}-isoprenyl pyrophosphate
IS	insertion sequence
KDG	2-keto-3-deoxy-D-gluconate
kdg$^+$	2-keto-3-deoxy-D-gluconate genotype
KDO	2-keto-3-deoxy-octonic acid; (3-deoxy-D-manno-octulo-sonic acid)
KDUD	ketodeoxygluconate dehydrogenase or 3-deoxy-D-glycero-2,5-hexodiulo-sonate dehydrogenase
kduD	ketodeoxygluconate dehydrogenase or 3-deoxy-D-glycero-2,5-hexodiulo-sonate dehydrogenase genotype
Km^r	kanamycin resistance
Km^s	kanamycin sensitivity
Lac^+	D-lactose utilization phenotype
lac$^+$	D-lactose utilization genotype
leu$^+$	leucine synthesis genotype
Lipid-P	carrier phospholipid
Lipid-P-P	carrier lipid pyrophosphate
LP	lipoprotein
lpp$^+$	lipoprotein production genotype
LPS	lipopolysaccharide
lss	genotype for lysis following exposure to DNA-damaging agents (mitomycin C, nalidixic acid or ultraviolet light)
lys$^+$	lysine synthesis genotype

Man	mannose
mcu$^+$	multiple carbohydrate utilization genotype
Met$^+$	methionine synthesis phenotype
met$^+$	methionine synthesis genotype
mob$^+$	mobility factor
MLO	mycoplasmalike organisms
MOI	multiplicity of infection
MPC	membrane protein complex
*m*RNA	messenger ribonucleic acid
MT	α-methyltryptophan
NADH	nicotinamide adenine dinucleotide (reduced)
nal^r	nalidixic acid resistance genotype
nic$^+$	nicotinic acid synthesis genotype
NMR	nuclear magnetic resonance
Noc	nopaline catabolism phenotype
Nos	nopaline synthesis phenotype
NPTII	neomycin phosphotransferase II
NTG	*N*-methyl-*N*′-nitro-*N*-nitrosoguanidine
nusA	*E. coli* gene locus for transcription termination
Occ	octopine catabolism phenotype
OCT	ornithine carbamoyl transferase
OGL	oligogalacturonate lyase; same as oligogalacturonate *trans*-eliminase (OGTE)
ogl	oligogalacturonate lyase genotype
OGTE	oligogalacturonate *trans*-eliminase; same as OGL
OM	outer membrane
OmpA	outer membrane protein
Onc	oncogenicity
Ori	origin of replication
orn$^+$	ornithine utilization genotype
Out$^+$	export of extracellular enzymes phenotype
out$^+$	export of extracellular enzymes genotype
P	plasmalemma
PALO	(phosphoacetyl)-L-ornithine
pan$^+$	pantothenic acid synthesis genotype
PAPS	3′-phospho-adenosyl 5′-phosphosulfate
PATE	polygalacturonic acid *trans*-eliminase, same as pectate lyase (PL)
P-C_{55}	undecaprenyl phosphate
PE	pectin esterase or pectin methyl esterase (PME)
peh	pectic enzyme-hydrolase genotype; previously *hpg*
Pel$^-$	pectic enzyme-lyase deficient phenotype
pel$^+$	pectic enzyme-lyase (=pectate lyase) genotype; previously *pat*$^+$
PEP	phosphoenolpyruvate
PG	polygalacturonase or pectic enzyme-hydrolyase (peh)
PGT	polygalacturonate
Pi	inorganic phosphate
pI	isoelectric point

PL	pectate lyase = PEL; formerly PATE
PME	pectin methyl esterase or pectin esterase (PE)
pme^{+}	pectin methyl esterase genotype
PMG	polymethylgalacturonase (pectin hydrolase)
pmi	mannosephosphate isomerase genotype
PML	polymethylgalacturonate lyase (pectin lyase)
PNL	pectin lyase
pnl	pectin lyase production genotype
PPi	inorganic pyrophosphate
pro^{+}	proline synthesis genotype
prt^{+}	protease production genotype
PS	periplasmic space
Psc	agrocinopine synthesis and catabolism
pSym	*Rhizobium* megaplasmids that specify nodulation, nitrogen fixation, and host-specificity
pur^{+}	purine synthesis genotype
pv.	pathovar
pyr^{+}	pyrimidine synthesis genotype
PYR	pyruvate
R	drug-resistant
rbs^{+}	D-ribose utilization genotype
rec^{+}	recombination-repair proficiency genotype
recA	general recombination genotype; repair of radiation damage
recB	recombination and repair of radiation damage genotype; exonuclease V subunit
recC	recombination and repair of radiation damage; exonuclease V subunit
rfa	the genes coding for the enzymes involved in lipopolysaccharide core biosynthesis
rfb	genes coding for enzymes involved in lipopolysaccharide biosynthesis
Rha	rhamnose
Ri	root-inducing
Rif^{+}	rifampicin resistance phenotype
rif^{+} (= *rif*^{r})	rifampicin resistance genotype
*r*RNA	ribosomal ribonucleic acid
rsi^{+}	race specific incompatibility genotype
sbc	exonuclease I genotype
SDG	saturated digalacturonate
SDS	sodium dodecyl sulfate
Ser^{+}	serine synthesis phenotype
ser^{+}	serine synthesis genotype
sid^{+}	siderophore production genotype
Sm^{r}	streptomycin resistance phenotype
Sp	spectinomycin
sp.	species
spc^{r}	spectinomycin resistance genotype
spp.	species
SR	syringomycin
ST	syringotoxin

str^{r}	streptomycin resistance genotype
Su^{r}	sulfonamide resistance phenotype
subsp.	subspecies
T	thymine
Tc^{r}	tetracycline resistant phenotype
Tc^{s}	tetracycline sensitive phenotype
T-DNA	Ti plasmid DNA fragment present in plant nuclear DNA
tet^{r}	tetracycline resistance genotype
tet^{+}	tetracycline synthesis genotype
thi^{+}	thiamine synthesis genotype
Thr^{+}	threonine synthesis phenotype
thr^{+}	threonine synthesis genotype
thy^{+}	thymine synthesis genotype
Ti	tumor inducing
Tm	thermal melting profile
Tn	transposon
TolG	see OmpA
Tox^{+}	toxin production phenotype
tox^{+}	toxin production genotype
Tox^{hyper}	toxin production greater than that of wild type
Tox^{leaky}	toxin production less than that of wild type
Tp^{r}	trimethoprim resistance phenotype
tra^{+}	transfer (sex factor) function genotype
*t*RNA	transfer ribonucleic acid
Trp^{+}	tryptophan synthesis phenotype
trp^{+}	trytophan synthesis genotype
ts	thermosensitive genotype
UDG	unsaturated digalacturonate
UDP	uridine diphosphate
ura^{+}	uracil synthesis genotype
UV	ultraviolet
uxa^{+}	genes for the enzymes involved in the catabolism of D-galacturonate (hexuronate)
uxu^{+}	genes for the enzymes involved in the catabolism of D-glucuronate (hexouronate)
V	vacuole
val^{+}	valine synthesis genotype
ves	vesicle
vir^{+}	virulence genotype
W	wall
wts	water-soaking genotype
xyl^{+}	xylose utilization genotype
XLB	Xylem-limited bacteria

GENETICS OF PATHOGENICITY FACTORS: APPLICATION TO PHYTOPATHOGENIC BACTERIA

A. K. Chatterjee and A. K. Vidaver

Department of Plant Pathology, Kansas State University, Manhattan, Kansas 66506 and Department of Plant Pathology, University of Nebraska, Lincoln, Nebraska 68583, USA

I. INTRODUCTION

During their normal life cycle, both terrestrial and aquatic plants are continually exposed to a large variety of microorganisms. Little is known about the bacterial pathogens of important aquatic crop plants (Andrews, 1976) so this article is confined to terrestrial plants. Of those microorganisms that come in contact with plants only a very limited number can colonize a particular host, and even fewer can interfere with normal growth and development by causing visible symptoms. Microorganisms, including bacteria, generally fall in broad categories with respect to their abilities to associate with plant hosts: (1) pathogens, those that can colonize plants and elicit pathological conditions, and (2) non-pathogens, those that may or may not colonize plant hosts but do not elicit pathological conditions. Some pathogens are opportunistic in that stressed hosts can be colonized and become diseased.

The symptoms of bacterial diseases of plants seem to result from a series of sequential events involving various factors of the host and the pathogen. Generally recognized events are recognition, multiplication, and translocation of the pathogen and, ultimately, stimulation or inhibition of the host. Multiple factors of the host and the pathogen seem to interact in these steps. For example, interactions between cell surface components of the host and the pathogen probably determine *compatibility*, host–pathogen interactions which normally result in disease. Bacterial multiplication and translocation within the host depend upon metabolic capabilities of the pathogen such as catabolism of substrates, production of detoxifying enzymes, and responses to chemotactic stimuli. The host influences pathogen activity by providing a conducive environment such as substrates utilizable by the pathogen and lack of toxic factors or lytic enzymes. Both cellular and subcellular plant components probably contribute to tissue specificity of a pathogen. Likewise, bacterial factors such as toxins, enzymes, or hormones that stimulate or inhibit the host must interact with a susceptible target(s); the absence of the target or its modification prevents such interactions and subsequent disease development. Thus, bacterial pathogenicity can be viewed as complementary interactions between various (macro)molecular components of the host and the pathogen. Modification of any of these components, resulting from fluctuations in environmental conditions or genetic changes, will influence the outcome of the interaction, i.e., disease development.

Genetic technologies have proved to be highly effective in understanding physiological and developmental processes in certain free-living as well as animal pathogenic bacteria. Analogous developments with plant pathogenic bacteria have been slow for various reasons. However, there has been a recent and growing interest to elucidate the genetic constitution of phytopathogenic bacteria and to apply genetic approaches in probing phytopathogenic mechanisms. Thus, we now know somewhat more about *extrachromosomal genomes* or plasmids of phytopathogenic bacteria, including the functions of some of the native plasmids. Genetic systems are being studied in several phytopathogenic bacteria and mapping of some chromosomal genes has been achieved, including genes that specify pathogenicity. Transposable elements (transposons; see page 16) are now known to interact with plasmid and chromosomal genomes of phytopathogenic bacteria, and these open up new possibilities of *in vivo* and *in vitro* genetic manipulations.

In this article we present certain aspects of what is currently known of bacterial determinants of plant pathogenicity and discuss in greater detail how genetic approaches can be effective in elucidating the structure and function of bacterial pathogenic determinants. In discussing areas of obvious or potential interest, we must often use the knowledge acquired with animal pathogens or free-living bacteria. Treatment of these areas, although largely speculative, should provide a perspective of approaches toward genetic analyses of bacterial phytopathogenic determinants. Since it is not our intent to present a comprehensive review on phytopathogenic bacteria, interested readers are urged to consult books (Elliott, 1951; Israilski, 1955; Dowson, 1957; Stapp, 1958; Gorlenko, 1965; Maas Geesteranus, 1972; Wood *et al.*, 1972; Leben, 1974; Dickinson and Preece, 1976; Heitefuss and Williams, 1976; Durbin, 1981; Goto, 1981; Lozano, 1981; Starr *et al.*, 1981; Bailey and Mansfield, 1982; Dickinson and Lucas, 1982; Rhodes-Roberts and Skinner, 1982; Mount and Lacy, 1982; Callow, 1983; Daly and Deverall, 1983; Fahy and Persley, 1983) and other relevant references (cited below) which summarize various facets of current knowledge of plant pathogenic bacteria. Also, the current literature on the tumor-inducing (Ti) plasmid of *Agrobacterium tumefaciens* has been extensively reviewed by Nester and Kosuge (1981), Kahl and Schell (1982), Merlo (1982), Ream and Gordon (1982), Roberts (1982), Van Montagu and Schell (1982), Hooykaas and Schilperoort (1984), and Nester *et al.* (1984), so that we will not consider this aspect in detail. Another area not treated in this article is that of biological control (Schroth and Hancock, 1981; Cook and Baker, 1983), a phenomenon which includes the use of some bacteria which may be related to plant pathogens.

II. PHYTOPATHOGENIC BACTERIA: AN OVERVIEW

Aside from their ability to associate with plants, most phytopathogenic bacteria are ordinary heterotrophic organisms. Plant pathogenicity is not unique to any particular genus, with the possible exception of *Xanthomonas*. Rather, phytopathogenic bacteria are diverse with respect to the hosts they infect and also in morphological, cultural, and biochemical properties such as Gram reaction, cell envelope components, cell surface appendages, pigmentation, catabolic and biosynthetic processes, energy metabolism, and bacteriophage and bacteriocin relationships and genetic constitution. Consequently, these bacteria can be differentiated from other taxa into the principal groups of *Agrobacterium*, *Corynebacterium*, *Erwinia*, *Pseudomonas*, *Spiroplasma*, *Streptomyces*, and *Xanthomonas*, and divisions of phloem-limited and xylem-limited bacteria. Mycoplasma-like organisms, minor pathogens, and anaerobes, about which little is known, are not discussed. Since the subject matter of this article requires some knowledge of some of these bacteria, a brief summary of major current taxonomic groups is appropriate along with a synopsis of pathogenic properties.

A. *Agrobacterium*

Agrobacterium and *Rhizobium*, the two genera within the family Rhizobiaceae, are differentiated physiologically by the ability to fix dinitrogen and elicitation of differential host reponses. Both are Gram-negative rods, motile by peritrichous flagella or a polar or lateral flagellum. Agrobacteria form nonpigmented colonies. Excess polysaccharide slime is generally produced upon growth on carbohydrate-containing media. Three species are recognized in a recent classification scheme (Holmes and Roberts, 1981), based upon numerical taxonomic analyses; these species correspond with three rather distinct biochemical groups (Moore and Cooksey, 1981). Revised descriptions

(Holmes and Roberts, 1981), without regard to phytopathogenic effect, were presented for the three species: (1) *A. tumefaciens*, the type species which produces galls on a wide variety of dicotyledonous plants; (2) *A. rhizogenes*, which causes hairy root or wooly knot; and (3) *A. rubi*, which causes galls on raspberries. The saprophytic, nonphytopathogenic state of *A. tumefaciens* strains is the descriptor used for strains previously classified as "*A. radiobacter*." This classification corresponds well with previous groupings into biotypes (see Holmes and Roberts, 1981). However, Kersters and DeLey (1984) argue for the retention of *A. radiobacter*.

The pathogenic properties of all *A. tumefaciens* and *A. rhizogenes* strains so far examined require plasmids (Hooykaas and Schilperoort, 1984; Nester *et al.*, 1984). Ti plasmids are either self-transmissible (Genetello *et al.*, 1977; Kerr *et al.*, 1977; Ellis *et al.*, 1982) or are mobilized by other conjugative plasmids (Chilton *et al.*, 1976; Van Larebeke *et al.*, 1977). The transfer and expression of Ti and Ri DNA segments in plant cells have been shown (Chilton *et al.*, 1982; Spanò *et al.*, 1982; White *et al.*, 1982; Willmitzer *et al.*, 1982).

In recent years agrobacteria have been extensively studied with respect to (1) the role of the tumor-inducing or Ti plasmids in tumorigenesis and the root-inducing or Ri plasmids in root proliferation, (2) the possible use of a segment of the Ti or Ri plasmid in genetic engineering, (3) the regulation of the catabolism of the opines and the sex factor activity associated with the Ti plasmid, (4) the function of the outer membrane (lipopolysaccharide layer) in pathogen establishment, and (5) the biological control of crown gall disease using a bacteriocin (agrocin 84)-producing strain of "*A. radiobacter*" (Drummond, 1979; Lippincott and Lippincott, 1980; Zambryski *et al.*, 1980; Nester and Kosuge, 1981; Bevan and Chilton, 1982; Kahl and Schell, 1982; Merlo, 1982; Van Montagu and Schell, 1982; Hooykaas and Schilperoort, 1984; Nester *et al.*, 1984). The implications of some of these findings in host: pathogen associations are discussed below.

B. *Corynebacterium* and Related Bacteria

Gram-positive, non-spore-forming bacteria included in the genus cause economically important diseases in many plants. In general, symptoms of plant diseases due to corynebacteria range from hypertrophy (*C. fascians*), gummosis (*C. iranicum*, *C. rathayi*, *C. tritici*), to wilts and/or leaf spots ("*C. betae*," "*C. flaccumfaciens*," "*C. insidiosum*," "*C. michiganense*," "*C. oortii*," "*C. poinsettiae*," "*C. sepedonicum*"). The most recent taxonomic analysis of Carlson and Vidaver (1982) showed that the species of *Corynebacterium* clustered into seven groups, from which basis the following species and subspecies combinations were proposed:

1. *C. michiganense* subsp. *michiganense*, *C. m.* subsp. *insidiosum*, *C. m.* subsp. *nebraskense*, *C. m.* subsp. *sepedonicum*, and *C. m.* subsp. *tessellarius*.
2. *C. flaccumfaciens* subsp. *flaccumfaciens*, *C. f.* subsp. *betae*, *C. f.* subsp. *oortii*, and *C. f.* subsp. *poinsettiae*.
3. The remaining species were retained, namely *C. ilicis*, *C. fascians*, *C. iranicum*, *C. rathayi*, and *C. tritici*.

Alternate schemes propose pathovar status for most species (Dye and Kemp, 1977) or species status in different genera (N. S. Mair, personal communication).

New taxa of corynebacteria proposed for the bacteria causing ratoon stunting disease of sugar cane and Bermuda grass stunting disease are *Clavibacter xyli* subsp. *xyli* and *C. xyli* subsp. *cyanodontis*, respectively (Davis *et al.*, 1984).

The species and subspecies can be differentiated by various biochemical tests (Dye and Kemp, 1977; Davis *et al.*, 1984), principal amino acids in cell walls (Yamada and Komagata, 1970; Davis *et al.*, 1984), isoprenoid quinones (Collins and Jones, 1980), DNA:DNA hybridization (Dopfer *et al.*, 1982), and the patterns of soluble proteins (Carlson and Vidaver, 1982; Davis *et al.*, 1984).

At present, the generic classification of phytopathogenic coryneform bacteria remains unsettled (Vidaver, 1982; Elia *et al.*, 1984). Proposals to assign most of them to other existing genera have been reviewed (Vidaver, 1982). Retention of *Corynebacterium* status seems less and less persuasive. Indeed, the forthcoming *Bergey's Manual* (Vol. 2) will transfer *C. fascians* to *Rhodococcus* or *Nocardia* and *C. flaccumfaciens* and related bacteria to *Curtobacterium* (N. S. Mair, personal communication). In addition, Davis *et al.* (1984) have proposed that the bacteria related to *C. michiganense* be transferred to the new genus of *Clavibacter*. Nevertheless, at present all these bacteria will be treated as members of *Corynebacterium* here for historical reasons and because there has not been time to determine the acceptance of these revisions by the scientific community.

C. *Erwinia*

The genus is a repository of phytopathogenic or plant-associated bacteria that are Gram-negative, fermentative, and usually motile with peritrichous flagella (Lelliott and Dickey, 1984). The members are extremely heterogeneous, but cluster into three broad phenotypic groups (Lelliott, 1974): the amylovora, the carotovora, and the herbicola groups.

The amylovora group contains such species as *E. amylovora*, *E. salicis*, *E.*

tracheiphila, *E. quercina*, *E. rubrifaciens*, and *E. nigrifluens*, which cause necrosis or wilts of plants. Some members of this group are pectolytic (Gardner and Kado, 1976; Schroth and Hildebrand, 1980). DNA hybridization data (Gardner and Kado, 1972; Murata and Starr, 1974; Brenner *et al.*, 1974) reveal considerable heterogeneity within the species of this group.

The carotovora group includes species and subspecies such as *E. carotovora* subsp. *carotovora*, *E. carotovora* subsp. *atroseptica*, *E. chrysanthemi*, *E. cyprepedii*, and *E. rhapontici*, all of which cause soft rot (tissue macerating) diseases in plants. Pectolytic activity has been detected in all species except *E. cyprepedii* and *E. rhapontici*. Homology among endopectate lyase genes of *E. carotovora* subsp. *carotovora* and *atroseptica* (Allen *et al.*, 1985) and *E. chrysanthemi* (Allen and Lacy, 1985) has been detected. A proposed reclassification of this group in *Pectobacterium* (Brenner *et al.*, 1977) is not acceptable, since other enterobacteria such as *Klebsiella* and *Yersinia* are known to be pectolytic (von Riesen, 1975, 1976; Starr *et al.*, 1977; Chatterjee *et al.*, 1979).

E. chrysanthemi strains were classified into six pathovars (Dickey, 1979; Dye *et al.*, 1980), the delineation of which was supported to some extent by DNA homology data (Brenner *et al.*, 1977). But more extensive studies of comparative reactions of 8 plant species to 353 strains of *E. chrysanthemi* led Dickey (1981) to conclude that usage of pathovar designations in *E. chrysanthemi* was premature.

The strains of *E. carotovora* subsp. *atroseptica*, *E. carotovora* subsp. *carotovora*, and *E. chrysanthemi* can be distinguished by serological and various biochemical properties (see Samson, 1973; De Boer *et al.*, 1979; Yakrus and Schaad, 1979; De Boer, 1980; Pérombelon and Kelman, 1980). Seemingly conflicting observations in some cases may simply reflect differences in cultural conditions, the procedures used to extract antigenic components, or the particular antigenic component used.

The herbicola group consists of species and subspecies *E. herbicola* subsp. *herbicola*, *E. herbicola* subsp. *ananas*, *E. stewartii*, and *E. uredovora* (Lelliott, 1974; Young *et al.*, 1978), all of which usually produce a yellow water-insoluble pigment (presumably carotenoid). While *E. herbicola* subsp. *herbicola* generally is associated with plants as an epiphyte, the other bacteria are known only as pathogens of plants (Gibbins, 1978). Members of this group can also be opportunistic pathogens of humans and animals (Ewing and Fife, 1972; Starr and Chatterjee, 1972; von Graevenitz, 1977; Gibbins, 1978; Starr, 1981b). The most recent taxonomic study by Brenner *et al.* (1984), using DNA hybridization and phenotypic tests, showed the inadequacy of the classification schemes presently used for this group of bacteria.

While the need for a taxonomic revision of the members of the genus is apparent (Starr, 1981b; Mergaert *et al.*, 1984) the classification recently proposed by Dye (1981) based on numerical taxonomy of phenotypic properties seems most useful for the present.

D. *Pseudomonas*

Phytopathogenic species of the genus infect a wide variety of plants, and produce symptoms ranging from necrosis, wilts and rots, to hyperplasia (galls and knots). Many species cause serious economic losses worldwide. Bacterial cells are Gram-negative, straight rods, and are motile by polar flagella. Energy metabolism is respiratory, never fermentative. Molecular oxygen is used as the terminal electron acceptor, except for certain strains of *P. solanacearum* which are capable of denitrification. Growth factors are not required.

Doudoroff and Palleroni (1974) recognized four phytopathogenic species within the genus; the species were differentiated by several biochemical tests including the production of fluorescent pigment, intracellular accumulation of poly-β-hydroxybutyrate, and absence of arginine dihydrolase. *P. syringae*, *P. cichorii*, and the opportunistic phytopathogen *P. aeruginosa* were placed with the fluorescent group, with most of the fluorescent phytopathogenic nomenspecies lumped under *P. syringae*. *P. marginalis*, another pathogen, was placed in biotype II of *P. fluorescens*. *P. solanacearum* and *P. cepacia* were the other two phytopathogenic species recognized under the nonfluorescent pseudomonads. This inadequate treatment of phytopathogenic pseudomonads was ascribed to the lack of distinguishing properties among previously named species or *nomenspecies*, except phytopathogenicity. Although 21 species of phytopathogenic pseudomonads were included in the Approved Lists of Bacterial Names (Skerman *et al.*, 1980), mostly in accord with the recommendations of Young *et al.* (1978), the status is still far from satisfactory (Palleroni, 1984) (*P. cepacia* was inadvertently left off the original Approved Lists). For example, within *P. syringae* are included a number of nomenspecies that are heterogeneous with respect to DNA homology (Pecknold and Grogan, 1973). The *r*RNA–DNA homology data (De Vos and De Ley, 1983) indicated the formation of three groups of *Pseudomonas* species with *P. syringae* strains clustering with the *P. fluorescens* *r*RNA group and *P. solanacearum* falling into a distinct group along with *P. cepacia*, *P. marginata*, *P. caryophylli*, and *P. lemoignei*. Considerable heterogeneity, however, was detected within each of the *r*RNA groups.

Despite extensive studies, the taxonomic status of *P. solanacearum* remains in a state of confusion. Heterogeneity among organisms included in this species is indicated by phenotypic or phenetic criteria, DNA homology values, DNA base composition, and phytopathogenicity. Various proposals have been put forward to subdivide *P. solanacearum* into groups based upon biochemical and serological properties, bacteriophage and bacteriocin sensitivity, and phytopathogenicity (Hayward, 1976). Despite the lack of uniformity, the three races described by Buddenhagen and Kelman (1964) and the four biotypes delineated by Hayward (1964) are generally used by most workers.

E. *Xanthomonas*

This genus of the family Pseudomonadaceae is currently composed entirely of phytopathogenic bacteria. The *r*RNA–DNA homology data revealed that the genus, despite its phenotypic similarity with the *Pseudomonas* group (see below), is a distinct taxon (De Vos and De Ley, 1983). Collectively, these bacteria infect a wide variety of plants and cause serious economic losses. Like pseudomonads, xanthomonads are Gram-negative, polarly flagellated (usually monotrichous), and obligately respiratory, using molecular oxygen as the terminal electron acceptor. Unlike certain pseudomonads, xanthomonads do not reduce nitrates. The majority of *Xanthomonas* spp. produce a unique, brominated, yellow pigment ("xanthomonadin") and copious slime on sugar-containing media. They are sensitive to triphenyl tetrazolium chloride and NaCl, relatively slow growing even in complex media and often lipolytic, proteolytic, pectolytic, and/or amylolytic (Starr, 1981a).

One fundamental taxonomic issue that remains unresolved is whether all or most nomenspecies (over 200 described thus far) are taxospecies (Van den Mooter *et al.*, 1985). Five taxospecies [*X. albilineans*, *X. ampelina*, *X. axonopodis*, *X. fragariae*, and *X. campestris* are currently recognized (Dye and Lelliott, 1974; Skerman *et al.*, 1980; Bradbury, 1984)] with 119 infra subspecific (pathovar) classifications of *X. campestris* (Young *et al.*, 1978). The proposal of Young *et al.* (1978), which accommodated nomenspecies differentiated by their pathogenic properties, is outside the purview of The International Committee of Systematic Bacteriology. Thus, while the taxonomic status of nomenspecies remains uncertain, the pathovar classification seems most appropriate at the present time.

F. Fastidious Prokaryotes

Both members of *Spiroplasma* and xylem-limited bacteria (XLB's) have been cultivated *in vitro* (Davis *et al.*, 1981; Tully and Whitcomb, 1981; Daniels, 1983; Davis *et al.*, 1984), but not mycoplasma-like organisms (MLO's) or other fastidious bacteria (McCoy, 1982). The cultivated bacteria are distinguished by differences in morphology, serology, and nutritional requirements (Tully and Whitcomb, 1981; Davis *et al.*, 1984) and *Spiroplasma* species by DNA:DNA hybridization (Lee and Davis, 1980).

It is apparent from the foregoing summary descriptions that the taxonomy of the major phytopathogenic bacteria is in a dynamic state. This is a healthy situation in that taxonomic problems are being assessed, reexamined, and sometimes resolved.

The lack of understanding of pathogenic mechanisms at the genetic or molecular level makes it difficult to justify the argument that phytopathogenic bacteria merit a taxonomic rank distinct from their nonphytopathogenic counterparts. Genetic evidence, although fragmentary, suggests otherwise. For example, *Rhizobium trifolii*, upon the acquisition of the Ti plasmid of *A. tumefaciens*, elicits a plant response akin to the latter (Hooykaas *et al.*, 1977) while retaining nodulation ability. Conversely, pSym plasmids carrying host-specific nodulation genes of *Rhizobium* spp. are expressed in *A. tumefaciens* (Hooykaas *et al.*, 1981, 1982b; Kondorosi *et al.*, 1982; Truchet *et al.*, 1984). Moreover, a significant DNA homology between *Rhizobium* plasmids involved in symbiotic nitrogen fixation and Ti plasmids suggested that plasmids share sequences derived from a common ancestral DNA (Hadley and Szalay, 1982; Prakash and Schilperoort, 1982). Homology is evident in the gene order of *Erwinia* chromosomes with nonphytopathogenic enterobacteria, such as *E. coli* and *Salmonella typhimurium* (Chatterjee, 1978; Pugashetti *et al.*, 1978; Chatterjee and Starr, 1980; Chatterjee *et al.*, 1981a; Kotoujansky *et al.*, 1982; Schoonejans and Toussaint, 1983) and in gene functions between *E. coli* and *Erwinia* spp. (Van Gijsegem and Toussaint, 1983; Chatterjee *et al.*, 1985; Van Gijsegem *et al.*, 1985a,b; Zink and Chatterjee, 1985; Zink *et al.*, 1985), thus suggesting their evolution from a common ancestor. Recombinational homology was detected between *P. syringae* pv. *glycinea* and *P. aeruginosa* (Leary, 1979) and between *E. amylovora* and *S. typhimurium* (Wu *et al.*, 1977). Plasmid transfer from nonphytopathogenic bacteria and their expression and maintenance in phytopathogenic bacteria shows some homology of both recipient and donor strains (Sanderson, 1976; Panopoulos *et al.*, 1978; Lacy and Leary, 1979; Chatterjee and Starr, 1980). Knowledge of epigenetic and genetic relatedness (Schleifer and Stackebrandt, 1983) among phytopathogens offers promise in resolving some of these taxonomic dilemmas.

III. PROGRESS IN GENETICS OF PHYTOPATHOGENIC BACTERIA

A. Introduction

Most phytopathogenic bacteria carry out physiological activities in a free-living state and in association with plant hosts. Thus, their genomes must code for functions required in host associations in addition to those that control vital physiological processes such as cell division, replication of DNA, and synthesis of other macromolecules. Genetic studies on phytopathogenic bacteria are therefore attractive for several reasons, the most obvious being to learn about gene functions required in host–pathogen interactions, and to assess the extent of overall genomic homology between phytopathogenic bacteria and their nonphytopathogenic counterparts. In probing host–pathogen interactions, genetic research should prove valuable in identifying pathogenic determinants, in elucidating their organization on chromosomes and/or plasmids, and in understanding functions of genes conferring pathogenicity and virulence traits. In addition, knowledge of the genetics of these pathogens should provide insight into rational chemical control, biological control with avirulent mutants and related bacteria, induction of resistance factors in plants, modification of plant hosts and, paradoxically, the potential production of useful compounds, the most notable of which is xanthan gum.

Pathogenic properties may change due to mutations or genetic rearrangements of chromosomes and/or extrachromosomal genomes. Such variations in conjunction with selection exerted by the host probably give rise to variants or so-called races of bacterial phytopathogens. An understanding of these mechanisms, including the identity of those genes that are more prone to modification and selection, may allow us to anticipate the genetic changes that are most likely to occur in the future and to advise plant breeders accordingly. Staskawicz (1983) reviewed the applications of molecular and genetic approaches in understanding plant disease.

Despite the recognition by phytobacteriologists of the potential benefits, genetic research on phytopathogenic bacteria has progressed rather slowly. Prior to and during the 1960s, a few studies of mutants (Kelman, 1954; Garber, 1960; Friedman, 1962; Jacobs *et al.*, 1965; Ercolani, 1970) and gene transformation in *X. campestris* pv. *phaseoli* (Corey and Starr, 1957a,b) and *A. tumefaciens* (Klein and Klein, 1953, 1956) constituted the full extent of genetic research. This slow development was primarily due to lack of personnel trained in areas of bacterial genetics and molecular biology.

During the 1970s, however, research on molecular aspects of plant pathogenicity and genetics developed rather rapidly. Knowledge was acquired of the biology of various phytopathogenic bacteria including the rule of the Ti plasmid in crown gall disease (Nester and Kosuge, 1981) and the production of such pathogenic determinants as toxins and hormones of phytopathogenic pseudomonads (Patil, 1974; Strobel, 1977, 1982; Durbin, 1981; Vidaver, 1982; Mitchell, 1984; Gross and Cody, 1985), and extracellular enzymes of soft rot erwinias (Bateman and Basham, 1976; Chatterjee and Starr, 1980; Collmer *et al.*, 1982a). High-molecular-mass wilting compounds produced by corynebacteria were also discovered (Strobel, 1977, 1982; Daly, 1981; Vidaver, 1982). Biochemical and ultrastructural studies (Goodman, 1978; Sequeira, 1979, 1983, 1984; Mazzucchi, 1983; Vance, 1983; Daly, 1984; Kolattukudy, 1984) began to show the significance of cell surface components in host–pathogen associations. Methods were developed that permitted scoring of virulence determinants (such as enzymes and toxins) and pathogenicity in model systems. The discovery of fatal infections in humans caused by presumed plant-associated bacteria (Starr and Chatterjee, 1972; von Graevenitz, 1977) renewed interest in phytopathogenic and epiphytic bacteria. Contemporary research with animal and human pathogenic bacteria with regard to the chemistry of capsules (Sutherland, 1977a, 1982; Troy, 1979), lipopolysaccharides (Wilkinson, 1977; Osborn, 1979) and outer membrane proteins (Inouye, 1979a,b; Osborn and Wu, 1980), toxinogenesis (Elwell and Shipley, 1980; Finn *et al.*, 1984), genetics (Hayes, 1968; Curtiss, 1969), plasmid biology (Falkow, 1975; Bennett and Richmond, 1978; Broda, 1979; Levy *et al.*, 1981), and recombinant DNA technology (Timmis, 1981; Hopwood and Johnston, 1984) provided further stimuli. As a result, during the past decade, significant progress has occurred in research on the genetics of bacterial plant pathogenicity (Chatterjee and Starr, 1980; Leary and Fulbright, 1982; Hooykaas and Schilperoort, 1984; Nester *et al.*, 1984; Chatterjee, 1985; Mills, 1985; Panopoulos and Peet, 1985). The structure of the Ti plasmid and its requirement for crown gall disease are now known (Nester and Kosuge, 1981; Kahl and Schell, 1982). The production of indoleacetic acid, required for virulence of *P. syringae* pv. *savastanoi* (Comai and Kosuge, 1980, 1982; Comai *et al.*, 1982) and coronatine in *P. syringae* pvs. *atropurpurea* (Sato *et al.*, 1983) and *glycinea* (Willis *et al.*, 1986), is now known to be determined by plasmids. Plasmids have also been detected in many Gram-positive and Gram-negative phytopathogenic bacteria and include some that are conjugative. Gram-negative phytopathogenic bacteria can accept and retain F and R plasmids from different donors. Conjugational and transductional systems are available in several bacterial species. Several chromosomal genes, including some that specify virulence, have been mapped and cloned.

The challenge for the future is to exploit genetic technologies in understanding bacterial pathogenic mechanisms and to better understand those genetic factors that alter bacterial pathogenic potential. In this section we briefly review the major accomplishments in genetic research on phytopathogenic bacteria but generally exclude agrobacteria (for details see Nester and Kosuge, 1981; Kahl and Schell, 1982; Merlo, 1982; Van Montagu and Schell, 1982; Hooykaas and Schilperoort, 1984; Nester *et al.*, 1984), and indicate some suggestions for using modern genetic technologies in investigating bacterial pathogenic mechanisms.

B. Strategy of Genetic Approaches

In probing pathogenic mechanisms, the strategy of genetic approaches would clearly differ depending upon the extent of our knowledge of host–pathogen interactions. For example, in those situations where biochemical findings implicate bacterial factor(s) in pathogenicity, genetic studies should seek verification of the biochemical evidence and elucidation of specific functions of those factors in pathogenic processes. With many phytopathogenic bacteria there is evidence for a requirement of cell envelope components in pathogenicity. Toxins, hormones, or enzymes produced by phytopathogenic bacteria are known to alter host physiology. In these instances, attempts are being made to modify genetically bacterial factors and then to examine the effect on pathogenic responses induced by cells carrying genetic changes (see below for the details of cell surface components, enzymes, toxins and hormones). Assessment of pathogenic responses must include data on survival and multiplication of the pathogen in plants, and whether or not bacteria are translocated from the site of inoculation. In those cases where a modified pathogen fails to elicit an overt symptom, ultrastructural and biochemical studies should provide further insights into infection processes. If an apparent genetic defect in a putative virulence factor does not alter bacterial virulence, it is desirable to establish by biological, biochemical, ultrastructural, and/or immunochemical criteria that the phenotype also persists in plants. Thus, it is possible to examine critically the consequences of quantitative and qualitative changes in putative determinants of pathogenicity.

In contrast, in those situations where there is no knowledge of bacterial determinants of pathogenicity, a somewhat different approach has to be used. In these instances, the only detectable change in pathogenic phenotype, resulting from a genetic change, would be altered virulence. We might recall that phenotypic changes in virulence could be due to alterations in many gene functions, including those that are specifically required in host associations (Table I). Thus, it is desirable to look for strains that are spontaneously avirulent and those that are avirulent due to transpositional events. Transposons (Tn elements) are useful in generalized mutagenesis as well as in creating gene fusions (see below). Such mutagenesis is particularly attractive since it provides a selectable phenotype. Avirulent mutants need to be examined carefully to rule out the possibility that avirulence results from defects in primary metabolic systems such as catabolism of substrates (i.e., in energy metabolism and formation of biosynthetic intermediates) or in biosynthesis of amino acids, purines, pyrimidines, vitamins, coenzymes, fatty acids, or phospholipids. Phenotypic changes should also be associated with Tn insertions. With the availability of genetically altered avirulent strains that are otherwise phenotypically similar to the parent, it should be possible to (1) identify the pathogenic event that is blocked; (2) determine, by complementation of the virulence trait in various independent mutants, if multiple gene products are required; and (3) ultimately characterize the gene product. Use of such approaches in conjunction with recombinant DNA technologies has

Table I. Examples of certain mutations that are associated with alteration of virulence of phytopathogenic bacteria[a]

Bacterial species	Phenotype of the primary mutation	Virulence property	References
Corynebacterium fascians	Auxotrophy	Avirulent	Jacobs *et al.* (1965)
C. michiganense subsp. *michiganense*	Auxotrophy	Avirulent	Ercolani (1970)
Erwinia amylovora	Noncapsular	Avirulent or reduced virulence	Ritchie and Klos (1977); Bennett and Billing, (1978); Ayers *et al.* (1979); Billing (1984)
	Auxotrophy	Avirulent[b]	Pugashetti and Starr (1975); Chatterjee and Starr (1978a)
	Streptomycin resistance	Avirulent	Bennett and Billing (1975)
	Altered in pathogenicity and hypersensitive reaction	—	Bauer and Beer (1985); Steinberger and Beer (1985)
E. aroideae (*E. carotovora* subsp. *carotovora*)	Auxotrophy	Avirulent	Garber *et al.* (1956)
E. carotovora subsp. *carotovora*	Deficiency in extracellular enzyme production	Avirulent	Beraha and Garber (1971); Beraha *et al.* (1974); Handa (1984); Zink and Chatterjee (1984a); Chatterjee *et al.* (1985)
	*c*AMP-deficient mutants altered in pectolytic enzyme production	?	Mount *et al.* (1979)
	Inability to catabolize lactose, arabinose, galactose and polygalacturonate in the presence or absence of *c*AMP	Avirulent	Handa (1984)
	General metabolic defect[c]	Reduced virulence	Friedman (1962)
E. chrysanthemi	Auxotrophy	Avirulent	Chatterjee and Starr (1978b)
	Pleiotrophic defect in carbohydrate utilization	Avirulent	Chatterjee and Starr (1977)
	Loss of extracellular pectolytic activity	Avirulent or reduced virulence	Chatterjee and Starr (1977, 1978b); Thurn and Chatterjee (1984)

Table I. (cont.)

Bacterial species	Phenotype of the primary mutation	Virulence property	References
E. chrysanthemi (cont.)	Bacteriocin resistant (outer membrane protein deficient)	Nonpathogenic	Expert and Toussaint (1985)
E. stewartii	Noncapsular	Reduced virulence or avirulent	Coplin *et al.* (1982)
	Non-watersoaking and/or capsular defect	Avirulent	Coplin *et al.* (1985)
Pseudomonas cepacia	Loss of extracellular pectinase activity	Avirulent	Vidaver (unpublished)
P. solanacearum	Auxotrophy	Avirulent	Coplin *et al.* (1974)
	Noncapsular	Avirulent	Kelman (1954)
	Acridine orange resistance	Avirulent; hypersensitive reaction negative	Message *et al.* (1978)
	Avirulence	—	Boucher *et al.* (1984)
P. syringae pv. *glycinea*	Auxotrophy	Avirulent	Thomas and Leary (1980)
P. syringae pv. *morsprunorum*	Bacteriophage resistance	Reduced virulence	Garrett *et al.* (1974)
P. syringae pv. *phaseolicola*	Nontoxinogenic	Reduced virulence	Patil *et al.* (1974)
	Nontoxinogenic	Absence or reduction of systemic chlorosis	Romeo and Patil (1984); Peet *et al.* (1985)
	Altered in pathogenicity and/or hypersensitive reaction	—	Anderson and Mills (1985); Panopoulos *et al.* (1985); Deasey *et al.* (1985); Lindgren *et al.* (1985)
P. syringae pv. *pisi*	Altered in pathogenicity	—	Malik *et al.* (1984)
P. syringae pv. *savastanoi*	Loss of indoleacetic acid production	Avirulent	Smidt and Kosuge (1978); Varvaro and Surico (1985)
P. syringae pv. *syringae*	Tryptophan requirement[d]	Avirulent	T. C. Currier (pers. comm.)
	Nontoxinogenic	Avirulent	Gonzalez and Vidaver (1978, 1979)
	Altered in virulence and/or hypersensitive reaction	—	Anderson and Mills (1985); Niepold *et al.* (1985a,b)
	Nontoxinogenic	?	Morgan and Chatterjee (1984a, 1985)
P. syringae pv. *tabaci*	Auxotrophy	Avirulent	Garber (1959); Vasil'eva and D'jakav (1970)

Table I. (cont.)

Bacterial species	Phenotype of the primary mutation	Virulence property	References
Xanthomonas campestris pv. *campestris*	Altered pathogenicity	—	Daniels *et al.* (1984a, 1985); Turner *et al.* (1985)
X. campestris pv. *oryzae*	Auxotrophy	Avirulent	Yamasaki *et al.* (1966)
	Pigment	Altered virulence	Tsuchiya *et al.* (1982)
X. campestris pv. *phaseoli*	Noncapsular	Avirulent	Corey and Starr (1957a)

[a] In Ti and Ri plasmids, induced mutations have been associated with avirulence. Mutations in genes affecting the pathogenicity of *Agrobacterium tumefaciens* and *A. rhizogenes* are not included. Examples of such mutations can be found in Lippincott and Lippincott (1966, 1969), Langley and Kado (1972), Liao and Heberlein (1979), Nester and Kosuge (1981), Douglas *et al.* (1982), Holsters *et al.* (1982), Hooykaas and Schilperoort (1984) and Nester *et al.* (1984).

[b] Genetic data (Pugashetti and Starr, 1975; Chatterjee and Starr, 1978a), strongly suggest that a mutation in a gene (*vir*) closely linked to *ser* caused avirulence.

[c] Mutants were selected on agar media containing tetrazolium; primary mutational phenotype not known.

[d] Most such mutants also fail to produce syringomycin. See Section IV.D. for additional details.

allowed identification of genes (DNA segments) for avirulence, pathogenicity and hypersensitive reaction (see Section III. G. for details).

C. Mutagenesis

1. *Chemical and Physical*

The induction of mutations is crucial for genetic and biochemical analyses of pathogenicity. The analysis of avirulent mutations should allow identification of virulence genes and the elucidation of biochemical events in pathogenicity. Thus, utmost care should be exercised in choosing the desired mutation (i.e., point mutations, deletions or insertions; Table II) and the use of mutagenic agents (Table III). Numerous reports (Clowes and Hayes, 1968; Drake, 1970; Roth, 1970; Miller, 1972; Drake and Baltz, 1976; Singer and Kusmierek, 1982) describe the properties of various mutagens, the mechanisms and efficiency of mutagenesis, and the procedures generally proved effective in

Table II. Functional and genetic properties of bacterial mutations[a]

Type of mutation	Nature of DNA alteration	Types of Genes involved	Effect on gene function	Revertible	Supressible by *t*RNA suppressors
Point					
Missense[b]	Base substitutions	Protein coding	Variable	Yes	Yes
Nonsense[c]	Base substitutions	Protein coding	Complete loss[e]	Yes[f]	Yes
Frameshift[d]	Base insertions or deletions	Protein coding	Complete loss[e]	Yes[g]	Yes
Other	Base substitutions	Regulatory elements; *t*RNA and *r*RNA genes[g]	Variable	Yes	No
Deletion	Loss of a DNA segment	All types	Complete loss	No	No
Insertion	Incorporation of a phage genome or transposable element	All types	Complete loss	Yes	No

[a] From Carlton and Brown (1981).
[b] Triplet base sequence is altered such that it codes for a different amino acid.
[c] Amino acid coding triplet is altered to code for no amino acid; causes premature termination of protein synthesis.
[d] Causes out-of-phase transcription of the base sequence.
[e] Unless very near the distal end of the gene.
[f] By second-site compensating mutations of the opposite type.
[g] *t*RNA, transfer RNA; *r*RNA, ribosomal RNA.

major groups of bacteria. For those contemplating chemical mutagenesis of phytopathogenic bacteria, Carlton and Brown's (1981) article and Miller's (1972) manual provide excellent discussion of the principles and procedures of chemical and physical mutagenesis.

2. *Transpositional*

As stated above, in probing pathogenic mechanisms, avirulent mutations resulting from insertions of a transposable element should prove extremely useful. Ordinarily, loss of virulence or pathogenicity due to mutation can be scored but cannot be selected. If the loss of gene function (i.e., virulence or pathogenicity) results from insertion of a transposable element along with a

Table III. Properties of some commonly used mutagenic agents[a]

Mutagen type	Presumed mechanism of mutagenesis	Type of mutation produced	Relative effectiveness	Special advantage or disadvantage
Radiations X rays, thermal neutrons	Primarily breakage of chromosomes	Deletions inversions	High	Equipment may be difficult to obtain
Ultraviolet	Pyrimidine dimerization	GC → AT transitions,[b] transversions,[c] deletions	Medium	Wide-spectrum mutagen, extent of killing must be carefully controlled, photoreactivation must be prevented
Chemicals Base analogs: 2-aminopurine, 5-bromouracil, 5-bromodeoxyuridine	Errors in DNA replication	Primarily AT→GC transitions	Low	Relatively inefficient
Hydroxylamine	Deamination of cytosine	GC→AT transitions	Low	Preferentially induces one-directional transitions
Nitrous acid	Deamination of cytosine and adenine	Bidirectional transitions, deletions	Medium	—
NTG[e]	Alkylation of bases at replication fork	Primarily GC→AT transitions	Very high	Highly mutagenic at low killing levels, induces multiple mutations in localized regions
EMS[e]	Alkylation of guanine	Primarily GC→AT transition, some AT→GT transitions and transversions	Medium	Similar mutational effect to that of NTG, but less likely to yield multiple mutations
Acridine half-mustards (ICR compounds)	Interacalation between bases during replication	Frameshifts—base insertions and deletions	High	Compounds may be difficult to obtain
Acridine dyes, ethidium bromide	Intercalation between bases during replication	Frameshifts, loss of extra-chromosomal elements	Low	Especially effective in curing cells of plasmids
Novobiocin	Blocks DNA replication	Curing of plasmids	High	Very effective in curing cells of plasmids

[a] From Carlton and Brown (1981).

[b] Transitions are single-base substitutions where a purine base (adenine or guanine) is changed to the other purine, or one pyrimidine base (cytosine or thymine) is exchanged for the other pyrimidine base.

[c] Transversions are base substitution mutations in which a purine base is exchanged for one of the two pyrimidine bases, or a pyrimidine base is changed to one of the two purine bases.

[d] NTG, *N*-Methyl-*N*′-nitro-*N*-nitrosoguanidine.

[e] EMS, ethyl methanesulfonate.

recognizable phenotype (such as drug resistance), mutation is usually subject to selection. In recent years transposable elements, principally coding for drug resistance, have proved useful in such genetic manipulations as mutagenesis, formation of Hfr donors, replicon fusions, and construction of F′ and R′ plasmids (Table IV; Kleckner *et al.*, 1977; Berg and Berg, 1983; de Bruijn and Lupski, 1984), especially in Gram-negative bacteria. In *A. tumefaciens*, analysis of transpositional mutations has allowed identification of some of the functions coded for by the Ti plasmid (Nester and Kosuge, 1981; Holsters *et al.*, 1982). Likewise, the use of transpositional mutagenesis has revealed plasmid-determined legume symbiosis and nitrogen fixation in *Rhizobium* spp. (Johnston *et al.*, 1978; Benyon *et al.*, 1980; Brewin *et al.*, 1980; Djordjevic *et al.*, 1982). Because of their numerous potential applications (Table IV), transposable elements are being used extensively in studies with phytopathogenic bacteria. It is therefore appropriate to summarize briefly some of their properties. The reader should consult review articles (Kleckner *et al.*, 1977; Calos and Miller, 1980; Starlinger, 1980; Kleckner, 1981; Saedler *et al.*, 1981; Berg and Berg, 1983) and book chapters (Starlinger and Saedler, 1976; Bukhari *et al.*, 1977;

Table IV. Potential uses of translocatable (Tn) genetic elements (transposons) in bacterial genetics[a]

I. Strain construction: transfer of otherwise unselectable markers by virtue of the drug resistance or other selectable marker (e.g., Hg^{2+} resistance) associated with insertion mutations
II. Isolation of mutants that have no readily scorable phenotype
III. Obtaining and exploiting selectable insertions near, but not within, regions of interest:
   A. Manipulation of markers with no selectable phenotype
   B. Localized mutagenesis
   C. Chromosomal mapping
   D. F′ mapping and complementation tests
   E. Cloning of genes
IV. Selectable insertions as a portable region of homology:
   A. Construction of Hfr strains with origin and direction of transfer specified
   B. Construction of new F′ or R′ plasmids
   C. Construction of specialized transducing phages
   D. Deletions and duplications with predetermined endpoints specified by positions of insertions
V. Generation of deletions with one endpoint at an insertion and the other endpoint unspecified
VI. Selection of chromosomal duplications with translocatable (selectable) elements and their use in chromosomal mapping

[a] Adapted from Kleckner *et al.* (1977).

Bennett and Richmond, 1978; Sherratt, 1981; Shapiro, 1983) for further details.

Transposable genetic elements or transposons are mobile DNA segments that can insert into a few or several sites in a genome. Such insertions, unlike classical recombination events, do not require DNA sequence homology and are independent of recombination systems. Based upon genetic organization, mechanistic properties, and DNA sequence homologies, three broad classes of transposable elements are recognized (Bukhari *et al.*, 1977; Calos and Miller, 1980; Kleckner, 1981): (1) IS (insertion sequence) elements that do not carry genes specifying a phenotype other than insertion function and are generally shorter than 2 kb, except for the gamma–delta sequence (Tn*1000*; Buxton and Drury, 1983); these are designated as IS*1*, IS*2*, IS*3*, etc. (Table V). (2)

Table V. Prokaryotic transposable elements[a]

Insertion sequence[b]	Normal occurrence in *E. coli*	Length (bp)	Inverted repeat (bp)	Comments	References[c]
IS*1*	Five–eight copies on chromosome	768	18/23	Full sequence for two different examples	Johnsrud (1979)
IS*2*	Five on chromosome, one on F	1327	32/41	Full sequence known	Ghosal *et al.* (1979)
IS*3*	Five on chromosome, two on F	1400	32/38	Sequences of 700 bp known	Sommer *et al.* (1979)
IS*4*	One or two copies on chromosome	1400	16/18	Sequence of ends known	Habermann *et al.* (1979)
IS*5*	Unknown	1250	Short	No sequence information	Chow and Broker (1978)
Gamma–delta[d]	One on F, one or more on chromosome	5700	35	Sequence of ends known	Reed *et al.* (1979)
pSC101 segment	On plasmid pSC101	200	30/36	Full sequence known, may be derived from gamma–delta or Tn*3*	Ravetch *et al.* (1979)

[a] Modified from Calos and Miller (1980).

[b] IS*51* was detected on plasmids and chromosomes of *P. syringae* pv. *savastanoi* (Comai and Kosuge, 1983). Similar sequences may also occur in *P. syringae* pv. *phaseolicola* (Szabo and Mills, 1984; Quant and Mills, 1984).

[c] Only a single reference is given. For additional references, see Calos and Miller (1980).

[d] Gamma–delta, Tn*1000*; Buxton and Drury (1983).

Transposons (Tn elements) that carry genes related to a phenotype such as drug resistance, lactose or raffinose utilization, arginine biosynthesis, mercury resistance, or enterotoxin production. Tn elements are generally longer than 2 kb, and a number of them possess IS elements at their ends (see Table VI; Kleckner, 1981). Kleckner (1981) proposed a slightly different subdivision of transposable elements to differentiate between transposons with and without known IS elements (her class II). (3) Self-replicating episome-like elements such as the entire bacteriophage genomes of Mu and ψ (*Agrobacterium* phage; Expert and Tourneur, 1982) (Table VI).

It is important to recognize that despite the presence of a large number of potential insertion sites on a DNA segment, these sites are not randomly distributed. On the contrary, insertions appear to be clustered in certain regions of the recipient genome referred to as "hot spots" (Kopecko *et al.*, 1976). Such clustering or "regional specificity" (Tu and Cohen, 1980) has been detected with a number of recipient plasmids and chromosomal genomes (for details see Botstein and Kleckner, 1977; Tu and Cohen, 1980; Kleckner, 1981). It should be noted that insertional specificity may be determined either by the transposon such as Tn*7* (Lichtenstein and Brenner, 1981; Turner *et al.*, 1984) or by the host. An example of the latter is the specificity of the insertion of Tn*5* in *Acinetobacter* sp. (Singer and Finnerty, 1984). While Tn*5* exhibits little site preference in many bacteria, including phytopathogens (see below and also Berg and Berg, 1983), insertion of the transposon occurred at a single site on the chromosome of an *Acinetobacter* sp. (strain H01-N) and at two sites on *A. calcoaceticus* (strain BD413).

In practice, transpositional mutagenesis usually involves the introduction into bacterial cells of Tn elements via a plasmid or bacteriophage vector. Plasmid vectors used in Tn mutagenesis usually carry two or more markers, one of which is coded by the transposon. Following mutagenesis, clones are selected that carry the Tn marker but not the plasmid marker. This differential phenotype must result from transposition of the Tn element into the recipient genome(s) and loss of the plasmid. Thus, it is critical to use a vehicle that either is naturally unstable in a desired host or could be rendered unstable by manipulating physiological conditions. Such bacteriophage and plasmid vehicles have been used in transpositional mutagenesis of several bacteria. For example, derivatives of bacteriophages P22 and λ are extremely efficient transpositional vehicles for *S. typhimurium* and *E. coli*, respectively (see Bukhari *et al.*, 1977; Kleckner *et al.*, 1977; B. D. Davis *et al.*, 1980; R. W. Davis *et al.* 1980, for additional details). Because of their limited host range these vehicles, with the possible exception of Pl::Tn*5* (see below), are not directly applicable in studies on phytopathogenic bacteria.

The broad host range R plasmid, pJB4JI (Beringer *et al.*, 1978; Hirsch and

Beringer, 1984), carries the transposon Tn*5* [kanamycin resistance, Km^r; it also confers streptomycin-resistance phenotype in *Rhizobium* spp. (Putnoky *et al.*, 1983; Selvaraj and Iyer, 1984) and in *Myxococcus xanthus* (Breton, 1984)] and the coliphage Mu sequence as well as other drug resistance markers. This plasmid has been quite effective in Tn mutagenesis of such phytopathogenic and plant-associated bacteria as *A. tumefaciens* (Garfinkel and Nester, 1980), *A. rhizogenes* (White and Nester, 1980), *E. amylovora* (Steinberger and Beer, 1983, 1985), *E. carotovora* subsp. *atroseptica* and *carotovora* (Zink *et al.*, 1982, 1984), *E. chrysanthemi* (Thurn and Chatterjee, 1982a; Chatterjee *et al.*, 1983), *E. herbicola* subsp. *herbicola* (Gantotti *et al.*, 1981), *P. solanacearum* (Boucher *et al.*, 1981, 1985a,b; Staskawicz *et al.*, 1982), *R. leguminosarum* (Beringer *et al.*, 1978), and *R. melitoti* (Duncan, 1981; Meade *et al.*, 1982). The plasmid was unstable in tested strains of these species and Tn*5* inserted either into the chromosome or into plasmid genome(s). The instability of pJB4JI apparently results from the presence of the Mu genome, which fails to replicate in these recipients. However, the suicidal property of pJB4JI may be strain dependent, since the plasmid was stable in certain strains of *E. amylovora* and *E. chrysanthemi* (Thurn and Chatterjee, unpublished data), and in a strain of *X. campestris* pv. *campestris* (Turner *et al.*, 1984).

Several other conjugative plasmids have been used in transposon mutagenesis in phytopathogenic pseudomonads. pAS8 Rep-1 was used to obtain Tn*7* insertions in genomes of several pathovars of *P. syringae* (Sato *et al.*, 1981). This is a hybrid plasmid of RP4 and ColE1, fused at their *Eco*RI sites and a Tn*7* (streptomycin resistance, Sm^r; trimethoprim resistance, Tp^r) insertion in a gene necessary for RP4-mediated replication. The plasmid cannot replicate in *Pseudomonas* spp., but is conjugation proficient and has a broad host range. The suicide plasmid pSUP1011 which carries Tn*5* (Simon *et al.*, 1983) produced auxotrophic mutations and attenuation of plant virulence in *P. syringae* pv. *phaseolicola* and *P. syringae* pv. *syringae* (Anderson and Mills, 1983, 1985) and in *X. campestris* pv. *campestris* (Daniels *et al.*, 1983; Turner *et al.*, 1984). Likewise, using pSUP1011, generalized Tn*5* mutagenesis was performed in a citrus strain of *P. syringae* pv. *syringae* (Morgan and Chatterjee, 1985). Therefore, this plasmid vehicle should prove useful in transpositional mutagenesis of species of Gram-negative phytopathogenic bacteria. Morales and Sequeira (1985c), by cloning into pBR325 the transfer genes of the Inc W plasmid, R388, constructed a conjugative plasmid which replicated by a colE1 replication system. Such a hybrid plasmid, pW1281, which also carried a copy of the transposon Tn*5*, failed to survive in *P. solanacearum* allowing the isolation of Tn*5* insertion mutants. This vector may also prove useful in generalized Tn*5* mutagenesis of *Xanthomonas* spp., as well as other *Pseudomonas* spp. Likewise, suicide vehicles for the delivery of Tn*1*, Tn*5* and Tn*9* to

Table VI. Prokaryotic transposons[a]

Transposon	Marker	Length (bp)	Inverted repeat	Comments[b]	References[c]
Tn*1*[d]	Ampicillin	4,957	38	Full sequence of Tn*3* known	Hedges and Jacob (1974)
Tn*2*[d]					Heffron *et al.* (1975a,b)
Tn*3*[d]					Kopecko and Cohen (1975); Cohen *et al.* (1979)
Tn*4*	Ampicillin, streptomycin, sulfonamide	20,500	Short	Contains Tn*3*	Kopecko *et al.* (1976)
Tn*5*	Kanamycin	5,400	1500	Sequence of ends known	Jorgensen *et al.* (1979); Schaller (1979)
Tn*6*	Kanamycin	4,200	Not detectable with electron microscopy		Berg *et al.* (1975)
Tn*7*	Streptomycin, trimethoprim	14,000	Not detectable with electron microscopy		Barth *et al.* (1976)
Tn*9*	Chloramphenicol	2,638	18/23	Full sequence known, flanked by direct repeats of IS*1*	Johnsrud *et al.* (1978); Alton and Vapnek (1979)
Tn*10*	Tetracycline	9,300	1400	Sequence of ends known	Kleckner *et al.* (1975)
Tn*204*	Chloramphenicol, fusidic acid	2,457	18/23	Full sequence known, flanked by direct repeat of IS*1*	Arber *et al.* (1978); Marcoli *et al.* (1980)
Tn*402*	Trimethoprim	7,500	Not detectable with electron microscopy		Shapiro and Sporn (1977)

Tn*501*	Mercuric ions	7,800	38	Sequence at ends known	Bennett *et al.* (1978)
Tn*551*	Erythromycin	5,200	35	In *Staphyloccus aureus*,[e] sequence of ends known	Pattee *et al.* (1977); Novick *et al.* (1979)
Tn*554*	Erythromycin, spectinomycin	6,200	ND[f]	In *S. aureus*[e]	Khan and Novick (1980)
Tn*732*	Gentamycin, tobramycin	11,000	ND		Nugent *et al.* (1979)
Tn*903*	Kanamycin	3,100	1050	Sequence of ends known	Oka *et al.* (1978); Young *et al.* (1979)
Tn*916*	Tetracycline	15,000	ND	In *Streptococcus faecalis*[e]	Franke and Clewell (1981)
Tn*917*	Erythromycin	5,100	Short	In *S. faecalis*[e]	Tomich *et al.* (1979)
Tn*951*	*lac*	16,600	Short		Cornelis *et al.* (1978)
Tn*1681*	Heat stable enterotoxin	2,088	768 (IS*1*)	Full sequence known	So *et al.* (1979); So and McCarthy (1980)
Tn*1721*	Tetracycline	10,900	Short	Sequence of ends known	Schmitt *et al.* (1979)
Bacterio-phaghe Mu		38,000	Absent	Sequence of ends known	Bukhari *et al.* (1977); Kahmann and Kamp (1979)
ψ (*Agrobacterium tumefaciens*)		ND	ND		Expert and Tourneur (1982)

[a] Modified from Calos and Miller (1980); additional transposons are continually being reported.

[b] The majority of transposons have been found in Gram-negative bacteria.

[c] Few representative references are cited. For additional citations see Calos and Miller (1980).

[d] Tn*1*, Tn*2*, and Tn*3* are very similar, if not identical. Other ampicillin-resistant transposons have also been described [see Tn*801* and Tn*802* (Bennett and Richmond, 1976, Bennett *et al.*, 1977), Tn*901* (Andreoli *et al.*, 1978), and Tn*1701* (Yamada *et al.*, 1979)] and are probably similar or identical to Tn*1*, Tn*2*, or Tn*3*. For nomenclature rules see Campbell *et al.* (1979).

[e] Gram-positive.

[f] Not determined.

Rhizobium spp. (Selvaraj and Iyer, 1983) may also work in these bacteria. The efficiency of Tn*5* mutagenesis with pGS9 (Selvaraj and Iyer, 1983) has already been demonstrated in *P. syringae* pvs. *tomato* (Cuppels, 1985), *syringae* (Kuykendall *et al.*, 1985; Xu and Gross, 1985) and *tagetes* (Kuykendall *et al.*, 1985).

Another efficient approach is to use plasmids that are thermosensitive (*ts*) in replication or maintenance, such as the F′ plasmid, F′*ts lac*$^+$::Tn*10*. Because of the *ts* mutation in plasmid replication, bacterial colonies that are Tc^r (Tn*10*) and Lac$^-$ at 40–42°C could result from the translocation of the Tn*10* element. On the other hand, those colonies that are both Lac$^+$ and Tc^r at 40–42°C could result from the integration of the plasmid into the chromosome, and produce a *high frequency* of *recombination* or Hfr state. The use of such plasmids yielded Hfr strains in *S. typhimurium* (Chumley *et al.*, 1979). Likewise, in a strain of *E. chrysanthemi*, Kotoujansky *et al.* (1982) obtained Hfr strains by chromosomal insertion of F′*ts lac*$^+$::Tn*10* (see below for additional details). Of 336 clones tested, those that were Tc^r at 40°C were also Lac$^+$, suggesting a low frequency of Tn*10* transposition, if at all (but see below for the regulation of transposition). Because of the restricted host range, the use of F′ plasmid vehicles is currently limited to *Erwinia* spp.

For most phytopathogenic bacteria, a useful vehicle would be one which has a broad host range and either will not replicate in the recipient (see above for the examples of such plasmids) or has a conditional mutation allowing counterselection of the plasmid vehicle under physiological conditions. For most, but not all, phytopathogenic bacteria, 35°C or higher is nonphysiological in that either growth does not occur (e.g., Billing, 1974; Chatterjee *et al.*, 1977) or variants including avirulent clones arise at a high frequency upon growth at 32–34°C or higher (Chatterjee and Gibbins, 1971; Garibaldi and Gibbins, 1975; Gantotti and Beer, 1982). Thus a plasmid that can be readily introduced into most phytopathogenic bacteria and that fails to replicate at 33–35°C would be an ideal vehicle. pMR5 (Robinson *et al.*, 1980), a temperature-sensitive derivative of the broad host range R plasmid, RP1, might prove useful. In *E. coli*, the frequency of loss of the plasmid was about 77% at 37°C and about 99.9% at 42°C. Replication of pMR5 may not occur in agrobacteria, erwinia, pseudomonads, and xanthomonads at lower growth temperatures (ca. 37°C). If so, this plasmid could serve as a transpositional vehicle in these Gram-negative phytopathogenic bacteria.

Mutation in a broad host range plasmid, pUT13, promoted its temperature-dependent segregation in such hosts as *E. chrysanthemi* and *E. coli* (Chatterjee *et al.*, 1981a). pUT13 is a Tn*406* (carbenicillin resistant; Cb^r) transpositional derivative of R751 (Meyer and Shapiro, 1980). Both the mutant and parent plasmids confer resistance to carbenicillin (Tn*406*) and trimethoprim (Tn*402*). In *E. chrysanthemi*, at 35–37°C the frequency of marker segregation of the mutant plasmid (pUT13 *ts-1*) was 100%; the parent plasmid (pUT13) was stable under similar growth conditions (Brown *et al.*, 1981). Preliminary data (Chatterjee *et al.*, 1981a) suggest that transposition of Tp^r or Cb^r from pUT13 and pUT13 *ts-1* occurs in *E. amylovora* and *E. chrysanthemi*. Future research should show whether or not these plasmids could serve as generalized transpositional vehicles for Gram-negative phytopathogenic bacteria.

McCammon *et al.* (1981) showed the potential usefulness of resident conjugative plasmids of phytopathogenic bacteria in transpositional mutagenesis. They transferred an *E. stewartii* conjugative plasmid into *E. coli*, inserted translocatable elements [Tn*10*, and Mu *cts* Km^r (pf7701)] into the plasmid genome, and then transferred the drug-resistant plasmids back into *E. stewartii*. The plasmids carrying Tn elements, apparently unstable in *E. stewartii*, gave rise to Tn*10* and Mu pf7701 transpositional mutants. (See below for the uses of coliphage Mu and its derivatives in creating gene fusions, chromosome mobilization, and *in vivo* gene cloning.) Since pathovars of *P. syringae* also harbor conjugative plasmid(s) (Gonzalez *et al.*, 1981a, 1984; Sato *et al.*, 1981; Currier and Morgan, 1983), a similar approach might prove useful in this bacterial species as well.

A Tn*5* derivative of the bacteriophage P1Cm*c1*.100, constructed by Berg *et al.* (1983), may prove useful in the mutagenesis of *Erwinia* spp. The bacteriophage, propagated in *E. coli*, was apparently restricted by the *S. typhimurium* restriction system and hence served as a suicide vector (Berg *et al.*, 1983). Since several *Erwinia* spp. can be infected with the P1 phage (Goldberg *et al.*, 1974; Jayaswal *et al.*, 1984) and since they probably have a restriction system different from *E. coli*, it would seem desirable to test if P1Cm*c1*.100::Tn*5* propagated in *E. coli* also acts as a suicide vector in various *Erwinia* species.

Several derivatives of the transposon Tn*5* have been constructed allowing either the selection of deletion mutants or the analysis of the regulatory genes. An example of the former is λ::Tn*5*–132 where a Tc^r gene is inserted into the Tn*5* DNA (i.e., IS*50*), thus the transposon codes for Tc^r (see Berg and Berg, 1983). Since positive selection for Tc^s clones is possible, use of this Tn*5* derivative allows the isolation of deletion strains resulting from imprecise excision of the element from genomes. Recently, Bellofatto *et al.* (1984) constructed a promoter probe derivative Tn*5*-VB32, which contains part of the IS*50*L sequence, the intact *NPT II* gene lacking its promoter but beginning with its translation initiation sequences, Tc^r gene of Tn*10*, and the complete IS*50*R sequence. This element when inserted into a genome can place the *NPT II* gene under the control of the genomic promoter sequence recognized by the expression of Km^r. Bellofatto *et al.* (1984), by placing Tn*5*-

VB32 into the P group R plasmid (pPH1JI) containing bacteriophage Mu sequences (pPH1JI::Mu = pJB9JI) constructed a plasmid pPV2 (pPH1JI ::Mu::Tn*5*-VB32). This plasmid, due to the presence of Mu sequences, was unstable in *Caulobacter crescentus* and gave rise to Tn*5*-VB32 insertion mutants that were auxotrophic or defective in motility. Evidence was presented for the expression of Km^r from *Caulobacter* promoter sequences. The use of Tn*5*-VB32 DNA either on pPV2 or on other suicide vehicles (such as pSUP1011) should prove extremely valuable in analyzing the regulation of virulence genes of many Gram-negative phytopathogenic bacteria. Recently, various derivatives of Tn*10* were also constructed by Way *et al.* (1984) that should allow gene fusions, higher efficiency, and stable transpositions, or generation of desired deletions. These transposons are currently available in λ vectors, thus their use is limited to *E. coli* K12 strains and DNA of other bacteria cloned into *E. coli*. However, placing these transposons on broad host range suicide vectors (some of which are described above), should facilitate genetic analysis of various phytopathogenic and other ecologically significant Gram-negative bacteria.

Insertion events can be controlled by repressors or inhibitors, at least in some cases (Biek and Roth, 1980; Berg and Berg, 1983; Rossetti *et al.*, 1984). Thus, in looking for Tn insertion mutants of a phytobacterial species not previously studied, it may be desirable to establish the kinetics of Tn transposition. In certain cases the efficiency of recovery of transpositional mutants might be enhanced by introducing an incompatible (= related) plasmid (Kleckner *et al.*, 1977), by selecting for drug-resistant clones that are also resistant to sex factor-specific phages (Hernalsteens *et al.*, 1980), or by growing plasmid-containing bacteria at low incubation temperatures (Kretschmer and Cohen, 1979; Cornelis, 1980).

Drug-resistant Tn elements are already powerful tools in the genetics of phytopathogenic bacteria. For example, using Tn*5* insertion mutations, Garfinkel and Nester (1980) have shown that chromosomal gene products are required in the attachment of the *A. tumefaciens* cells to the host cell wall, in addition to functions encoded by the Ti plasmid. The use of Tn*5* and Tn*7* also has permitted the elucidation of several Ti plasmid-coded functions (see Nester and Kosuge, 1981; Holsters *et al.*, 1982, for further details). In *E. herbicola*, Gantotti *et al.* (1981) found that Tn*5* insertion in a plasmid resulted in loss of bacteriocin activity. They deduced that bacteriocin production was determined by a plasmid genome. In *P. solanacearum*, Tn*5* insertions (Boucher *et al.*, 1981, 1985; Staskawicz *et al.*, 1982) have yielded an array of mutants, including some that were avirulent. Physical and genetic analysis of Tn*5* induced avirulent mutants suggests clustering of phytopathogenicity genes in *P. solanacearum* (Boucher *et al.*, 1985a,b). Tn*5* insertions into the *E. chrysanthemi* chromosome resulted in mutants that were auxotrophic, unable to catabolize galacturonate or polygalacturonate, defective in the production of protease or unable to export pectate lyase, polygalacturonase and cellulase (Chatterjee *et al.*, 1983, 1985b; Thurn and Chatterjee, 1985a). The mutants have facilitated studies on the catabolism of polygalacturonate and the export of various extracellular enzymes. Likewise, in *P. syringae* pv. *syringae*, the analysis of syringotoxin-negative Tn*5* insertion mutants revealed a clustering of some of the *tox* genes on the chromosome (Morgan and Chatterjee, 1985). In addition, the availability of mutants with Tn*5* insertions in genes for toxinogenesis in *P. syringae* pv. *phaseolicola* (Romeo and Patil, 1984; Peet *et al.*, 1985) and for pathogenicity and/or hypersensitive reaction in *E. amylovora* (Bauer and Beer, 1985; Steinberger and Beer, 1985), *P. syringae* pv. *phaseolicola* (Bertoni *et al.*, 1985; Lindgren *et al.*, 1985) and *P. syringae* pv. *syringae* (Anderson and Mills, 1985) have facilitated the isolation of corresponding wild-type genes (DNA segments) and their subsequent analysis (see below).

As stated above (also see Table IV), the construction of Hfr strains can be facilitated by the use of Tn elements. The homology between the Tn sequences present in the chromosome and in a transmissible plasmid promotes the integration of the plasmid. Such integration of a transmissible plasmid into the chromosome usually results in an Hfr state. Tn directed Hfr formation has been reported in *S. typhimurium* (Chumley *et al.*, 1979), and with an R factor in *E. coli* (Danilevich *et al.*, 1978). Transposon-facilitated recombination was noted in *Vibrio cholerae* (Johnson and Romig, 1979) and *A. tumefaciens* (see Section III.E.1.; Pischl and Farrand, 1983). In *E. chrysanthemi*, Chatterjee *et al.* (1982) have constructed donor strains by exploiting the homology between Tn*5* sequences introduced into the chromosome and an F′ plasmid (F′*ts lac*$^+$::Tn*10*::Tn*5*). The genetic evidence suggested that the direction of chromosome transfer in such donor strains was determined by chromosomal Tn*5* insertion sites. Further studies under way should verify if the Hfr state occurred in donor strains. Tn*5* homology also produced putative Hfr-type donors in *X. campestris* pv. *campestris* (Turner *et al.*, 1984).

Transposons coding for drug resistance have also been effective in elucidating functions specified by indigenous plasmids of several phytopathogenic bacteria, besides the Ti plasmid of *A. tumefaciens*. For example, by using Tn*7* mutagenesis, Sato *et al.* (1981) associated a Tra$^+$ function with a 30-MDa plasmid (pBPW1) of *P. syringae* pv. *tabaci*. These results were later confirmed by Tn*3* labeling of pBPW1 (Obukowicz and Shaw, 1983). Likewise, the application of Tn*1* mutagenesis allowed the detection of sex factor activity in an indigenous plasmid of *P. syringae* pv. *syringae* (Gonzalez *et al.*, 1981a, 1984). Obukowicz and Shaw (1985) marked the plasmids in *P. syringae* pvs. *angulata* and *tabaci* with Tn*3* and then isolated strains that had lost the Tn*3*-carrying plasmids. Studies with the cured strains indicated that the plasmids were not required in "host association" or production of the phytotoxin, tabtoxin. Comai and Kosuge (1982) cloned a fragment of a pIAA plasmid of *P. syringae* subsp. *savastanoi* into *E. coli*, and used Tn*1* insertion mutations to determine the orientation and location of *iaaM* (a gene locus that specifies tryptophan monooxygenase; see Section IV.E. for the details of genes and enzymes of the indoleacetic acid biosynthetic pathway) on a chimeric plasmid. Thus, the combined use of transposons and recombinant DNA technologies (see Section III.F.; Timmis, 1981) have allowed the identification of functions specified by several plasmids, as well as chromosomal segments of various Gram-negative phytopathogens. A similar approach should prove valuable in analyzing pathogenicity genes on the chromosomes or plasmids of Gram-positive phytopathogenic bacteria, e.g. through the use of Tn*916* (Tc^r) and Tn*917* (Em^r) (Gawron-Burke and Clewell, 1984; Ike and Clewell, 1984).

Transposons have also been used in marker-exchange mutagenesis. This procedure generally requires recombinant DNA technology (see below), in which cloned DNA fragments are first mutagenized by insertions of transposon(s) into restriction fragments. The mutagenized DNA is introduced into the bacterial cell, homologous recombination occurs, and the vector is eliminated either by selection or suicide (Ruvkun and Ausubel, 1981; Comai *et al.*, 1983; White *et al.*, 1983). This process simplifies enormously the physical and genetic characterization of mutant alleles in the desired bacterial host. This procedure has been used in *A. tumefaciens* (Garfinkel *et al.*, 1981; White *et al.*, 1983), *E. carotovora* subsp. *carotovora* (Zink *et al.*, 1985a,b), *E. chrysanthemi* (Roeder and Collmer, 1985a,b), and *X. campestris* pv. *campestris* (Turner *et al.*, 1985). Site-specific mutagenesis of cloned DNA can be performed by oligonucleotide-directed mutagenesis or by misincorporation via error-prone polymerization (Zakour and Loeb, 1982; Lathe *et al.*, 1983; Kunkel, 1985). Myers *et al.* (1985) have also described a general and efficient method for saturation mutagenesis of cloned DNA fragments.

3. Gene Fusions

Directed fusion of genes, facilitated by the use of transposons, is a powerful tool in examining the regulation of gene expression as well as gene function. Of the various available approaches, the specialized transducing phage, Mu *c* ts dl (Ap^r, *lac*), constructed by Casadaban and Cohen (1979), has been widely used in creating gene fusions and then analyzing (i) the expression of various genes of *E. coli* and other enterobacteria (Débarbouillé and Schwartz, 1979; Smith and Gots, 1980; Mulligan *et al.*, 1982), and *P. putida* (Shapiro, 1984) and (ii) the genetics of the transport of periplasmic and outer membrane proteins (Silhavy *et al.*, 1979; Michaelis and Beckwith, 1982). This specialized phage carries the structural genes of the *E. coli* lactose operon without the lactose promoter, hence lactose genes are not expressed in this state. Upon integration of the phage in proper orientation into a transcribed region, lactose genes are hooked to the adjacent promoter regulatory region which then allows the expression of the lactose genes, a product of which, such as β-galactosidase, can be measured readily. The insertion of the phage into a gene causes a mutation which, because of its acquired Lac$^+$ phenotype, can be selected readily in a population of Lac$^-$ cells.

In addition to the Mu-*lac* transcription fusion forming phage [Mu dI1 (Ap *trp'CBA' lac'ZYA*); 37.2 Kb], Casadaban and Chou (1984) constructed a *lac* transcription-*lacZ* translation fusion forming phage [Mu dII301 (Ap *lac'ZYA*); 35.6 Kb]. While these bacteriophages are defective (d) for Mu phage growth, they are capable of intracellular transposition (Castilho *et al.*, 1984). These two fusion forming phages differ in *lac* sequences on one side of *lacZ* allowing either transcription fusion or transcription and translation fusion. The transcription of Mu dI1 (Ap *trp'CBA lac'ZYA*) fusion phages starts from the target gene and the translation of *lacZYA* appears to be initiated from or within the *trpA* gene. In contrast, transcription and translation in Mu dII301 proceeds from the target gene in the *lacZ* gene to produce enzymatically active hybrid proteins. Such *lac* fusions which produce hybrid proteins with β-galactosidase (the product of the *lacZ* gene) at the carboxyl terminal end, have many applications including studies of gene regulation and protein localization and in purifying polypeptide segments of interest.

Casadaban and associates (Castilho *et al.*, 1984) also subsequently constructed a series of mini-Mu-*lac* gene fusion elements which offer several advantages over the Mu dI1 and Mu dII301 fusion forming phages described above: (1) the new elements (viz. Mu dII1681; Mu dII1734; Mu dII4041) carry an additional drug-resistant marker (e.g. Km^r); (2) some (Mu dI1734 and Mu dII1734) lack the Mu transposition genes but the function can be complemented by, for example, resident Mu *c*ts prophage, producing insertions that remain stable in the absence of transposases; and (3) the relatively small size of the new elements (e.g. Mu dII1681, 14.2 Kb; Mu dII1734, 9.7 Kb; and Mu dII4041, 7.5 Kb) facilitates mutagenesis of plasmids. Castilho *et al.* (1984) have also described in detail an efficient procedure for selecting mini-Mu insertions into plasmids. Since Mu-*lac* transposition and gene fusion have been detected in *A. tumefaciens* and in *P. aeruginosa* (Castilho *et al.*, 1984) and since Mu-receptive (sensitive) strains are already available in various *Erwinia* spp. (see below), these gene fusion elements should have a general application in many Gram negative phytopathogenic bacteria.

Mu-derived transposons that utilize *lux* (genes for light production) in creating gene fusions are also available. Engebrecht *et al.* (1985) constructed mini-Mu-*lux* transposons which carry *lux* without the native promoter, the transposase genes (*A* and *B*), the gene for repressor of transposition (*c*, a temperature sensitive allele) of Mu dII1681, and either Km^r or Tc^r. In these transposons, the *lux* DNA (*luxE luxB luxA luxD luxC*) is oriented to align the target gene transcription with that of *lux*. Thus, insertion of this transposon within a gene inactivates the gene function and, when inserted in the proper orientation, fuses the target gene transcription with that of *lux* resulting in light production. Light produced by bacterial cells carrying fused *lux* genes can be quantitatively measured by means of X-ray or photographic film or by a scintillation counter (Engebrecht *et al.*, 1985). This simple yet elegant gene fusion procedure, already proven effective in *E. coli* and *Vibrio fischeri* (Engebrecht *et al.*, 1985), may facilitate studies of gene regulation in bacteria that are Mu-sensitive or of phytobacterial genes cloned into *E. coli*. A preliminary report suggests the usefulness of a *lux* delivery vector usable in *Agrobacterium*, *Erwinia*, *Pseudomonas* and *Xanthomonas* spp. (Shaw *et al.*, 1985).

Until recently, the application of the Mu-derived gene fusion elements was restricted to *E. coli* because of the limited host range of the specialized phage. However, considering its potential uses, the host range has been broadened by using various genetic approaches and bacteriophage derivatives (Csonka *et al.*, 1981; Gowrishankar and Pittard, 1982; Holley and Foster, 1982; Lee *et al.*, 1980; Rosenfeld and Brenchley, 1980).

Two approaches of the application of Mu-mediated gene fusions in the genetics of *Erwinia* spp. merit discussion. The procedure developed by Csonka *et al.* (1981) entailed the construction of an *E. coli* double lysogen carrying a defective Mu *c* ts dl (Ap^r, *lac*) phage (with deletions in host range genes) and a Mu-P1 helper phage (Mu hP1) that conferred the P1 host range, preparation of phage lysate from the double lysogen, infecting a P1-sensitive (*galE*), restriction-deficient, modification-proficient strain of *S. typhimurium*, and isolating a double lysogen carrying Mu d1 and Mu hP1 phages. Induction of such strains yielded lysates which generated gene fusions (for example *pyr*::*lac*) in *S. typhimurium*. Using a similar approach Jayaswal *et al.* (1984) isolated an array of auxotrophic mutants and demonstrated the occurrence of *his*::*lac* fusions in *E. carotovora* subsp. *carotovora*. Subsequently, Handa (1984, 1985) obtained Mu-insertion mutants of *E. carotovora* subsp. *carotovora* that were pectolytic but nonpathogenic (see the section on pectolytic enzymes for additional details). While the stability of Mu-insertion mutations, the linkage between Mu(Ap^r) and mutational phenotypes, and the physical state of Mu insertions are not yet known, the findings nevertheless suggest the potential use of the gene fusion approach in studies of the pathogenicity factor of this and possibly other *Erwinia* spp.; for example, *E. chrysanthemi*, where Mu-sensitive strains are available (Faelen *et al.*, 1981; Expert and Toussaint, 1985). An alternative approach entails the use of the plasmid, F$^+$::Mu d(Ap^r *lac*) constructed by Lee *et al.* (1980). The procedure involves transfer of the plasmid to appropriate recipients by selecting for Ap^r, partial induction of Mu d(Ap^r *lac*) by growing at 37°C, and then looking for appropriate mutants that can utilize lactose. Since various *Erwinia* spp. are receptive to F plasmids (primes), the use of F$^+$::Mu d(Ap^r *lac*) should allow construction of gene fusions and thereby facilitate studies on the expression of virulence genes in infected plants or plant tissue.

The collection of transpositional mutations, detected by loss of one or more functions, constitutes only the first step in genetic analysis of a phenotype. Subsequent studies should characterize the mutation both genetically and physically. It is essential to establish the linkage between the transposon marker and the gene function lost due to the insertion of the Tn element. This can be accomplished by recombination experiments to show the inheritance of the Tn or selected marker and the concomitant loss of gene function producing the mutational phenotype. Near isogenic lines constructed in this way should also prove useful in pathophysiological investigations.

The importance of physical characterization of putative Tn-insertion mutations is illustrated by the findings with *R. meliloti* and *E. chrysanthemi*. In *R. meliloti*, Meade *et al.* (1982) obtained Tn*5* insertion mutations using the R plasmid, pJB4JI (see page 21). The use of Mu and Tn*5* radioactively labeled probes showed that while certain mutations carried the Tn*5* sequence (bona fide Tn*5* mutants), others possessed both Mu and Tn*5* sequences. Similar observations have been noted in *E. chrysanthemi* (A. K. Chatterjee and K. K. Thurn, unpublished). Since the bacteriophage Mu behaves as a transposon (Calos and Miller, 1980; also see Table VI), the findings are not too surprising. In *E. chrysanthemi*, certain Tn*5* insertion mutants carried two copies of Tn*5* DNA separated by intervening *E. chrysanthemi* DNA which lacked the *Eco*RI site but had *Kpn*I and *Sal*I sites (Chatterjee *et al.*, 1985b) (*Eco*RI and *Kpn*I do not cut Tn*5*, while *Sal*I cuts Tn*5* once; Jorgensen *et al.*, 1979). In this case, erroneous conclusions concerning the physical state of Tn*5* could have been drawn based solely upon the characterization of *Eco*RI genomic fragments. The results show that the genetic constitution of Tn-insertion mutants needs to be analyzed rigorously prior to their application in physiological, phytopathological, or recombinant DNA research.

D. Resident Plasmids

As stated above, various pathogenic determinants of animal and human pathogenic bacteria are located on extrachromosomal DNA capable of autonomous replication. For example, in pathogenic *E. coli*, enterotoxin and hemolysin production and cell surface properties are plasmid determined;

likewise, enterotoxin production in *Staphylococcus aureus* is controlled by a plasmid (Elwell and Shipley, 1980). Such findings raised the question whether certain virulence determinants of plant pathogens might also be specified by plasmids. Intensive investigations with *A. tumefaciens* have shown that the Ti plasmid determines tumorigenicity, synthesis and catabolism of opines, exclusion of the bacteriophage AP1, sensitivity to bacteriocin (agrocin 84), and bacterial host specificity and sexuality (see Nester and Kosuge, 1981; Kahl and Schell, 1982; Zambryski *et al.*, 1983, for details). Comai and Kosuge (1980) demonstrated an association between the presence of a 34-MDa plasmid in a strain of *P. syringae* pv. *savastanoi* and bacterial production of the growth hormone, indoleacetic acid (IAA). The convincing evidence for this role included the inability of cured strains to produce IAA, and the restoration of the IAA^+ phenotype upon the reintroduction of the plasmid DNA. Comai and Kosuge (1982) subsequently cloned a fragment of pIAA1 encoding tryptophan monooxygenase (*iaaM*) into the plasmid vector, RSF1010, and it was expressed in *E. coli*. Additional plasmids were later found to code for IAA; these were larger (48 MDa; Comai *et al.*, 1982). All oleander strains contained IAA plasmids; olive and privet strains did not (Comai *et al.*, 1982). In the latter case, with the exception of an olive strain (EW1017) (Surico *et al.*, 1984), it seems likely that the genes are chromosomally located. Studies in progress (T. Kosuge, personal communication) should reveal whether or not additional genes required in IAA synthesis such as *iaaH* are located on plasmids and if they are organized in an operon. Recent studies of T. Kosuge and associates implicate plasmid(s) in the host range of *P. syringae* pv. *savastanoi* and the involvement of cytokinin in abnormal tissue proliferation (Surico *et al.*, 1984). If confirmed, these findings would parallel those in *A. tumefaciens* where host range genes and auxin–cytokinin ratio (which in turn determines the tumor morphology) are specified by the Ti plasmid (Nester and Kosuge, 1981; Kao *et al.*, 1982; Akiyoshi *et al.*, 1983).

Stall *et al.* (1984) have shown that a plasmid in *X. campestris* pv. *vesicatoria* confers resistance to copper, host specificity, and sexual proficiency. Aside from explaining pathogenic potential of this bacterium, the observation is an example of the evolution of ecologically desirable traits. In addition to learning about the structure and function of these genes, it would also be of interest to determine whether or not the gene for Cu^{2+} resistance and host specificity function as transposons.

Plasmids have been detected in various phytopathogenic species of bacterial groups: *Agrobacterium* (Nester and Kosuge, 1981), *Corynebacterium* (Gross *et al.*, 1979; Murai *et al.*, 1980; Lawson *et al.*, 1982; Hendrick *et al.*, 1984; Mogen *et al.*, 1985), *Erwinia* (Chatterjee and Starr, 1973b; Gibbins *et al.*, 1976; Coplin and Rowan, 1978; Panopoulos *et al.*, 1978; Schukin *et al.*, 1980; Sparks and

Lacy, 1980; Coplin *et al.*, 1981a,b; Gantotti *et al.*, 1981; Kado and Liu, 1981; Gantotti and Beer, 1982; Thiry and Faelen, 1984), *Pseudomonas* (Curiale and Mills, 1977; Gonzalez and Vidaver, 1978, 1979, 1980; Panopoulos *et al.*, 1978; Gantotti *et al.*, 1979; Comai and Kosuge, 1980; Currier and Morgan, 1981, 1983; Jamieson *et al.*, 1981; Kado and Liu, 1981; Quant and Mills, 1981; Sato *et al*, 1981; Szabo *et al.*, 1981; Vidaver et al., 1981; Zischek *et al.*, 1981; Piwowarski and Shaw, 1982; Rosenberg *et al.*, 1982; Malik *et al.*, 1984; Sato, 1984; Bender and Cooksey, 1985; Morales and Sequeira, 1985a; Willis *et al.*, 1986), *Xanthomonas* (Kado and Liu, 1981; Stall *et al.*, 1984; Civerolo, 1985; Haas *et al.*, 1985; N. W. Schaad, personal communication), and *Spiroplasma* (Ranhand *et al.*, 1980; Archer *et al.*, 1981). Thus, the presence of plasmid(s) in phytopathogenic bacteria is a common rather than a rare occurrence. In cases where plasmid(s) have not been readily detected, utmost care needs to be exercised to rigorously eliminate the possibility of the presence of unusually large species of extrachromosomal DNA, as detected in *A. tumefaciens* (Zaenen *et al.*, 1974; Currier and Nester, 1976) and *P. solanacearum* strains (Zischek *et al.*, 1981; Rosenberg *et al.*, 1982; Morales and Sequeira, 1983). In *P. solanacearum*, Currier and Morgan (1981) failed to detect plasmid DNA in 19 of 27 strains examined; plasmids of variable mass (3.5 to 250 MDa) were detected in the remainder. In general, similar results were obtained by Morales and Sequiera (1985a), except that they detected >500 MDa plasmids in several strains. They also raised the possibility of using a small (7.5-Kb) plasmid, pWI 15, after appropriate modifications, as a cloning vector for *P. solanacearum*.

As indicated previously, there is conclusive evidence for plasmid involvement in bacterial phytopathogenicity in some cases, but not in others. However, the presence of a 18-MDa plasmid in virulent strains and its absence in avirulent strains implicates it in the plant pathogenicity of *X. campestris* pv. *glycines* (Haas *et al.*, 1985) Initial studies with both *P. syringae* pv. *syringae* (Gonzalez and Vidaver, 1979) and *P. syringae* pv. *phaseolicola* (Gantotti *et al.*, 1979) implicated plasmids in the production of syringomycin and phaseolotoxin, respectively, but more recent investigations (Gonzalez *et al.*, 1981b, 1984; Jamieson *et al.*, 1981; Currier and Morgan, 1983) have shown evidence to the contrary (see Section IV.D. on phytotoxins for additional details).

Despite lack of evidence for plasmid involvement in pathogenicity in these cases, the plasmids of these and other phytopathogenic bacteria may well be important in their ecological behavior, including plant pathogenicity. For instance, a phenotype that could directly affect pathogenicity is the bacterial cell surface (see Section IV.A.) Altered bacteriophage and bacteriocin susceptibility in strains of *A. tumefaciens* and *P. syringae* pv. *syringae* "cured" of resident plasmid(s) tentatively suggest cell surface changes due to plasmids. Plasmid-mediated changes in LPS and OM proteins have been noted in *E. coli*,

R. trifolii, and *S. typhimurium* (Iyer *et al.*, 1978; Hoffman *et al.*, 1980; Russa *et al.*, 1982). Plasmid acquisition by members of the *C. michiganense* group changes EPS structure and decreases virulence (Vidaver *et al.*, 1985). The colonization factor of *E. coli* (Elwell and Shipley, 1980), form I cell surface antigen of *Shigella sonnei* (Kopecko *et al.*, 1980), virulence antigens of *Yersinia* spp. (Gemski *et al.*, 1980; Ben-Gurion and Shafferman, 1981; Portnoy *et al.*, 1983, 1984), virulence (invasion of epithelial cells) of *S. flexneri* 2a (Maurelli *et al.*, 1984), and increased cell surface charge and hydrophobicity of *Y. enterocolitica* (Lachica and Zink, 1984), are also plasmid determined.

In examining plasmid-specified phenotypes for ecological significance in phytopathogenic bacteria, attention should be paid to traits that may determine bacterial survival under seemingly adverse conditions. After all, most phytopathogenic bacteria are facultative pathogens, and they must exist in nature in the free-living state, i.e., in the absence of the plant host. Plasmid-specified Cu^{2+} resistance in *X. campestris* pv. *vesicatoria* (see above) and *P. syringae* pv. *tomato* (Bender and Cooksey, 1985) may protect the bacterium against the commonly used copper-based biocides; likewise arsensate/arsensite resistance in *C. flaccumfaciens* subsp. *oortii* may protect the bacterium against commonly used arsenicals (Hendrick *et al.*, 1984). In addition, plasmids of *Spiroplasma* and other fastidious bacteria might code for properties relative to survival in insects.

Many plasmids code for the transformation of various substrates, such as unusal hydrocarbons (Chakrabarty, 1976), pesticides (Pemberton and Fisher, 1977; Chakrabarty, 1980), and heavy metals (Summers and Silver, 1978; Robinson and Tuovinen, 1984; Williams and Silver, 1984). Thus, the possibility should be considered that certain resident plasmids of phytopathogenic bacteria might determine the metabolism of substrates available to these bacteria in their natural habitats, such as opines, the biosynthesis and catabolism of which are coded for by Ti plasmids of *A. tumefaciens* (Nester and Kosuge, 1981), or heavy metals, such as arsenic (Hendrick *et al.*, 1984) and copper (Stall *et al.*, 1984).

Plasmid-specified resistance to UV light also remains an attractive possibility, particularly in those phytopathogens (for example *P. syringae* pv. *syringae*) that are predominantly epiphytic. Certain enterobacterial and pseudomonad plasmids have been known to confer resistance to UV light (Falkow, 1975). However, plasmids can also confer UV sensitivity (Lehrbach *et al.*, 1978).

Plasmids of phytopathogenic bacteria might code for bacteriocins or represent genomes of temperate bacteriophages. In nontumorigenic *A. tumefaciens* (syn. *A. radiobacter*), *E. coli*, *E. carotovora* subsp. *carotovora* strain 379 and other bacteria, bacteriocin production is determined by plasmids (Konisky, 1978; Slota and Farrand, 1982; Ward and Copeman, 1985). The genome of the temperate phage, P1, exists in an extrachromosomal state in *E. coli* (Ikeda and Tomizawa, 1968). Many phytopathogenic bacteria are known to be lysogenic and producers of bacteriocins or bacteriocin-like substances. Vidaver (1976) has reviewed the ecological significance of bacteriocinogeny. Temperate phages of phytopathogenic bacteria could alter the phenotype of cells by altering cell surface properties as in *Salmonella* spp. (see Section IV.A.2.a.v.) code for virulence factor(s) as in *C. diphtheriae* (Pappenheimer, 1977) or alter the genetic constitution by mediating gene transfer.

Resident plasmids can determine sexuality, i.e., mediate conjugation (see below for the details of this mode of gene transfer) in several phytopathogenic bacteria. These include *A. tumefaciens* (Genetello *et al.*, 1977; Kerr *et al.*, 1977; Ellis *et al.*, 1982), *C. flaccumfaciens* subsp. *oortii* (Hendrick *et al.*, 1984), *E. herbicola* (Chatterjee and Starr, 1973b), *E. stewartii* (Coplin and Rowan, 1978; Coplin *et al.*, 1981a), *E. uredovora* (Thiry and Faelen, 1984), *P. syringae* pv. *mori* (Sato, 1984), *P. syringae* pv. *phaseolicola* (Poplawsky and Mills, 1983), *P. syringae* pv. *syringae* (Gonzalez *et al.*, 1981), *P. syringae* pv. *tabaci* (Sato *et al.*, 1981), *P. syringae* pv. *tomato* (Bender and Cooksey, 1985) and *X. campestris* pv. *vesicatoria* (Stall *et al.*, 1984). Sexuality in the genetic evolution of phytopathogenic bacteria is probably of considerable significance. Since IS or IS-like elements occur in plasmids and chromosomes of several bacteria (see above and Table V), it is likely that plasmids of phytopathogenic bacteria contain similar genetic elements. The T-DNA region of the Ti plasmid can be considered a transposon (Zambryski *et al.*, 1983) but whether an IS-like element is present is not yet certain. The presence of insertion sequences (IS*51*) was detected on plasmids as well as chromosomes of *P. syringae* pv. *savastanoi* (Comai and Kosuge, 1983). Similar DNA sequences may also be present in the genomes of *P. syringae* pv. *phaseolicola* (Szabo and Mills, 1984). Such translocatable elements, by causing insertions, inversions, or deletions, could increase the potential variability of pathogens. Such a potent mechanism of genetic variation, primarily by reassortment of the existing gene pool, would confer a high degree of plasticity onto an organism, with the potential to counteract an altered inimical host environment.

In this context, Szabo *et al.* (1981) found that a 98-MDa resident plasmid in a strain of *P. syringae* pv. *phaseolicola* integrated into and excised from the bacterial chromosome. Moreover, the data (Szabo *et al.*, 1981; Curiale and Mills, 1982) strongly suggested that the excision of the plasmid was imprecise. While the phenotype(s) encoded by the 98-MDa plasmid is not yet known, the findings nevertheless reveal interactions between the chromosomal and plasmid genomes that result in genomic rearrangements. Such reassortment of genetic material could have profound consequences on bacterial physiology and ecology. Campbell (1981) has discussed the evolutionary significance of extrachromosomal (viral as well as plasmid) DNA, transposons, and insertion sequences.

Many phytopathogenic as well as animal pathogenic bacteria lose virulence upon repeated cultivation on artificial media and regain virulence upon passage through natural or surrogate hosts. Thus, the pathogenic property of many bacteria is extremely variable; bacteria possessing virulence traits are selected by the host, whereas such traits, apparently of little consequence on rich artificial media, can be counterselected under this condition. Of the various possible genetic mechanisms, alteration of gene sequence (hence expression) by genome rearrangement (e.g., site-specific recombination; Simon *et al.*, 1980; Silverman and Simon, 1983; Plasterk and Van de Putte, 1984) mediated by IS or IS-like elements in plasmid or chromosomal genomes could account for such variability, as well as other phenotypic changes noted frequently, such as the occurrence of capsular and noncapsular forms in *P. solanacearum* (Kelman, 1954). Indeed, the observations of Schell and Vinson (1985) are consistent with the notion that the virulence of *P. solanacearum* may be associated with a cluster of genes that were coordinately controlled by a genetic switching mechanism. A model (Fig. 1), proposed by Weiss and Falkow (1984) for the regulation of virulence-associated genes in the human pathogen, *Bordetella pertusis*, may also apply to *P. solanacearum* and other phytopathogenic bacteria. An attractive feature of this model is that it takes into account genetic changes as well as environmental signals in the expression of virulence.

While the plasmids of phytopathogenic bacteria have been studied in some detail, relatively little or no attention has been paid to chromosomal genes. Plasmid studies have been valuable in elucidating the structure and functions coded by their genomes, but both plasmid and chromosomal genomes should be studied to understand complementary interactions between these genomes and their physiological and ecological effects. Plasmids of phytopathogenic bacteria probably coevolved with the chromosomal genome (Mills and Gonzalez, 1982). One implication of such coevolution is that the replicons have adapted to a common replicative machinery and complement some cell functions. Several lines of evidence support such a hypothesis. For example, in *A. tumefaciens* and *P. syringae* subsp. *savastanoi*, plasmids seem to carry out specialized functions that complement functions of chromosomal genes. Bacterial attachment, an early stage in the development of crown gall disease, seems to be determined by a chromosomal gene(s) of *A. tumefaciens* (Nester and Kosuge, 1981; Douglas *et al.*, 1982). In *P. syringae* pv. *savastanoi*, tryptophan, the precursor of IAA, is apparently produced by enzymes specified by chromosomal genes.

Another hypothesis of plasmid evolution is that certain phytopathogenic

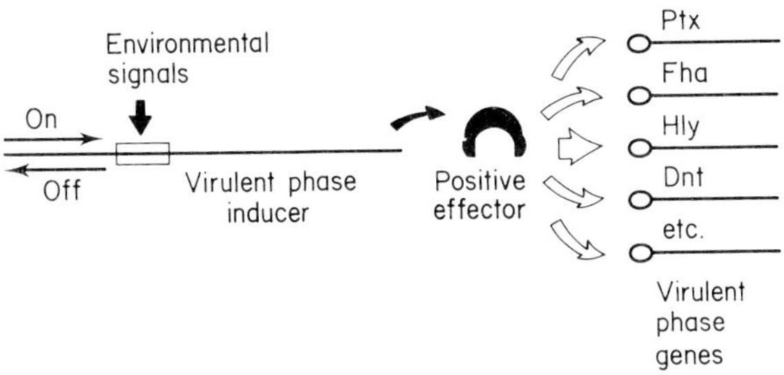

Fig. 1. Model for regulation of the virulence-associated genes by a positive effector. The lines represent the DNA coding regions for the regulatory gene and the virulence-associated genes. The circles represent the site of action for the common regulatory element. The *trans*-acting positive effector could be a protein encoded by the virulent-phase inducer gene, or this gene could encode an activity which synthesized a product required for the expression of the virulence-associated genes, analogous to cyclic AMP. Environmental signals, such as temperature, Mg^{2+} ions, or nicotinic acid, could modulate the expression of the positive effector. The molecular mechanism of ON to OFF switch of phase variation is unknown, but is diagrammed here as an invertible sequence as a working hypothesis. Abbreviations: Ptx, pertussis toxin; Fha, filamentous hemagglutinin; Hly, hemolysin; Dnt, dermonecrotic toxin; etc., other virulence factors. (Reproduced from Weiss and Falkow, 1984.)

bacteria acquired extrachromosomal DNAs that escaped the host's restriction mechanisms, and have since persisted in cells as "endoparasites." Nevertheless, it seems more likely that resident plasmids are important in the biology of phytopathogenic bacteria, coding for factors required in host associations or for functions that enable better survival in nature. In contrast to the ecology of R factors in animal and human pathogenic bacteria (Bennett and Richmond, 1978), no studies, with the possible exception of studies of certain Ti plasmids, have been made to compare the ecological behavior (i.e., survival, dissemination, and pathogenicity in nature) of phytopathogenic bacteria harboring or lacking resident plasmids. The segregation and structural modification(s) of resident plasmids in bacterial cells growing within plant tissues and on artificial media also remain unknown.

E. Genetic Systems

Gene transfer systems are essential in analyzing the genetic and biochemical bases of bacterial plant pathogenicity and other functions. In bacteria, gene transfer is brought about by three processes: (1) *conjugation*, where DNA transfer occurs following a physical contact between the donor and the recipient cells; (2) *transformation*, in which the naked DNA (of the donor) is taken up by the recipient; and (3) *transduction*, where a bacteriophage incorporates donor DNA and transfers it to the recipient bacterium. The mechanics of these processes, and methodological details, have been discussed by Bennett and Richmond (1978) and Curtiss (1981), respectively. All known modes of gene transfer have been reported in various phytopathogenic bacterial species (Table VII). While conjugation and transformation have been reported in several species, transduction has been convincingly reported only in *E. chrysanthemi* (Chatterjee and Brown, 1980; Resibois *et ai.*, 1984) and *P. syringae* pv. *syringae* (Nordeen and Currier, 1983).

Protoplast fusion has proved to be a useful tool for the study of genomic interactions (recombination) in various Gram-positive bacteria (such as *Bacillus* and *Streptomyces* species) as well as in a few Gram-negative bacteria (such as *E. coli* and *Providencia alcalifaciens*) (Hopwood, 1981). While the success of protoplast fusion in plant pathogenic bacteria is unknown, it may prove valuable in genetic studies on Gram-positive plant pathogens of *Corynebacterium* and *Streptomyces* species. Interested readers should consult articles by Ferenczy (1981) and Hopwood (1981).

1. Conjugation

Gene transfer by conjugation requires the presence of a sex (fertility) factor in one of two states: "free" as a plasmid (extrachromosomal), or integrated into the chromosome. All conjugative plasmids carry in their genomes *tra* (transfer) genes. Aside from self-transfer, some conjugative plasmids can mobilize chromosomal markers; this property is known as *chromosome mobilization ability* (Cma; Holloway, 1979). If conjugative plasmids integrate into the chromosome, they may produce a strain that can bring about a high frequency of recombination; such strains are designated as Hfr strains (Hayes, 1968).

In matings between an Hfr (male) donor and the recipient (female) bacterium, chromosome transfer occurs in a polarized manner with a fixed origin of gene transfer (located in the sex factor) and a terminus. Since the mating pair formed between the donor and recipient cells is rather fragile, they break spontaneously and DNA transfer then stops; this gives rise to a gradient in the frequency of transfer of chromosomal markers (Hayes, 1968; B. D. Davis *et al.*, 1980). This makes it possible to deliberately interrupt matings and determine the kinetics of entry of markers into the recipient as a function of time of mating (Fig. 2). This allows mapping of the chromosome in time units, i.e., in minutes (Bachmann *et al.*, 1976; Bachmann, 1983; Sanderson and Roth, 1983). The entire chromosome of the Hfr donor, however, can be transferred to the recipient depending upon mating being

Table VII. Indigenous characters (genomes) transferred and modes of transfer in phytopathogenic bacteria

Bacterial species	Plasmid genome, chromosomal loci, or bacteriophage DNA transferred[a]	Mode of transfer	References
Agrobacterium tumefaciens	Ti plasmid[b]	Conjugation (RP4)	Chilton *et al.* (1976)
		Conjugation	Genetello *et al.* (1977); Kerr *et al.* (1977); Ellis *et al.* (1982)
	ade$^+$, *chl*^{r}, *cys*$^+$, *glu*$^-$, *his*$^+$, *ile*$^+$, *ilv*$^+$, *leu*$^+$, *met*$^+$, *pan*$^+$, *phe*$^+$, *pyr*$^+$, *rif*^{r}, *spc*^{r}, *str*, *thi*$^+$, *trp*$^+$, *ura*$^+$, *val*$^+$	Conjugation (R68.45, pDP37 and pDP38)	Hamada *et al.* (1979); Bryan *et al.* (1982); Hooykaas *et al.* (1982a); Pischl and Farrand (1982, 1983, 1984)
	Ti plasmid	Transformation	Holsters *et al.* (1978)
	Virulence	Transformation	Klein and Klein (1953, 1956)
	LR-4	Transfection	Milani and Heberlein (1972)
	pS8cc186	Transfection	Holsters *et al.* (1978)
Corynebacterium flaccumfaciens subsp. *oortii*	*asi*^{r}, *asa*^{r}	Conjugation (pDG101)	Hendrick *et al.* (1984)
Erwinia amylovora	*arg*$^+$, *cys*$^+$, *gal*$^+$, *his*$^+$, *ilv*$^+$, *leu*$^+$, *met*$^+$, *rbs*$^+$, *ser*$^+$, *thr*$^+$, *trp*$^+$, *vir*$^+$	Conjugation (Hfr strains; pULB113)	Chatterjee and Starr (1973a, 1978a); Pugashetti and Starr (1975); Pugashetti (1976); Pugashetti *et al.* (1978); Ross *et al.* (1984); Chatterjee *et al.* (1985a)
E. carotovora subsp. *atroseptica*	*arg*$^+$, *gua*$^+$, *leu*$^+$, *lys*$^+$, *pur*$^+$, *trp*$^+$	Conjugation (pULB113)	Ross *et al.* (1984); Chatterjee *et al.* (1985a)
E. carotovora subsp. *carotovora*	*arg*$^+$, *gua*$^+$, *his*$^+$, *leu*$^+$, *lys*$^+$, *out*$^+$, *pur*$^+$, *trp*$^+$	Conjugation (R68.45; pULB113)	Pérombelon and Boucher (1978); Ross *et al.* (1984); Zink and Chatterjee (1984a); Chatterjee *et al.* (1985a); also see Addendum

Table VII. (cont.)

Bacterial species	Plasmid genome, chromosomal loci, or bacteriophage DNA transferred[a]	Mode of transfer	References
E. chrysanthemi	*ade*$^{+}$, *arg*$^{+}$, *cys*$^{+}$, *exu*$^{+}$, *gal*$^{+}$ *gtu*$^{+}$, *his*$^{+}$ *idg*$^{+}$, *ile*$^{+}$, *ilv*$^{+}$ *kdg*$^{+}$, *leu*$^{+}$ *lys*$^{+}$, *mcu*$^{+}$, *met*$^{+}$ *pel*$^{+}$ *pro*$^{+}$, *pur*$^{+}$ *rif*^{r}, *ser*$^{+}$, *str*^{r}, *thr*$^{+}$, *thy*$^{+}$, *trp*$^{+}$, *ura*$^{+}$, *uxa*$^{+}$ *uxu*$^{+}$, *xyl*$^{+}$	Conjugation (Hfr strain; R plasmids R68.45, R100*drd*-56, Rts1, RK2, RP4, RP11–42, RP4::Tn*7*); pULB113	Chatterjee and Starr, (1977, 1978b); Guimares *et al.* (1978); Lacy (1978); Chatterjee (1980); Chatterjee *et al.* (1981a, 1985a); Prokulevich and Fomichev (1981); Kotoujansky *et al.* (1982); Schoonejans and Toussaint (1983); Ross *et al.* (1984)
	ade^{+}, *arg*$^{+}$, *his*$^{+}$, *idg*$^{+}$, *ilv*$^{+}$ *leu*$^{+}$, *rif*^{r}, *ser*$^{+}$, *thr*$^{+}$, *trp*$^{+}$, *ura*$^{+}$	Transduction (Erch-12)	Chatterjee and Brown (1980, 1981)
	arg^{+}, *gau*$^{+}$, *his*$^{+}$, *ile*$^{+}$, *leu*$^{+}$, *thr*$^{+}$, *thy*$^{+}$, *trp*$^{+}$	φEC2	Resibois *et al.* (1984)
E. stewartii	*arg*$^{+}$, *cys*$^{+}$, *his*$^{+}$, *ilv*$^{+}$, *lu*$^{+}$, *lys*$^{+}$, *pro*$^{+}$, *rif*^{r}, *ser*$^{+}$	Conjugation (pDC252–1 and Mu *c* ts62 interaction)	McCammon and Coplin (1982)
	pDC151[b]; pDC250[b]	Conjugation	Coplin and Rowan (1978)
Pseudomonas solanacearum	*met*$^{+}$, *nal*^{r}, *rif*^{r}, *str*^{r}, *trp*$^{+}$	Transformation	Liu (1973); Coplin *et al* (1974); Boucher and Sequeira (1978); Lê *et al.* (1978)
P. syringae pv. *glycinea*	*arg*$^{+}$, *his*$^{+}$, *ilv*$^{+}$, *leu*$^{+}$, *lys*$^{+}$, *nalA*$^{+}$, *orn*$^{+}$, *phe*$^{+}$, *pro*$^{+}$, *pur*$^{+}$	Conjugation (RP1, R6886)	Lacy and Leary (1976); Fulbright and Leary (1978)

P. syringae pv. *lachrymans*	*str*^{r}	Conjugation	Yano *et al.* (1979); Yano and Fujii (1983)
P. syringae pv. *morsprunorum*	*ilv*$^+$, *met*$^+$, *nic*$^+$	Conjugation (indigenous fertility system)	Errington and Vivian (1981)
P. syringae pv. *phaseolicola*	*met*$^+$, *phe*$^+$)	Conjugation (R18–1, RK2	Guimares *et al.* (1978)
	pEX8080	Conjugation	Poplawsky and Mills (1983)
P. syringae pv. *savastanoi*	*iaa*$^+$ (pIAA1) [c]	Transformation (RSF1010; pIAA1)	Comai and Kosuge (1980)
P. syringae pv. *syringae*	*arg*$^+$, *his*$^+$, *leu*$^+$, *met*$^+$, *phe*$^+$, *trp*$^+$	Conjugation (R68.45)	T. C. Currier (pers. comm.)
	pCG131::Tn*1* [b]	Conjugation	Gonzalez *et al.* (1984)
	arg$^+$, *aro*$^+$, *his*$^+$, *met*$^+$ *ser*$^+$ *trp*$^+$	Transduction (PSSY 15)	Nordeen and Currier (1983)
	pRO161 [c]	Transformation (RSF1010; pRO161)	Gross and Vidaver (1981)
	asp$^+$, *cys*$^+$, *leu*$^+$, *met*$^+$, *thr*$^+$, *val*$^+$	Transformation	Twiddy and Liu (1972)
P. syringae pv. *tabaci*	pBPW1::Tn*7* [b]	Conjugation	Sato *et al.* (1981)
P. syringae pv. *tomato*	*cup*$^+$	Conjugation	Bender and Cooksey (1985)
Spiroplasma citri	Bacteriophage	Transfection	McCammon *et al.* (1985)
Xanthomonas campestris pv. *campestris*	*gliv*$^+$	Conjugation (pRK2073::Tn*5*)	Turner *et al.* (1984)
X. campestris pv. *oryzae*	*leu*$^+$	Transformation	Yamasaki *et al.* (1966)
X. campestris pv. *phaseoli*	*cap*$^+$, *str*^{r}	Transformation	Corey and Starr (1957a,b)
X. campestris pv. *vesicatoria*	*cup*$^+$, host specificity	Conjugation (pXvCu)	Stall *et al.* (1984); Stall (1985)

[a] See Appendix I for marker designations. (Plasmids or phage are designated by letters and numbers.)
[b] Plasmids are conjugation proficient (*tra*$^+$).
[c] Plasmids were cotransformed with RSF1010.

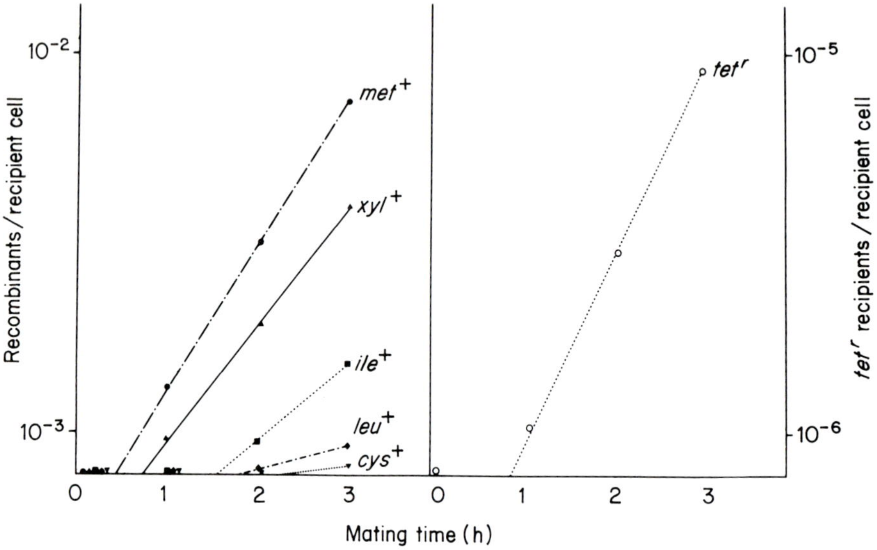

Fig. 2. The times of entry of *met* ⁺, *xyl* ⁺, *ile* ⁺, *leu* ⁺, *cys* ⁺, and *tet* ʳ from *Erwinia chrysanthemi* donor strain Hfrq to several auxotrophic recipient strains. For details see text and Kotoujansky *et al.* (1982). (Reproduced from Kotoujansky *et al.*, 1982.) Authors' designation for the donor was *E. carotovora* subsp. *chrysanthemi*.

carried out for an appropriate length of time and the efficiency of DNA transfer. In *E. coli* (Bachmann and Low, 1980), the entire chromosome can be transferred to the recipient in a 100-min mating period, while it requires about 300 min for an Hfr strain of *E. chrysanthemi* (Chatterjee and Starr, 1977) to transfer its chromosome.

Stable recombination between a naturally occurring conjugative plasmid and the host chromosome is a rare phenomenon, except between F-like plasmids belonging to incompatibility groups FI and FII and enterobacterial chromosomes. (*Incompatibility groups* are defined on the basis of interactions between plasmids—related plasmids fail to coexist in the same cell, hence are incompatible; Novick *et al.*, 1976.) Compared to other conjugative plasmids, F factor or F′ plasmids (*prime* [′] *plasmids* contain genomic segments) are unusual in that they can integrate into the chromosomes of a variety of enterobacteria such as *E. coli* (Low, 1972; Bachmann *et al.*, 1976), *S. typhimurium* (Sanderson *et al.*, 1972; Sanderson and Hartman, 1978), *E. amylovora* (Chatterjee and Starr, 1973a; Pugashetti, 1976; Pugashetti *et al.*, 1978), *E. chrysanthemi* (Chatterjee and Starr, 1977; Chatterjee, 1978; Prokulevich and Fomichev, 1981; Kotoujansky *et al.*, 1982), *Citrobacter freundii* (de Graaff *et al.*, 1974), and *Y. pseudotuberculosis* (Lawton *et al.*, 1968). However, the stability of the resulting Hfr strains is variable depending upon the bacterial species and strain. Recent studies have shown IS elements (Table V) on the F genome as well as on the *E. coli* chromosome. Integration of F most likely involves recognition between these sequences followed by reciprocal recombination (Curtiss, 1981). The formation of Hfr strains in *E. amylovora* and *E. chrysanthemi* suggests that their chromosomes have insertion sequence(s) recognized by the F genome. The presence of at least three different insertion sequences in the F genome (Table V) may account for the insertion of F or F-prime plasmids into chromosomes of bacteria that are taxonomically distant to *E. coli*.

As discussed above, homology between transposon sequences present in a plasmid and in the chromosome was exploited in the isolation of Hfr strains (Chumley *et al.*, 1979). Such Hfr-type donor strains in *E. chrysanthemi* were obtained by creating Tn*5* homology between an F′ plasmid (F′*ts lac*$^+$::Tn*10*::Tn*5*) and the chromosome (Chatterjee *et al.*, 1982; see Section III.C.2. for further details). Broad host range R plasmids, which also contain transposons, can also be used in the isolation of Hfr-type donors. Danilevich *et al.* (1978) observed that the insertion of Tn*1* into the *E. coli* chromosome sharply increased the frequency of the chromosomal integration of a temperature-sensitive replication mutant (pEG1) of the R plasmid, RP4. The *recA*, (e.g., a mutation causing a deficiency in the host recombination system) or recombination-dependent integration appeared to be directed by Tn*1* sequences present in the chromosome and the plasmid. Tn*5*-facilitated chromosome mobilization in *A. tumefaciens* was accomplished (Pischl and Farrand, 1983, 1984) by constructing kanamycin sensitive (Km^s) derivatives of R68.45 by cleavage with restriction enzymes, inserting Tn*5* into the Km^s plasmid, and introducing the Tn*5*-carrying plasmids into bacterial strains containing chromosomal Tn*5* insertions. Such plasmid-bearing strains mobilized chromosomal markers at a 100-fold higher frequency than previously noted with the R plasmid, R68.45 (Hamada *et al.*, 1979). One Tn*5*-carrying plasmid, pDP37, mediated polarized transfer from a single origin near *ilv* (Pischl and Farrand, 1984). However, the transmission gradients of a number of pDP37-containing donor strains, each differing at the site of the chromosomal Tn*5* insertion, revealed the occurrence of only two classes. The gradient of marker transfer for one class of donors was Ilv$^+$ > Rif^r > His$^+$ > Met$^+$, while for the other class it was His$^+$ > Rif^r > Ilv$^+$ > Met$^+$ (Pischl and Farrand, 1984). The apparent interaction of the plasmid with only two chromosomal segments implicates factors other than the mere presence of Tn*5* copies in plasmid–chromosomal interactions. Despite the uncertainty of the molecular aspects of interaction, it is apparent that a similar approach could be used in *Erwinia*, *Pseudomonas*, and *Xanthomonas*, as previous studies have shown that many of these phytopathogenic bacteria are receptive to RP4 and other P group plasmids (Panopoulos *et al.*, 1978; Lacy and Leary, 1979; Chatterjee and Starr, 1980; Vincent and Fulbright, 1983; Walsh *et al.*, 1984).

Since IS elements occur in various R plasmids including R68.45, the Cma of plasmids could be ascribed to transient cointegrate formation between the plasmid and host chromosome. The number of chromosomal genes transferred and the frequency of transfer would depend upon the relative stability of the cointegrates formed and the number of sites at which cointegrate formation can occur. In *E. chrysanthemi* strain EC16, R68.45 transferred markers located in different regions of the chromosome at about equal frequency and transferred small chromosomal segments (Chatterjee, 1978, 1980). Thus, R68.45 probably interacts with the *E. chrysanthemi* chromosome at various sites. Naturally occurring deletions of R68.45 in various *Erwinia* spp. (Currier and Morgan, 1982) have confirmed earlier observations of Leemans *et al.* (1980) and Riess *et al.* (1980), that duplicated regions of R68.45 are responsible for the Cma of the plasmid. Willets *et al.* (1981) suggest that the duplicated region of R68.45 constitutes a highly transposable insertion sequence, IS*21*, which could mediate cointegrate formation between R68.45 and the bacterial chromosome and account for Cma.

In various enterobacteria, including *Erwinia* spp., chromosome mobilization has been observed with coliphage Mu and plasmid cointegrates. Mu is known to behave as a transposon (Table VI), and mediate various types of chromosomal rearrangements such as deletions, inversions, fusions of two replicons, and translocation of genes into a plasmid (Toussaint *et al.*, 1977). Thus, Mu is a powerful tool in the genetics of *E. coli* and other enterobacteria such as *C. freundii* (de Graaff *et al.*, 1973) and *K. pneumoniae* (Bachhuber *et al.*, 1976; Rao, 1976). As a prelude to the realization of such possibilities, including gene isolation and gene mapping, bacteriophage Mu and R plasmid cointegrates (RP4::Mu *c*ts62, R68::Mu c^+ 445–447, or RK2::Mu *c*ts62) have been introduced into several phytopathogenic bacteria such as *E. carotovora* subsp. *carotovora*, *E. stewartii*, and *P. solanacearum* (Boucher *et al.*, 1977; Pérombelon and Boucher, 1978; Coplin, 1979; Murooka *et al.*, 1981) with the production of Mu phage. Mu-mediated chromosome transfer was noted in *E. carotovora* subsp. *carotovora* (Pérombelon and Boucher, 1978) and *E. stewartii* (McCammon and Coplin, 1982). In *E. stewartii*, McCammon and Coplin (1982) observed chromosome mobilization in a donor which was lysogenic for Mu *c*ts62 and carried the plasmid, pDC252.1, an indigenous *E. stewartii* plasmid derepressed for transfer and labeled with transposons Tn*1* and Tn*10*. Chromosomal markers were transferred at a 100- to 1000-fold higher frequency upon heat induction of the donor than in the absence of heat induction. The frequency of recombinant formation suggested multiple origins of transfer. Upon heat induction Mu may have transposed into the transmissible plasmid and into various sites within the *E. stewartii* chromosome. The presence of Mu sequences both in the plasmid and chromosome could then promote replicon interactions, resulting in an enhanced Cma. This hypothesis predicts interactions of Mu with genomes of this enterobacterial phytopathogen, which is further supported by the expression of RP4::mini-Mu plasmids in various *Erwinia* spp.

The availability of mini-Mu derivatives (Faelen *et al.*, 1979) and the RP4::mini-Mu plasmids (Van Gijsegem and Toussaint, 1982) now greatly facilitates the applications of Mu technologies (Faelen *et al.*, 1977) in the genetics of various bacterial species, including phytopathogens. In mini-Mu derivatives, the viral genomic segment, except for Mu ends and the Mu gene *A*, was deleted, thereby removing all the viral functions lethal to the bacterial host but retaining the property of most Mu-mediated chromosomal rearrangements (Faelen *et al.*, 1979). Subsequently, Van Gijsegem and Toussaint (1982) constructed RP4::mini-Mu derivatives, thereby expanding the host range of mini-Mu. RP4::mini-Mu plasmids mobilized chromosomal markers in *Alcaligenes eutrophus*, *E. amylovora*, *E. carotovora* subsp. *atroseptica*, *E. carotovora* subsp. *carotovora*, *E. chrysanthemi*, *E. coli*, and *P. fluorescens* (Van Gijsegem and Toussaint, 1982; Lejeune *et al.*, 1983; Schoonejans and Toussaint, 1983; Ross *et al.*, 1984; Chatterjee *et al.*, 1985a). The RP4::mini-Mu plasmid, pULB113, was particularly useful in the construction of primes (i.e., *in vivo* gene cloning) in all of these bacteria as well as in *Enterobacter cloacae*, *K. penumoniae*, *P. mirabilis*, and *S. typhimurium* (Van Gijsegem and Toussaint, 1982, 1983; Kadam *et al.*, 1984; Riley *et al.*, 1984; Ross *et al.*, 1984; Seeberg and Wiedemann, 1984; Sprenger and Lengeler, 1984; Van Gijsegem *et al.*, 1984; Chatterjee *et al.*, 1985a). (Some examples of phytobacterial genes cloned *in vivo* are given in Table VIII.) Since pULB113 primes can carry about 50- to 150-kb pairs of insert DNA, this system of *in vivo* cloning is functionally similar, and in some respects (viz. in complementation studies) even superior, to *in vitro* cosmid gene cloning (see Section III.F. for *in vitro* gene cloning).

It is also feasible to mobilize chromosomal markers by creating homology between a conjugative (R) plasmid and the chromosome. By using recombinant DNA technology (see below), Barth (1979) inserted small segments (from 0.05–0.34%) of the *E. coli* chromosome into the broad host range plasmid, RP4. Such recombinant plasmids, but not the parent plasmid, mobilized the *E. coli* chromosome in a polarized manner. The origin of transfer, deduced from transfer frequency, was different for each of the recombinant plasmids. The Cma of the plasmids was abolished when the donor strain carried a *recA*

Table VIII. Some examples of phytobacterial genes (genomes) cloned by *in vivo* and *in vitro* genetic technologies.

Bacterial species	Genes or plasmid genome clones[a]	Mode of cloning	References
Agrobacterium tumefaciens and *A. rhizogenes*	Ti plasmid and chromosomal fragments	*In vitro*	Nester and Kosuge (1981); Hooykaas and Schilperoort (1984); Nester *et al.* (1984)
Erwinia amylovora	*arg*$^+$, *gal*$^+$, *his*$^+$, *leu*$^+$, *met*$^+$, *thy*$^+$	*In vivo*[b]	Chatterjee *et al.* (1985a)
	lpp$^+$, pathogenicity/HR	*In vitro*	Yamagata *et al.* (1981); Bauer and Beer (1985)
E. carotovora subsp. *atroseptica*	*arg*$^+$, *gal*$^+$, *his*$^+$, *leu*$^+$, *met*$^+$, *thy*$^+$	*In vivo*[b]	Chatterjee *et al.* (1985a)
E. carotovora subsp. *carotovora*	*arg*$^+$, *gal*$^+$, *his*$^+$, *leu*$^+$, *met*$^+$, *thy*$^+$	*In vivo*[b]	Chatterjee *et al.* (1985a)
	gal$^+$, *lac*$^+$, *leu*$^+$, *peh*$^+$, *pel*$^+$, *pro*$^+$, *prt*$^+$ *recA*$^+$, *xyl*$^+$	*In vitro*	Allen *et al.* (1984); Roberts *et al.* (1984); Zink and Chatterjee (1984a,b); Lei *et al.* (1985); Roberts *et al.* (1985); Willis and Chatterjee (1985); Willis *et al.* (1985); also see Addendum
E. chrysanthemi	*cel*$^+$, *clb*$^+$, *kdg*$^+$, *pel*$^+$, *uxa*$^+$, *uxu*$^+$	*In vivo*	Van Gijsegem and Toussaint (1983); Ross *et al.* (1984); Van Gijsegem *et al.* (1985)
	arg$^+$, *cel*$^+$, *clb*$^+$, *gal*$^+$, *leu*$^+$, *pel*$^+$, *pme*$^+$, *pro*$^+$, *prt*$^+$, *xyl*$^+$	*In vitro*	Barras *et al.* (1984a,b, in prep); Keen *et al.* (1984); Collmer *et al.* (1985a,b); Kotoujansky *et al.* (1985)

E. herbicola	*ina*$^{+}$; gene for yellow pigment	*In vitro*	Orser *et al.* (1983, 1985); Perry *et al.* (1985)
E. stewartii	*gal*$^{+}$, *leu*$^{+}$, *pro*$^{+}$, *recA*$^{+}$, *cps*$^{+}$, *wts*$^{+}$	*In vitro*	Coplin *et al.* (1984, 1985); Dolph and Coplin (1985)
Pseudomonas syringae pv. *glycinea*	Gene for race specificity (*rsi*$^{+}$); plasmid specifying coronatine production [c]	*In vitro*	Staskawicz *et al.* (1984a); Napoli *et al.* (1985); Willis *et al.* (1986)
P. syringae pv. *phaseolicola*	Gene for ornithine carbamoyltransferase; pathogenicity/HR; phaseolotoxin	*In vitro*	Panopoulos *et al.* (1984); Clements *et al.* (1985); Lindgren *et al.* (1985); Peet *et al.* (1985)
P. syringae pv. *savastanoi*	pIAA (*iaaM*$^{+}$; *iaaH*$^{+}$; *iaaL*$^{+}$)	*In vitro*	Comai and Kosuge (1982); Glass and Kosuge (1985); Yamada *et al.* (1985)
P. syringae pv. *syringae*	*flu*$^{+}$, *ina*$^{+}$, *rec*$^{+}$, *sid*$^{+}$; pathogenicity/HR	*In vitro*	Loper *et al.* (1983); Orser *et al.* (1983); Loper *et al.* (1984); Niepold *et al.* (1985b); Panopoulos *et al.* (1985)
Xanthomonas campestris pv. *campestris*	Pathogenicity, enzyme export (?)	*In vitro*	Daniels *et al.* (1984b, 1985)
X. campestris pv. *vesicatoria*	Plasmid genome [d]	*In vitro*	R. E. Stall (pers. comm.)
X. campestris pv. *malvacearum*	Avirulence	*In vitro*	Gabriel (1985)

[a] See Appendix I for marker designations.

[b] *In vivo* gene cloning was done by using the RP4::mini-Mu plasmid, pULB113 (Van Gijsegem and Toussaint, 1982). See text for the properties of this plasmid.

[c] An 8.3-kb plasmid appears to specify coronatine production in *P. syringae* pv. *glycinea*. See Table XIV for some of the properties of this toxin.

[d] This conjugative plasmid specifies host specificity as well as resistance to copper.

mutation. Thus, Cma appeared to result from interactions between DNA segments in common between the plasmid and chromosomal genomes. By using a similar approach, it should be possible to construct recombinant plasmids carrying segments of the chromosomes of phytopathogenic bacteria; such recombinant plasmids, depending on the size of DNA insertion, should prove useful not only in developing genetic systems but also in analyzing gene expression and gene function.

In phytopathogenic agrobacteria, erwinia, and pseudomonads, mobilization of chromosomal markers has been brought about by various R plasmids, and Hfr strains have been obtained in *E. amylovora* and *E. chrysanthemi* (see text and Table VII). Hfr strains have proved extremely useful in mapping chromosomal genes of *E. amylovora* and *E. chrysanthemi* by time of entry, by gradient of transmission, and by linkage data (Chatterjee, 1978; Chatterjee and Starr, 1980; Prokulevich and Fomichev, 1981; Kotoujansky *et al.*, 1982). However, the formation of Hfr strains in these *Erwinia* spp. was a rare event (Chatterjee and Starr, 1980) and did not show different directions of chromosome transfer, as in *E. coli* (Bachmann and Low, 1980) or in *Salmonella typhimurium* (Sanderson and Hartman, 1978). Thus, gene loci that were distal to the origin of transfer were difficult to map. This problem has now been alleviated to a great extent by the use of R plasmids that mobilize the *E. chrysanthemi* chromosome (Chatterjee *et al.*, 1981a, 1985a; Schoonejans and Toussaint, 1983; Van Gijsegem *et al.*, 1985b), the isolation of different Hfr strains (Kotoujansky *et al.*, 1982; Chatterjee *et al.*, 1982), and the use of generalized transducing bacteriophages (Chatterjee and Brown, 1980; Resibois *et al.*, 1984).

The phenomenon of *integrative suppression*, whereby inserted genes can dominate or control a prior phenotype, has not yet been exploited in the isolation of Hfr type donors in phytopathogenic bacteria. In *E. coli*, temperature-sensitive (*ts*) mutations in replication of the chromosome can be phenotypically suppressed due to integration of a plasmid (Nishimura *et al.*, 1971, 1973; Chandler *et al.*, 1977); chromosome replication is thus controlled by replication genes of the integrated plasmid. This approach with conjugative plasmids (such as F′ and R plasmids) enables a strong selection for Hfr type donors (Bennett and Richmond, 1978). Essentially similar results can be obtained with *ts* mutations affecting replication of conjugative plasmids, such as F′*ts* lac^+, F′*ts* gal^+, or R*ts*1. For example, Hfr strains in *E. coli* were obtained by selection of the plasmid-specified phenotype under nonpermissive conditions of 42°C (Beckwith *et al.*, 1966; Terawaki *et al.*, 1975; Yoshimoto and Yoshikawa, 1975). R*ts*1, a naturally occurring R plasmid thermosensitive in replication, enabled Guimares *et al.* (1978) to obtain *E. chrysanthemi* clones that apparently mobilized chromosomal markers in a polarized sequence.

Whether or not a stable integration of the plasmid (R*ts*1) occurred in the *E. chrysanthemi* chromosome was not clear. However, a similar approach should prove useful in studies of phytopathogenic bacteria that can grow at a temperature which prevents replication (maintenance) of an introduced plasmid.

The use of temperature-sensitive plasmids in the isolation of Hfr strains of *E. chrysanthemi* was clearly documented by Kotoujansky *et al.* (1982). Starting with an F′ plasmid (F′*ts lac*$^+$::Tn*10*) and a strain of *E. chrysanthemi* (isolated from *Saintpaulia*) capable of growth at 40°C, they obtained donor strains by selecting for clones that were Tc^r at the nonpermissive temperature (reflecting Tn*10* insertion). Data from interrupted mating experiments, gradient of marker transfer, and linkage between markers indicated that this Hfr strain had a different origin of transfer than a previously reported *E. chrysanthemi* Hfr strain resulting from the integration of F′*lac*$^+$ plasmids (Chatterjee and Starr, 1977; Prokulevich and Fomichev, 1981). The data now available (Chatterjee *et al.*, 1981a; Kotoujansky *et al.*, 1982) suggest a high homology in chromosomal gene order of *E. chrysanthemi* strains originating on such plant hosts as *Chrysanthemum*, *Dieffenbachia*, and *Saintpaulia*. Moreover, genetic homology also is evident between *E. chrysanthemi* and *E. coli* chromosomes. It will be of considerable interest to determine if recombinational homology also exists between strains of *E. chrysanthemi* originating from other plant hosts, and if the organization of the genes for plant virulence is similar in these strains.

R plasmids with Cma (Holloway, 1979; Van Gijsegem and Toussaint, 1982) are undoubtedly powerful tools in the genetics of Gram-negative phytopathogenic bacteria. Their use allows construction of donor strains of desired genetic background and a better estimate of linkage by performing reciprocal crosses (Chatterjee, 1980). In those phytopathogenic bacteria in which generalized transduction is not yet known, linkage data may be obtained by using R plasmids or other plasmids with a readily selectable phenotype. Moreover, R plasmids generally have a broad host range which would facilitate genetic analysis of strains that differ in host specificity (races) and other pathogenic properties. For example, linkage maps and the circularity of chromosomes of beneficial *Rhizobium* spp. have been delineated by the exclusive use of R plasmids (Meade and Signer, 1977; Beringer *et al.*, 1978). Linkage between various chromosomal markers of *A. tumefaciens* (Bryan *et al.*, 1982; Hooykaas *et al.*, 1982a; Pischl and Farrand, 1983, 1984), *E. amylovora* (Pugashetti *et al.*, 1978; Chatterjee and Starr, 1980), *E. carotovora* subsp. *carotovora* (Zink and Chatterjee, 1984a), *E. chrysanthemi* (Chatterjee and Starr, 1977; Chatterjee, 1980; Chatterjee and Brown, 1980; Chatterjee *et al.*, 1981a; Kotoujansky *et al.*, 1982; Prokulevich *et al.*, 1983; Schoonejans and Toussaint, 1983), and *P. syringae* pv. *glycinea* (Fulbright and Leary, 1978) has been estimated. Circularity of the *A. tumefaciens* and *E. chrysanthemi* chromosome has now been established (Hooykaas *et al.*, 1982a; Pischl and Farrand, 1984; Prokulevich *et al.*, 1984; G. R. K. Sastry, personal communication).

Further advances in chromosomal genetics of *A. tumefaciens*, *E. carotovora* subsp. *carotovora*, and *E. stewartii* are expected. In these cases, plasmids are known to mobilize chromosomal markers (Table VII). An indigenous fertility system detected in *P. syringae* pv. *morsprunorum* (Errington and Vivian, 1981; Table VII) should allow genetic studies of this phytopathogen. Genetic analysis of pathogenic factors in such important pathogens as *P. syringae* pv. *savastanoi*, *P. syringae* pv. *syringae*, and *P. syringae* pv. *phaseolicola* and probably several *X. campestris* pathovars would be facilitated with the development of efficient conjugational genetic systems. This would probably require testing the expression, including Cma, of various R plasmids in strains of these bacterial species and/or isolating plasmid derivatives with enhanced Cma. In *E. stewartii*, *P. syringae* pv. *syringae*, *P. syringae* pv. *tabaci*, *P. syringae* pv. *tomato* and *X. campestris* pv. *vesicatoria*, the finding of sex factor activity of indigenous plasmids needs to be extended to determine the interactions of the plasmids with host chromosomes and whether or not Hfr-type donor strains can be obtained. The recent finding (Hendrick *et al.*, 1984) of a conjugative plasmid in *C. flaccumfaciens* subsp. *oortii* suggests the possibility of mobilizing chromosomal segments in this and other phytopathogenic *Corynebacterium* spp.

2. *Transduction*

Bacteriophages active against most phytopathogenic bacteria are readily obtained from nature (Baigent *et al.*, 1963; Okabe and Goto, 1963; Vidaver, 1976; Chatterjee and Starr, 1980) with the possible exception of certain members of the "carotovora" group of the genus *Erwinia* (viz. *E. chrysanthemi*; Paulin and Nassan, 1978; Chatterjee and Brown, 1980). Most such bacteriophages are virulent or lytic in that infected cells die, liberating progeny bacteriophage particles. Many phytopathogenic bacteria are also lysogenic in that cells carry viral genomes in a prophage stage which replicates in synchrony with the bacterial chromosome. Many lysogenic bacteria can be induced to produce large numbers of bacteriophage particles by such treatments as UV irradiation or mitomycin treatment (Borek and Ryan, 1973). Small numbers of bacteriophage particles are normally released spontaneously from lysogenic cells during growth.

Bacteriophages produced by lysogenic cells are designated as *temperate phages*. The genomes of temperate phages, in the prophage state, can exist as a plasmid, for example P1 phage in *E. coli*, or integrated into the bacterial chromosome as in P22 (*S. typhimurium*) or λ (*E. coli*). Depending upon the expression of viral genes, temperate phages can alternate between two states:

prophage (latent) state or lytic (virulent) state. If most of the viral functions are repressed, lysogeny prevails, whereas if phage functions are expressed, the lytic state, or vegetative production, occurs. The expression of a temperate phage in a host depends upon the genetic constitution of the virus and the host as well as the intracellular environment (Hershey, 1971; Bukhari *et al.*, 1977; Birge, 1981).

Two general types of transduction are recognized: (1) *specialized transduction*, and (2) *generalized transduction*. In generalized transduction any chromosomal or plasmid gene has an equal chance of being packaged within the bacteriophage coat. In contrast, in specialized transduction only those gene loci flanking the prophage attachment site are incorporated into the phage genome and transferred.

The detection of specialized transducing phages for phytopathogenic bacteria poses a problem, since a specialized transducing phage can only transduce one or two of the several thousand genes present on the bacterial chromosome. An obvious approach involves testing for the transfer of any number of gene(s) by an induced phage lysate. Alternatively, one looks for the transduction of a specific desired trait; the probability of success, however, is much reduced in such trials.

With a generalized transducing phage such as P1, segments of the host genome are picked up upon the induction of the lysogenic donor or after lytic growth of the virus in the donor strain. In contrast, specialized transduction apparently occurs only when the prophage of the specialized transducing phage is induced and not when the phage is obtained through a lytic cycle. Certain phages, such as P22 (which infects *S. typhimurium*), can bring about both specialized or generalized transduction (Sanderson and Hartman, 1978).

Both virulent and temperate phages can mediate transduction. The recovery of cells infected with transducing particles that are in a minority, compared to cells infected with normal particles, requires the prevention of later infection (adsorption) and subsequent killing of transductants by normal phages (produced by productive cells). Thus, transducing particles do not necessarily render cells immune to infection by normal particles. With temperate transducing phages, recovery can be accomplished by using a recipient that is lysogenic for the normal phage. In this case, advantage is taken of the phenomenon called *superinfection immunity*, whereby the incoming bacteriophage DNA will not be expressed due to its sensitivity to the repressor produced by the prophage of the lysogenic cell. Killing of transductants can also be prevented by using a multiplicity of infection of less than one to minimize multiple infections, and by adding antiviral antibody after adsorption of the virus by bacterial cells. In those cases where bacteriophage adsorption requires the presence of divalent cations, such as Ca^{2+} for P1 adsorption (Roth, 1970), the use of chelators after adsorption prevents subsequent infection by normal viral particles and the killing of recipient cells. It may not be feasible to apply any of the above conditions in screening virulent (lytic) phages of phytopathogenic bacteria for transducing activity. An alternate approach that might prove useful involves the use of UV-irradiated bacteriophage lysates in transductional crosses. In *E. chrysanthemi* (Chatterjee and Brown, 1980), *R. leguminosarum* (Buchanan-Wollaston, 1979), and *Myxococcus xanthus* (Campos *et al.*, 1978), UV irradiation stimulated the frequency of transductants, presumably by neutralizing the killing effect of virus. In *P. syringae* pv. *syringae*, however, UV did not increase the number of transductants (Nordeen and Currier, 1983). Since stimulation of host recombination systems following the use of irradiated phage lysates has also been reported (Benzinger and Hartman, 1962), it is difficult to interpret transductional frequencies.

The donor DNA (exogenote), introduced into a recipient by generalized or specialized transducing phage, frequently pairs with the region of homology within the resident genome (endogenote), and is subsequently incorporated into the endogenote by crossing over. The partial zygote (merozygote) then segregates to yield a haploid recombinant cell. If the exogenote persists and replicates as part of the prophage, a partially diploid heterozygous clone results.

In specialized transduction, the exogenote is sometimes found to persist and function but not to replicate. In such an event one cell in a clone will harbor the exogenote at any given time. This condition, known as *abortive transduction*, has proved extremely useful in analyzing complementation between mutant alleles (see Hayes, 1968).

Mapping of chromosomal loci, i.e., measurement of genetic distances between genes, by generalized transduction is done by determining the cotransductional frequency in two-factor and three-factor crosses (Clowes and Hayes, 1968; Roth, 1970; Miller, 1972; Chatterjee and Brown, 1980). As with other gene mapping systems, transductional mapping is also based upon the premise that the frequency of recombination between two markers is directly proportional to the distance between them. However, the size of the transducing DNA fragment, determined by the phage coat, affects the frequency of joint transduction between markers. The coliphage P1 can transduce a fragment of the bacterial genome equivalent to 2 min (unit) of map distance (Bachmann *et al.*, 1976), while P22 can transduce genomic fragments of *S. typhimurium* corresponding to one unit length (Sanderson and Roth, 1983). Thus, markers separated by more than 1.0 map unit distance will not be cotransduced by P22. In *E. chrysanthemi*, among Erch-12 transductants, Chatterjee and Brown (1980, 1981) detected transductional linkage between *ser-thr*, *arg-idg* [a locus that specifies indigoidine (pigment production)], and *rif-ade*. In conjugational recombinants, these pairs of markers were also coinherited at a high frequency, suggesting a tight linkage between markers. Transductional linkage between several pairs of markers (e.g., *arg-trp*; *arg-met*) has also been detected in *P. syringae* pv. *syringae* (Nordeen and Currier, 1983).

In *E. chrysanthemi*, as with other bacteria, differences were observed in frequencies of coinheritance of nonselected markers in reciprocal crosses (Chatterjee and Brown, 1980). For example, the frequency of the inheritance of the *thr* marker among Ser^+ recombinants was consistently higher than the occurrence of *ser* in Thr^+ recombinants; transduction by Erch-12 of certain markers (*ser*, *ura*) was also higher relative to others (*ilv*, *trp*). Perhaps a site of phage (Erch-12) encapsulation or DNA packaging on the *E. chrysanthemi* genome is close to the left of the *ser* locus and another farther away to the right of the *thr* locus. In other bacteriophage–host systems (Susskind and Botstein, 1978), specific sites for preferential initiation of encapsulation seem to exist on the host genome. Moreover, the putative existence of specific sites of encapsulation might also explain the high frequency of transduction of certain markers. In other generalized transducing systems, such as P22 in *S. typhimurium* (Ebel-Tsipis *et al.*, 1972; Schmieger, 1972), certain chromosomal markers are also transduced at a much higher frequency relative to others.

Kemper (1974) discussed the factors that can affect recombinational frequency between markers, and derived an equation which transforms transduction frequency into linear distance between markers. A similar equation was proposed by Wu (1966) and used to determine map distances between *E. coli* genes (Bachmann *et al.*, 1976). Both equations may be useful in gene mapping of plant pathogens.

Despite certain limitations (Bachmann *et al.*, 1976), generalized transduction is a powerful tool in fine-structure mapping of bacterial genes. In both *E. coli* and *S. typhimurium*, transductional data have been valuable in establishing marker location and estimating distances between markers on chromosomal linkage maps (Sanderson and Hartman, 1978; Bachmann and Low, 1980; Bachmann, 1983; Sanderson and Roth, 1983). Up to now, among phytopathogenic bacteria, transductional linkage has been established only in *E. chrysanthemi* (Chatterjee and Brown, 1980, 1981) and *P. syringae* pv. *syringae* (Nordeen and Currier, 1983). However, even in these bacterial phytopathogens more knowledge is required in order to establish "*linkage groups*," i.e., delineation of groups containing genes that are continuously linked to each other in overlapping transducing fragments carried by phages. This will require introduction of many additional genetic markers and mapping the gene loci by transductional (and conjugational) crosses.

In developing transducing systems in phytopathogenic *Erwinia* spp., the possibility of sensitizing bacteria, or causing bacteria to be infected with coliphages (such as P1), should be explored. With a heat-inducible, kanamycin-resistant derivative of P1 phage P1 Km *c1r*100, Goldberg *et al.* (1974) isolated lysogens of *E. amylovora* and *E. carotovora* subsp. *carotovora*. The bacteriophage (P1) titer was low in heat-induced lysates of Km^r *E. amylovora*. Similar "lysates" of *E. carotovora* subsp. *carotovora*, however, did not yield P1 phage (Goldberg *et al.*, 1974). In similar studies, but using different strains of these bacteria, we could not obtain P1 lysogens (A. K. Chatterjee, unpublished data). In contrast, Jayaswal *et al.* (1984) detected P1-sensitivity in a wild-type pectolytic strain (Brig-P1A) of *E. carotovora* subsp. *carotovora*. This apparent discrepancy might reflect differences in phage receptor sites and in restriction of P1 DNA among *Erwinia* strains. An alternate approach, used in *S. typhimurium* (Enomoto and Stocker, 1974), might prove fruitful. This would entail isolation in *Erwinia* spp. of lipopolysaccharide-defective (rough) mutants (see Section IV.A.2.a.v.), testing for P1 sensitivity (including the capacity to adsorb the phage) of such LPS-deficient mutants, and introducing restriction deficiency in strains that adsorb P1 particles but do not support the replication of the phage. Based upon the contribution of the P1 transductional system in the genetic analysis of *E. coli*, it is reasonable to predict that the availability of P1-sensitive clones would facilitate genetic research on such important pathogens as *E. amylovora* and *E. carotovora* subspecies.

It may now also be possible to apply lambda technology in genetic analysis of various Gram-negative phytopathogenic bacteria including *Erwinia* spp. normally insensitive to this coliphage. This insensitivity probably results from the inability to adsorb λ due to the lack of the receptor, i.e. the product of *malB* (= *lamB*) gene. While the absence of some or all of the maltose genes in *Erwinia* has not been documented, this appears to be the case since *Erwinia* strains that have been tested cannot utilize maltose as a carbon source (A. K. Chatterjee, unpublished). If the absence of *lamB* product is the limiting factor, it should be possible to construct λ-adsorbing strains by introducing *lamB* via F′112 or pTroy plasmids. Palva *et al.* (1981) first sensitized *Salmonella typhimurium* by introducing F′112. Subsequently, Harkii and Palva (1984), by introducing *E. coli nusA* (Holowachuk and Friesen, 1982) (*nusA* = An *E. coli* gene locus for transcription termination; Bachmann, 1983), constructed derivatives of *S. typhimurium* that supported lytic and lysogenic modes of growth of various λ–derivatives. Because *Erwinia* spp. are related to *E. coli*, a similar approach should be applicable in these bacteria, opening new possibilities of *in vivo* genetic manipulations.

Ludwig *et al.* (1984) constructed broad host range plasmids carrying a *lamB*^{c} [a constitutive *lamB*; pTroy], a *lamB-lacZ* fusion (pTroy 2041) or its derivative. While λ-adsorbing strains of *S. typhimurium* and *Klebsiella pneumoniae* resulted from the introduction of pTroy, those of *R. meliloti* and *A. tumefaciens* were obtained only with pTroy 2041 derivatives. These developments raise the possibility of the application of λ-technology to nonenterobacterial phytopathogens such as pseudomonads and xanthomonads. (For the use of Mu phage and its derivatives see the section on Gene Fusion.)

In a number of bacteria, certain traits are determined exclusively by genes present in the phage genome. Thus, such traits are expressed in lysogenic or phage-infected but not in phage-free bacteria. For example, certain temperate phages possess genetic information that modifies cell surface properties (lipopolysaccharide composition; see Section IV.A.2.a.v.) of *Salmonella* spp. and toxinogenesis of *C. diphtheriae* (Barksdale and Arden, 1974). The β phage of *C. diphtheriae* carries structural genes of diphtheria toxin (Barksdale, 1955; Uchida *et al.*, 1971), and converts nontoxinogenic (Tox^-) strains to Tox^+ upon infection (Freeman, 1951; Pappenheimer and Gill, 1973). This process, whereby the bacterium acquires new trait(s) as a result of bacteriophage infection, is known as *phage conversion* (Barksdale and Arden, 1974), and the bacteriophages responsible are designated as converting phages.

Phage conversion differs from specialized transduction in that (1) converting phages are not defective as are specialized transducing phages (see above), and (2) converting phages transfer a gene(s) not normally found in the bacterial genome.

In at least certain *Xanthomonas* spp. (Goto and Starr, 1972; Wu, 1972a,b), preliminary data indicate changes in bacterial phenotypes due to lysogeny. However, these observations have not been pursued, so that it would be premature to ascribe the phenotypic changes to phage conversion. Whether or not the genes for specific phenotypes are integral parts of bacteriophage genomes is not known, much less the gene products that modify bacterial phenotypes. Such determinations could be made by modifying phage genomes with transposons (see above) or by constructing *in vitro* recombinant DNA molecules (see below) carrying segments of the phage genomes. The expression of modified phage genomes in an appropriate recipient could be examined after infection or transfection (see below and Benzinger, 1978).

3. *Transformation*

Transformation of chromosomal genes has been detected in several genera of both Gram-positive and Gram-negative bacteria, including plant pathogens of *Agrobacterium*, *Pseudomonas*, and *Xanthomonas* (see Table VII and Lacy and Leary, 1979). Other bacteria in which such transformation has been reported include human pathogens and saprophytes (Tomasz, 1969; Hotchkiss and Gabor, 1970; Juni, 1972; Cosloy and Oishi, 1973; Notani and Setlow, 1974; Biswas *et al.*, 1977; Bergmans *et al.*, 1980; Henner and Hoch, 1980; Smith *et al.*,

1981). Transformation of plasmid DNA has been accomplished in many bacterial species including such phytopathogens as *A. tumefaciens* (Holsters *et al.*, 1978), *P. syringae* pv. *savastanoi* (Comai and Kosuge, 1980), *E. amylovora* (Bauer and Beer, 1983), *E. carotovora* subsp. *atroseptica* (Hinton *et al.*, 1985), *E. carotovora* subsp. *carotovora* (Berman *et al.*, 1983; Hinton *et al.*, 1985; Zink *et al.*, 1985), *P. solanacearum* (Morales and Sequeira, 1985b), *P. syringae* pv. *syringae* (Gross and Vidaver, 1981), and *P. syringae* pv. *phaseolicola* (Gantotti *et al.*, 1979). *Transfection*, that is transformation of bacteriophage genomes, is known in *A. tumefaciens* and a number of other bacteria (Milani and Heberlein, 1972; Notani and Setlow, 1974; Benzinger, 1978; Holsters *et al.*, 1978).

Compared to conjugation and transduction, transformation appears to be less efficient for chromosomal mapping because cotransformation can be detected for only closely linked markers. Nevertheless, cotransformation was used in fine-structure mapping of genes in *Bacillus subtilis* (Henner and Hock, 1980). In recombinant DNA technology (see below), transformation is an essential step in the introduction of DNA molecules cloned into nonconjugative plasmids. Thus, transformation is a potentially useful tool in research involving "*cryptic*" *plasmids* (i.e., those of unknown function) of phytopathogenic bacteria (see above).

Studies with nonphytopathogenic bacteria have shown that transformation of chromosomal genes requires binding of transforming DNA to special sites on the cell surface of *competent bacteria* (i.e., cells that have the ability to take up transforming DNA; see below), enzymatic cleavage of the externally bound DNA (to yield fragments of variable size depending upon the species), penetration into the cell of DNA fragments, conversion of the double-stranded DNA to a single-stranded form, and integration of transforming DNA by recombination with the bacterial chromosome. The specificity of DNA binding and the temporal sequence of some of the steps differ depending upon the bacterial species (for details see Tomasz, 1969; Hotchkiss and Gabor, 1970; Notani and Setlow, 1974; Low and Porter, 1978).

Various factors such as competence of recipient cells, physical state of the transforming DNA, and recombination proficiency of the recipient affect transformation frequency. Of these, competence of recipient cells is of paramount importance. The proportion of competent cells varies during the cell cycle, and is affected by cultural conditions as well as by the genetic constitution of the recipient (Notani and Setlow, 1974; Curtiss, 1981; Hanahan, 1983). For example, competence of *E. coli* cells is achieved by Ca^{2+}-heat shock treatment (Mandel and Higa, 1970), although the basis for this induction is not clearly understood. However, Hoekstra and associates (Reijenders *et al.*, 1979; Hoekstra *et al.*, 1980) obtained 100-fold higher transformation efficiencies with chromosomal markers of *E. coli* by eliminating washing of recipient cells with Ca^{2+}, conducting the heat shock in the presence of both Ca^{2+} (30 m*M*) and Mg^{2+} (26 m*M*), and selecting transformants on media containing Ca^{2+}, Mg^{2+} and a low concentration (2×10^{-4} m*M*) of phosphate. In a comprehensive study, Hanahan (1983) determined the effects on the efficiency of plasmid transformation in *E. coli* of various factors such as culture conditions and pretreatment of cells before DNA exposure with Mn^{2+}, Ca^{2+}, Rb^{2+} or K^{+}, dimethyl sulfoxide, dithiothreitol, and hexamine cobalt. Thus, it could be important to examine the effects of various conditions in optimizing transformation frequences in phytopathogenic bacterial species.

As stated above, transformation of plasmid DNA has been detected in a wide variety of bacterial species, including several phytopathogens. The conditions affecting transformation of plasmid DNA appear less stringent than those for chromosomal genes. This might be ascribed to the physical state of transforming DNA (*ccc*-DNA of plasmids vs linear DNA of chromosomal genes) and the autonomous replication of plasmid DNA upon its penetration into the recipient cell, in contrast to recombination of the incoming chromosomal genes with the recipient chromosome.

Different conditions are required in various phytopathogenic bacteria to induce competence for the uptake of plasmid or bacteriophage DNA. In *A. tumefaciens*, freezing and thawing of recipient cells stimulated the uptake of bacteriophage, Ti plasmid, or R plasmid DNAs (Holsters *et al.*, 1978). The use of polyethylene-glycol (Klebe *et al.*, 1983) allowed an efficient transformation of plasmids (pWI297 and pVCD2) in *P. solanacearum* (Morales and Sequeira, 1985b). In contrast, in *E. herbicola*, the Ca^{2+}-heat shock conditions recommended for transformation of *E. coli* with plasmid DNA (Cohen *et al.*, 1972) yielded good results (Lacy and Sparks, 1979). Hinton *et al.* (1985), by modifying the procedure of Hanahan (1983), obtained a high frequency of transformation of ColE1-derived plasmids in *E. carotovora* subsp. *atroseptica* and *carotovora*. The Ca^{2+}-based procedures (Mandel and Higa, 1970; Chakrabarty *et al.*, 1975) were slightly modified to obtain transformation of R plasmids (RSF1010 and pBR322) into *P. syringae* pv. *phaseolicola* (Gantotti *et al.*, 1979), and cotransformation of a resident plasmid and RSF1010 into *P. syringae* pv. *savastanoi* (Comai and Kosuge, 1980). Replacement of Ca^{2+} by Mg^{2+} stimulated the frequency of RSF1010 transformants in *P. syringae* pv. *savastanoi* (Comai and Kosuge, 1980). Gross and Vidaver (1981) found that washing of *P. syringae* pv. *syringae* with a $CaCl_2$ (150 m*M*)–glycerol (10%) mixture significantly enhanced the transformation of R plasmids (pRO161 and RSF1010). These findings collectively suggest that while a Ca^{2+}-heat shock treatment is generally effective in inducing competence for transformation of plasmid DNA in phytopathogenic bacteria, the specific steps may require modification depending upon the recipient bacterial species or strain.

In analyzing phenotypes specified by "cryptic" plasmids of phytopathogenic bacteria it is essential to transform the plasmid genetically marked with a transposon (see above) or to cotransform such plasmids (for which there is no selection) along with a plasmid which can be readily selected. For example, Kretschmer *et al.* (1975) found that when transformation was performed with excess DNA of the nonselectable plasmid in the presence of nonsaturating concentrations of DNA of the selectable plasmid, a high proportion (between 50 and 85%) of transformants also received the nonselectable plasmid. A similar approach allowed Comai and Kosuge (1980) to cotransform pIAA1 (a plasmid that controls IAA production) and RSF1010 DNAs into *P. syringae* pv. *savastanoi* by selecting for Sm^r; the frequency of cotransformation increased with an increase in the molar ratio of pIAA1 to RSF1010 DNAs. Gross and Vidaver (1981), however, did not detect cotransformation of pCG131 (a "cryptic" plasmid of *P. syringae* pv. *syringae*; Gonzalez and Vidaver, 1979) with RSF1010, although the R plasmid was readily transformed. Since compatibility between pCG131 and RSF1010 is not yet known, the possibility of some interaction between these plasmids remains.

In transformation of "cryptic" plasmids of phytopathogenic bacteria, the technique of cotransformation of plasmid DNA with linear chromosomal DNA might prove useful (Bergmans *et al.*, 1980). By treating an auxotrophic (*leu*) strain of *E. coli* deficient in exonuclease V and exonuclease I (*recB21*, *recC22*, *sbc15*) with a nonsaturating concentration of chromosomal (leu^+) DNA and saturating concentration (excess) of plasmid (pBR322 or ColE1) DNA, Bergmans *et al.* (1980) obtained Leu^+ transformants, a proportion (5–10%) of which had also received plasmid DNA. Thus, screening of 100 Leu^+ transformants allowed identification of at least five clones that also contained plasmid DNA. In contrast to cotransformation with R plasmids, this procedure should eliminate potential problems of incompatibility between plasmids. Moreover, since the primary selection is for chromosomal gene(s), the transformants should carry a single desired plasmid. This would also eliminate steps that might be required to remove the selectable plasmid from double transformants following the primary selection. Whether or not this approach would be applicable to phytopathogenic bacteria depends upon successful transformation with linear chromosomal DNA.

F. Recombinant DNA Technology

The technology of *in vitro* DNA (and RNA) manipulations, developed during the past decade, has revolutionized research on genome structure and gene function. Thus all the genetic accomplishments with *E. coli* can now conceivably be done in less well-studied bacteria as well as in complex eukaryotes; additional genetic "tricks" are also possible. The development of recombinant DNA technologies is built upon advances in genetics and enzymology in bacterial systems, particularly *E. coli*. The knowledge of structure, replication, restriction, and modification of plasmids and bacterial viruses has been instrumental in the evolution of recombinant DNA technologies. The principles, ethical considerations and methodological aspects of recombinant DNA have been discussed extensively (Cohen, 1975; Beers and Bassett, 1977; Sinsheimer, 1977; Boyer and Nicosia, 1978; Hahn, 1978; Collins, 1979; Garland and Williamson, 1979; Timmis and Pühler, 1979; Wu, 1979; Setlow and Hollaender, 1979, 1980; B. D. Davis *et al.*, 1980; R. W. Davis *et al.*, 1980; Leisinger *et al.*, 1981; Timmis, 1981; Gryczan, 1982; Kreft *et al.*, 1982; Maniatis *et al.*, 1982). Timmis (1981) particularly has provided not only a comprehensive treatment of the subject, but also includes the applications of recombinant DNA technologies to agriculture and other disciplines. Hopwood and Johnston (1984) discuss the limitations, as well as potential, of engineering microbes useful in industry and agriculture. We present here some of the highlights of recombinant DNA technology including suitable references.

The construction of recombinant DNA molecules, or the molecular cloning of genes, requires (1) a means of breaking and joining DNA molecules from different sources; (2) a suitable gene carrier (*vector*) that can replicate along with the foreign DNA covalently linked to it; (3) a method for the introduction of hybrid DNA molecules into an organism; and (4) a means of selection for clones that have acquired the hybrid DNA molecule in the presence of a large population of recipient cells (Cohen, 1975; Timmis, 1981). A typical outline of the procedure is given in Table IX and Fig. 3.

Table IX. Procedural outline for generating chimeric molecules *in vitro*[a]

1. Digest DNA from two sources separately with endonuclease; run control gels to ensure that digestion is complete
2. Mix the two types of digested DNA; incubate at 30°C for 15 min to open up self-circularized molecules
3. Dilute DNA to final concentration of about 15 μg/ml. Incubate at 0°C for 24 hr
4. Add ligase "cocktail" and ligase. Incubate at 10°C for 3–6 hr, and then hold at 0°C. Purity and activity of the ligase must be carefully checked
5. Introduce molecules into host strain by transformation. This involves holding cells in 0.1 *M* CaCl at 0°C for 30 min, giving a heat shock (e.g., 2 min at 42°C), and holding for a further 90 min at 0°C, before plating
6. If possible, there should be a selection for chimeric transformants or some simple screening method for detecting them. A period of incubation before selection may be necessary to allow expression of the chimeric functions

[a] Procedures particularly effective for *E. coli*. (Adapted from Broda, 1979.)

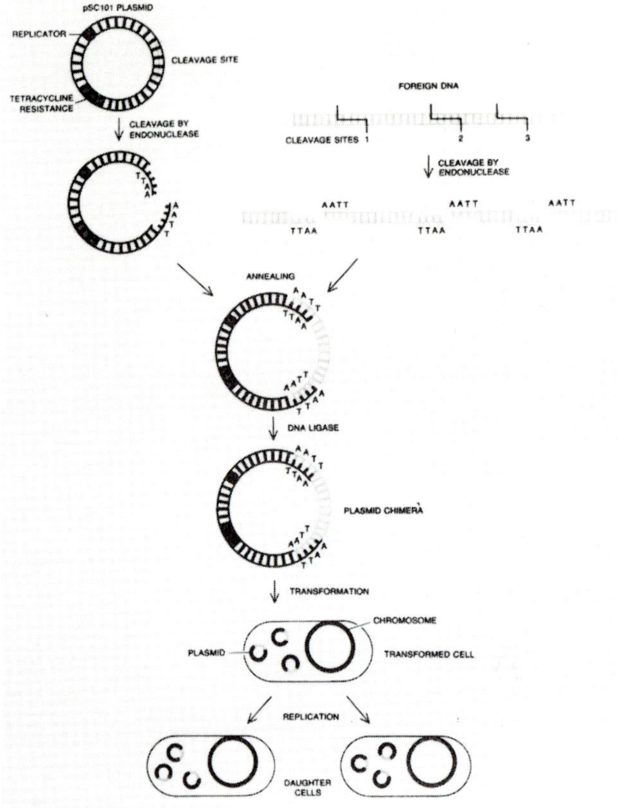

Fig. 3. Recombinant DNA technique. Foreign DNA is spliced into the pSC101 plasmid and introduced with the plasmid into the bacterium *Escherichia coli*. The plasmid is cleaved by the endonuclease *Eco*RI at a single site that does not interfere with the plasmid's genes for replication or for resistance to tetracycline (top left). The nucleotide sequence recognized by *Eco*RI is also present in the other DNA, so that a foreign DNA exposed to the endonuclease is cleaved about once in every 4000 to 16,000 nucleotide pairs on a random basis (top right). Fragments of cleaved foreign DNA are annealed to the plasmid DNA by hydrogen bonding of the complementary base pairs, and the new composite molecules are sealed by DNA ligase. The DNA chimeras, each consisting of the entire plasmid and a foreign DNA fragment, are introduced into *E. coli* by transformation, and the foreign DNA is replicated by virtue of the replication functions of the plasmid. (Reproduced from Cohen, 1975.)

DNA fragments are produced by restriction enzymes, a special class of endonucleases that recognizes specific sites of variable length within double-stranded DNA (restriction enzymes normally function to destroy foreign DNA entering a cell). One type of restriction enzyme (type I endonucleases) binds to specific targets but cleaves DNA randomly at some distance from the recognition site. Hence, these are not useful in the construction of hybrid DNA molecules. In contrast, enzymes of the other class (type II) generally cleave DNA at the binding or target site. DNA fragments are produced, the numbers and sizes of which are fixed, depending upon the frequency of the occurrence of the target site in a given genome. [Several plant pathogens are commercial sources for specific restriction enzymes (Bukhari *et al.*, 1977; Roberts, 1980a,b).] For example, the restriction enzyme *Eco*RI cleaves the fertility factor, F DNA, into 19 fragments whereas *Bam*HI produces 11 fragments (Skurry *et al.*, 1978). The cleavage sites of several restriction endonucleases for the Ti plasmid are shown in Fig. 4. Pomponi *et al.* (1983) have shown cleavage sites for three restriction endonucleases in the Ri plasmid.

Recognition sites of most type II enzymes possess *twofold rotational symmetry*, i.e, they are palindromic sequences that have a central axis of symmetry and are read identically in both directions from that axis (also see Smith, 1979; Roberts, 1978, 1980a,b). Cleavage at these sites can be clean or aligned or staggered, thereby generating fragments with single-stranded termini with either 3′ or 5′ ends.

Joining two DNA ends is critical to obtaining *in vitro* recombinants. Three strategies are basically used for such joining: (1) ligation of sticky ends generated by restriction endonuclease digestion; (2) ligation of blunt ends generated by certain enzymes or mechanical shearing and either digestion with single strand-specific S_1 nuclease or treatment with *E. coli* DNA polymerase; or (3) addition of linkers or adapters and then joining the adapters together either by ligation or annealing to form stable bonds (Schleif and Wensink, 1981). These methods are considered below.

Cohesive or sticky ends are obtained if the enzymes produce termini that can join by hydrogen bonding. If the base sequence of the single-stranded ends, generally two to five nucleotides in length, is perfectly complementary, the termini can join because of the twofold symmetry of the target site. Under these conditions, a fragment may be converted into a circle, provided it is 250–300 base pairs or longer; two or more fragments may pair (by hydrogen bonding), giving rise to fragment oligomers; or DNA fragments from different sources possessing complementary base pairs at single-stranded termini may join. Two unrelated DNA fragments can therefore join, provided that they have matching base pairs (A–T; G–C) in the single-stranded ends (Fig. 3). Such hydrogen-bonded fragments can then be covalently joined or ligated by enzymes from *E. coli* or the coliphage T4 (Weiss *et al.*, 1968; Modrich *et al.*, 1973).

The ligating step is generally carried out at 10–15°C so as to prevent the "melting" or breakage of hydrogen bonds of base pairs at the ends. For

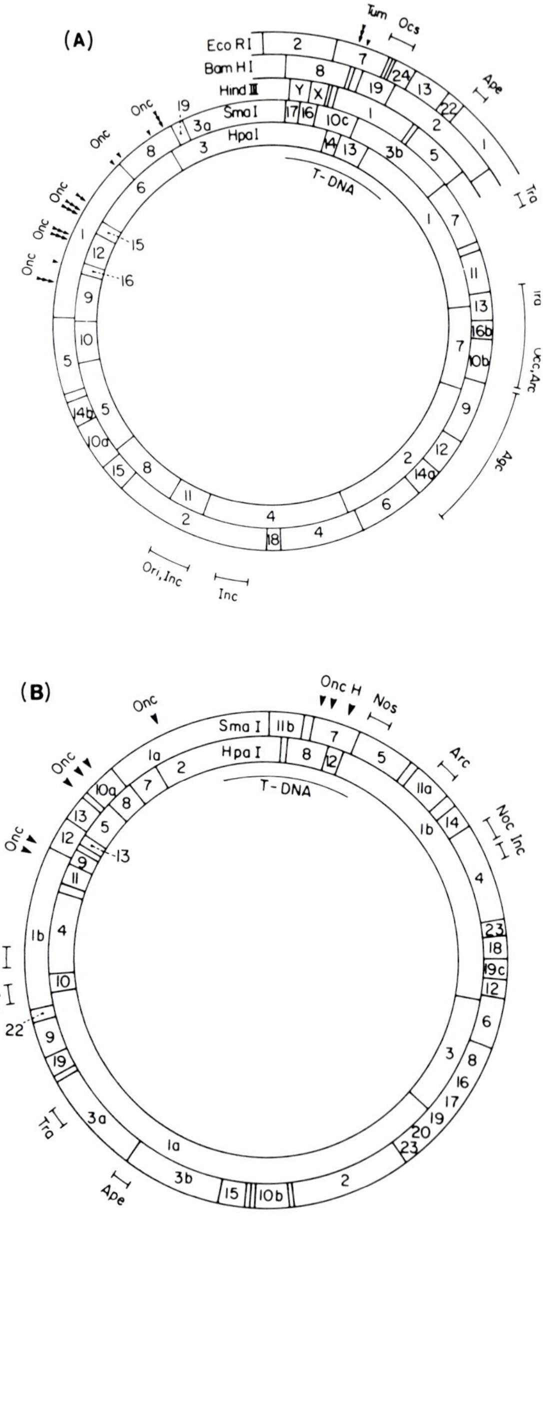

(A)
Eco RI
Bam HI
Hind III
Sma I
Hpa I
Tum
Ocs
Ape
Tra
Onc
T-DNA
Occ, Arc
Agc
Ori, Inc
Inc
(B)
Sma I
Hpa I
Onc
Onc H
Nos
Arc
Noc
Inc
T-DNA
Psc
Agr^s
Tra
Ape

instance, the midpoint of the melting curve (T_m value) of ends formed by *Eco*RI, which would possess four A–T base pairs, is 5–6°C. Thus incubation of the *Eco*RI-generated fragments at 10°C in the presence of DNA ligase ensures both the catalytic activity of the enzyme and the stability of hydrogen-bonded fragments. Ligation at a higher temperature can be done in cases where single-stranded termini possess more G–C pairs and consequently form more stable complexes.

Since DNA structure is symmetrical, insertion of a DNA fragment could occur in one of two orientations. The expression of the inserted DNA could depend upon the orientation of insertion.

As noted earlier, certain restriction enzymes produce clean cuts, usually referred to as blunt or flush ends or butts. A DNA ligase, coded by the coliphage T4, can join flush ends under appropriate conditions (see Sgaramella, 1972, for the technical details).

Another approach involves the addition of adapters or deoxyribonucleotides (dA or dT) to 3′ ends of DNA strands to facilitate annealing between two different DNA species by generating complementary base pairs (A–T or G–C) at the ends. For example, dA residues are added to 3′ ends of strands of restriction fragments of one DNA species (DNA I), and dT residues are linked to 3′ ends of strands of the fragments of another DNA species (DNA II). It is apparent that the strands possessing terminal dA residues (DNA I) or dT residues (DNA II) cannot self-anneal to produce circles or oligomers but would anneal when fragments of DNA I (dA tails) and DNA II (dT tails) are mixed because of the pairing between A and T.

The joining of dA or dT residues is brought about by terminal deoxynucleotidyl transferase, which normally adds nucleotides to a single-stranded 3′ terminus. Hence, the restriction fragments are treated with a 5′ exonuclease coded by λ bacteriophage, which exposes 3′ ends of restriction fragments; dA or dT residues are then added by the terminal deoxynucleotidyl transferase. Alternatively, advantage is taken of the finding (Roychoudhury *et al.*, 1976) that in the presence of cobalt ions, terminal deoxynucleotidyl transferase adds deoxynucleotides to 3′ ends in the presence of the complementary strand, thus eliminating the step involving exonuclease treatment. The *in vitro* conversion of the annealed fragments into covalently closed forms is brought about by

---

Fig. 4. Restriction endonuclease digest maps of *Ti plasmids*. (A) An octopine (pTi-$B_6$806) and (B) a nopaline (pTI-C58)-type plasmid are shown. For details of the map locations of markers see Nester and Kosuge (1981). Phenotypic designations: Agc, agropine catabolism; Agr^s, agrocin sensitivity, Apc, exclusion of phage API; Arc, arginine catabolism; Inc, incompatibility; Noc, nopaline catabolism; Nos, nopaline synthesis; Occ, octopine catabolism, Onc, oncogenicity; Ori, origin of replication; Psc, agrocinopine synthesis and catabolism; Tra, plasmid transfer; T-DNA, plasmid DNA fragment present in plant nuclear DNA. (Reproduced from Nester and Kosuge, 1981.)

filling in the gaps by deoxynucleotides and the enzyme DNA polymerase I, and then covalent joining of the strands by DNA ligase. However, these steps are generally not required since the transformed host cell, such as *E. coli*, can convert annealed fragments to covalently closed forms.

The joining of DNA fragments via dA and dT tails offers certain advantages in that inter- and intramolecular annealing between the same DNA species is prevented, annealed molecules are quite stable, an increased yield of hybrid DNA molecules is obtained, and the ligation step is eliminated (Clarke and Carbon, 1975). This method has proved highly efficient in cloning *E. coli* chromosomal genes into plasmids (Clarke and Carbon, 1975). A similar approach could be used in cloning chromosomal segments of Gram-negative phytopathogenic bacteria.

An additional approach in cloning of DNA fragments with flush ends involves the use of linkers, short synthetic nucleotides with one or more endonuclease recognition sites (Timmis, 1981). This involves (1) ligating linkers to DNA fragments; (2) cleavage of the linker covalently attached to DNA fragments with an appropriate endonuclease to produce the specific termini; (3) joining of such DNA fragments to a vector having complementary termini; and (4) introducing hybrid molecules into host bacteria. The availability of a number of synthetic linkers and adapters with recognition sites for various restriction endonucleases now allows an investigator to choose appropriate linkers and endonucleases in gene cloning ventures. One potential problem in this cloning strategy may arise if the gene(s) to be cloned contains one or more cleavage sites for the endonuclease used in generating specific termini on the linker. This problem can be overcome by using a linker having a site for an endonuclease that does not cleave the desired DNA fragment. Alternatively, DNA fragments can be treated with a specific modification methylase prior to the addition of linkers; the modification of bases may prevent the cleavage of DNA fragments during the step involving endonuclease treatment. The construction of asymmetric linkers containing single preformed specific cohesive termini has been reported (Bahl *et al.*, 1978; Norris *et al.*, 1979). The use of such linkers in producing hybrid DNA molecules would eliminate the step involving endonuclease treatment and hence the potential problem of the internal cleavage of DNA fragments.

In selecting vectors of recombinant DNA molecules for phytopathogenic bacteria several factors should be taken into consideration: (1) ease of introduction of the vector and the expression and stability of the vector-specific phenotype; (2) potential of amplification (increased number of molecules) and isolation of hybrid DNA; (3) presence on the vector of a single or a limited number of cleavage (cloning) site(s) for various endonucleases; and (4) amount of DNA that can be joined to the vector.

An important strategy in cloning experiments is the ability to rapidly select and/or identify clones carrying hybrid DNA molecules. This is accomplished in various ways (Timmis, 1981). An example is the use of plasmid vectors, such as pBR322 (Bolivar *et al.*, 1977), where the drug-resistant markers are subject to *insertional inactivation*, i.e., the insertion of a DNA fragment within the drug-resistant gene results in its inactivation, and hence cells carrying the hybrid plasmid are susceptible to the drug. Thus, the clones carrying hybrid DNA can be recognized by differential susceptibility to drugs using replica plating techniques. Likewise, insertional inactivation of a conditional lethal effect (such as killing by temperate bacteriophages or bacteriocins) of certain vectors on normally sensitive bacterial hosts allows selection of clones carrying hybrid DNA molecules (see Bolivar and Backman, 1979; Kahn *et al.*, 1979; Timmis, 1981, and commercial catalogs for additional details).

Large (32–45 kb) pieces of DNA can be cloned by using *cosmids* [small plasmids that contain the λ cohesive end site (*cos*) and carry antibiotic resistance gene(s) or genes for other phenotypes] in conjunction with *in vitro* packaging procedures, i.e., packaging of DNA into viable phage particles using various bacteriophage proteins (Becker and Gold, 1975; Hohn and Murray, 1977; Sternberg *et al.*, 1977; Hohn and Collins, 1980; Ish-Horowicz and Burke, 1981). In this case advantage is taken of the finding that only DNA molecules within a certain size class can be packaged into a viable virion and transduced. Thus a size selection is imposed so hybrids are obtained that fill a phage head with approximately 37–50 kb DNA.

Since the availability of a suitable cloning vector is critical in the selection and subsequent studies on hybrid DNA molecules, much effort has been directed toward the development of vectors useful in *E. coli* and other bacteria (Tables X and XI; see also Casadaban *et al.*, 1980; Prentki and Krish, 1982; Brosius, 1984a,b; Kahn and Timblin, 1984; Zagursky and Berman, 1984; and commercial catalogs). Some vectors are useful only in *E. coli* or related bacteria, such as *Erwinia* spp., while others, because of their broad host range, such as those derived from P, Q, and W groups of plasmids, have been and can be used in various Gram-negative phytopathogenic bacteria (Tables X and XI; see also Close *et al.*, 1984; Selvaraj *et al.*, 1984; Priefer *et al.*, 1985). In addition, both narrow and broad host range transcriptional/translational gene fusion vectors are available (Casadaban *et al.*, 1980; Minton, 1984; Nano *et al.*, 1984) which can be useful in investigating gene regulation, protein export and isolation of gene product(s). The development of "shuttle vectors," i.e., those that can replicate in both Gram-negative and Gram-positive bacteria (Ehrlich, 1978; Macrina *et al.*, 1982, 1983; Larson and Hershberger, 1984; Ozaki *et al.*, 1984; Santamaría *et al.*, 1984; Sullivan *et al.*, 1984), increase the potential of cloning genes from Gram-positive phytopathogens. Cloning vectors developed for saprophytic streptomycetes (Chater *et al.*, 1982; Richardson *et al.*, 1982) or bacilli (Gryczan, 1982; Kreft *et al.*, 1982) may prove useful in cloning

Table X. Some general purpose high copy number plasmid cloning vectors for *E. coli*[a]

Plasmid vector	Parent replicon[b]	Cloning sites	Selection marker[c]	(1) Insertional inactivation[d] or (2) positive selection[e]	Size (kb)	Copy number	Amplification[f]
pBR322	pMB1	*Pst* I	Tc^r	1, Ap^r	4.4	40–50	Yes, Cm
		*Cla*I, *Hin*dIII, *Bam*HI, *Sal* I	Ap^r	1, Tc^r			
		*Eco*RI, *Ava*I, *Pvu*II, *Bal* I	Ap^r, Tc^r	—			
pBR325	pMBI	*Pst* I	Tc^r, Cm^r	1, Ap^r	5.4	40–50	Yes, Sp
		*Hin*dIII, *Bam*HI, *Sal* I	Ap^r, Cm^r	1, Tc^r			
		*Eco*RI	Ap^r, Tc^r	1, Cm^r			
pACYC177	p15A	*Pst* I	Km^r	1, Ap^r	3.7	20	±, Cm
		*Hin*d III, *Xho* I	Ap^r	1, Km^r			
		*Bam*HI, *Hin*d II	Ap^r, Km^r	—			
pACYC184	p15A	*Eco*RI	Tc^r	1, Cm^r	4.0	20	Yes, .Sp
		*Hin*d III, *Bam*HI, *Sal* I	Cm^r	1, Tc^r			
pKT235	ColD	*Eco*RI	Ap^r, Km^r	1, Cm^r	8.5	20	Yes, stationary phase
		*Hin*d III, *Xho*I	Ap^r, Cm^r	1, Km^r			
		Pst I	Km^r, Cm^r	1, Ap^r			
		*Bst*E II, *Sst*I, *Bam*HI, *Kpn* I	Km^r, Cm^r, Ap^r	—			
pKY2289	ColEI	*Xma* I, *Eco*R I	Ap^r	2, MC	11.0	20	Yes, Cm
pKN80	ColEI	*Hin*d II, *Hpa* I	Ap^r	2	15.9	20	Yes, Cm

[a] Adapted from Timmis (1981).
[b] Plasmids from which the replication genes were incorporated into a plasmid vector.
[c] Tc^r, tetracycline resistance; Ap^r, ampicillin resistance; Cm^r, chloramphenicol resistance; Km^r, kanamycin resistance.
[d] Insertional inactivation of the indicated property.
[e] Positive selection, in the case of pKY2289, on mitomycin C-containing medium, MC.
[f] By growth of host cells in chloramphenicol (Cm), spectinomycin (Sp), or stationary phase.

genes of phytopathogenic streptomycetes and possibly nocardia and coryne-bacteria. The progress in this area has been so rapid that new listings and properties of vectors are continually being published; the reader is alerted to these developments.

As discussed above, transfer of plasmid DNA by conjugation or transformation has been accomplished in several phytopathogenic bacterial species. In *Erwinia*, *Pseudomonas* or *Xanthomonas* spp., recombinant plasmids were introduced by transformation and by conjugation using an intermediate host (Daniels *et al.*, 1984b, 1985; Coplin *et al.*, 1985; Staskawicz *et al.*, 1984, 1985; Neipold *et al.*, 1985; Roeder and Collmer, 1985a,b; Zink *et al.*, 1985a,b).

Several approaches have been used in detecting the presence of the desired cloned genes. For those genes which have a selectable phenotype, such as prototrophy or RecA$^+$ (resistance to such DNA damaging agents as UV light, methyl methane sulfonate or nitroquinoline oxide), transformants or transductants were selected directly for the phenotype (Keener *et al.*, 1984; Ehrenshaft and Mills, 1985; Zink *et al.*, 1985b). In those instances (e.g., the genes for extracellular enzymes) where the phenotypes can be scored but cannot be selected, cells containing recombinant DNA were first selected for the vector phenotype and then scored for the expected phenotypes or products of the desired genes (Barras *et al.*, 1984a; Keen *et al.*, 1984; Collmer *et al.*, 1985a,b; Lei *et al.*, 1985; Van Gijsegem *et al.*, 1985a; Zink and Chatterjee, 1985). A plaque assay procedure was also used by Kotoujansky *et al.* (1985) for the detection of *pel*$^+$ genes cloned into a λ vector. When the expression of cloned DNA in *E. coli* was suspect or a specific gene of a multigene phenotype (viz. toxin production; host specificity and/or pathogenicity) was sought, the desired recombinant DNA was obtained by complementation of appropriate strains or mutants by mobilizing recombinant plasmids via conjugation (Daniels *et al.*, 1984b; Gabriel, 1985; Staskawicz *et al.*, 1984, 1985; Coplin *et al.*, 1985). Another approach that is specific but relatively laborious, is to detect sequences by using a colony hybridization technique as outlined in Fig. 5. The use of transposon (Tn) insertion mutants, however, facilitated the identification of the desired recombinant clones. The procedure entails isolating genomic DNA sequences flanking the Tn DNA and using the DNA fragment as a hybridization probe in the detection of the recombinant DNA (Neipold *et al.*, 1985a,b).

G. Isolation of Host Association Genes

Using the approaches discussed in the previous section, a number of phytobacterial genes have been cloned (Table VIII), including several that specify plant pathogenicity, virulence or hypersensitive reaction (HR). The findings

Table XI. Some cloning vectors proven useful or potentially useful in Gram-negative phytopathogenic bacteria

Plasmid/ cosmid	Cloning sites	Selection	Size (kb)	Remarks	References
pBR329	*Hind* III, *Bam* HI, *Sal* I *Pvu* I, *Pst* I *Eco*R I, *Pvu* II, *Bal* I	Ap^r, Cm^r (Tc^s) Tc^r, Cm^r, (Ap^s) Ap^r, Tc^r (Cm^s)	4.15	May prove useful in *Erwinia* spp.; like pBR322, pBR329 allows cloning by insertional inactivation of the drug-resistant markers, and can be amplified	Covarrubias and Bolivar (1982)
pHP34	As for pBR322 (Table X) except for two sites for *Eco*R I; also has a unique *Sma*I site immediately flanked by two *Eco*R I sites	Ap^r and/or Tc^r, depending on the restriction enzyme	4.37	May prove useful in *Erwinia* spp.; allows cloning of blunt-ended DNA fragments; particularly useful in the sequencing of small cloned fragments and in *in vitro* mutagenesis	Prentki and Krisch (1982)
RSF1010	*Eco*R I	Sm^r, Su^r	8.9	Broad host range; pIAA fragments were cloned into the plasmid by Comai and Kosuge (1982)	Nagahari and Sakaguchi (1978)
pKT231	*Eco*R I, *Sst* I *Hind* III, *Xma* I, *Xho* I *Cla* I	Km^r (Sm^s) Sm^r (Km^s) Km^r, Sm^r	13.0	Broad host range	Bagdasarian *et al.* (1981)
pKT247	*Eco*R I, *Sst* I	Ap^r (Sm^s)	11.5	Broad host range; carries λ *cos* site	Bagdasarian *et al.* (1981)
pRK2501	*Hind* III, *Xho* I *Sal* I *Bgl* II, *Eco*R I	Tc^r (Km^s) Km^r (Tc^s) Km^r, Tc^s	11.1	Broad host range; derived from RK2	Kahn *et al.* (1979)
pRO1614	*Bam* HI, *Hind* III	Cb^r (Tc^s)	3.9	Broad host range	Olsen *et al.* (1982)
pLAFR1	*Eco*R I	Tc^r	21.6	Broad host range; derived from pRK290, carries λ *cos* site; can be conjugally transferred by the helper plasmid pRK2013	Friedman *et al.* (1982)
pKT240	*Eco*R I	Ap^r, Km^r (Sm^r)	12.5	Broad host range promoter-probe cloning vehicle; derived from RSF1010; DNA carrying promoter(s) inserted into the *Eco*R I site allows the expression of *aph* gene and hence resistance to Sm	Bagdasarian *et al.* (1983)
pMMB6	*Eco*R I	Ap^r	11.6	Broad host range expression vector; derived from pKT240; gene(s) without its promoter inserted into the *Eco*R I site expressed from the upstream *lac* promoter	Bagdasarian *et al.* (1983)

pMM22	*Eco*R I	Ap	12.7	Broad host range ... pKT240; utilizes *lac*^{Q} gene and *lac* promoter in controlled expression of gene(s) inserted into the *Eco*R I site	... *al.* (1983)
pHK17	*Eco*R I *Sal* I *Hind* III, *Xho* I	Km^{r}, Tc^{r} Km^{r} (Tc^{s}) Tc^{r} (Km^{s})	12.8	Broad host range; carries λ *cos* site; the plasmid is Tra^{-} but Mob^{+} which allows its conjugal cotransfer with the helper plasmid, pRK2013. The cosmid has been used to make a clone bank of the *A. tumefaciens* pTiA6 plasmid	Klee *et al.* (1982)
pRK290	*Eco*R I, *Bgl* II	Tc^{r}	20	Broad host range; derived from RK2; can be conjugally transferred by the helper plasmid, pRK2013.	Ditta *et al.* (1980)
pVK100	*Eco*R I *Hind* III, *Xho* I *Sal* I	Tc^{r}, Km^{r} Tc^{r} (Km^{s}) Km^{r} (Tc^{s})	23	Broad host range; derived from pRK290; carries λ *cos* site; can be conjugally transferred by the helper plasmid, pRK2013	Knauf and Nester (1982)
pVK101	As for pVK100; also carries a unique *Bgl* II site	As for pVK100	21.3	As for pVK100 but lacks the λ *cos* site	Knauf and Nester (1982)
pVK102	As for pVK100 but lacks the *Eco*R I site	As for pVK100	23	As for pVK100. The cosmid was used to generate a clone bank of *A. tumefaciens* plasmid pTiA6	Knauf and Nester (1982)
pSCC31	*Hind* III, *Bgl* II	Ap^{r}	5.15	Positive selection cloning vehicle; derived from pKC30 and pAN4; contains *Eco*R I endonuclease gene downstream from λ p_L; transformants are obtained only when the structural integrity is destroyed by cloning at the *Hind* III or the *Bgl* II site; allows regulated overproduction of hybrid proteins; may prove useful in *Erwinia* spp.	Cheng and Modrich (1983)
pUN121	*Eco*R I, *Xma* I, *Sma* I, *Hind* III, *Bcl* I	Ap^{r} (Tc^{r})	4.4	A positive selection vector dreived from pTR262 (Roberts *et al.*, 1980) and pBR322 by recombinant techniques and oligonucleotide-mediated mutagenesis; tetracycline resistance gene is under transcriptional control of the repressor protein encoded by phage λ *cI* gene; insertion of DNA at the restriction sites in the repressor gene allows the expression of tetracycline resistance; may prove useful in *Erwinia* spp.	Nilsson *et al.* (1983)
pGV1106	*Pst* I, *Bgl* II, *Bam* HI, *Sac* II *Eco*R I	Km^{r}, Sm^{r}/Sp^{r} Km^{s} (Sm^{s}/Sp^{s})	8.78	Broad host range cloning vector; derived from W-plasmid SA; can be mobilized by an N-type plasmid	Leemans *et al.* (1982)

Table XI. (cont.)

Plasmid/ cosmid	Cloning sites	Selection	Size (kb)	Remarks	References
pGV1113	*Hind* III, *Bam HI*, *Pst* I, *Bgl* II	Sm^r/Sp^r, Su^r Sm^r/Sp^r (Su^s)	8.48	As for pGV1106	Leemans *et al.* (1982)
pGV1122	*Pst* I *Hind* III, *Bam* HI, *Sal* I	Sm^r/Sp^r, Tc^r Sm^r/Sp^r (Tc^s)	10.9	Broad host range cloning vector; derived from pBR325 and pGV1113; contain *bom* site of *colE1, hence efficiently mobilizable by I-type plasmids; carries the col*E1-type origin in addition to the W-type origin of replication, thus the plasmid can be amplified in *E. coli* by chloramphenicol; Tc^s clones can be enriched by cycloserine treatment	Leemans *et al.* (1982)
pGV1124	*Hind* III, *Bam* HI, *Sal* I *Eco*R I	Sm^r/Sp^r, Cm^r Sm^r/Sp^r (Cm^s)	10.9	Broad host range cloning vector; derived from pGV1113 and pACYC184; amplifiable; mobilizable by I-type plasmids; Cm^s clones can be enriched by cycloserine treatment	Leemans *et al.* (1982)
pUC8 and pUC9	*Eco*R I, *Sma* I, *Xma* I, *Bam* HI, *Sal* I, *Acc* I, *Hind* II, *Pst* I, *Hind* III	Ap^r	2.7	A general purpose cloning vehicle for *E. coli* and possibly for *Erwinia* spp.; cloning in any one of the nine sites can be monitored by the appearance of colonies on an indicator medium; allows rapid sequencing of the cloned fragments	Vieira and Messing (1982)
pKEN037 (pIN-I-A1)	*Eco*R I, *Hind* III, *Bam* HI	Ap^r	4.9	An expression vehicle derived from pBR322; carries *lpp* promoter and a transcription termination signal from the *lpp* gene; allows constitutive expression of the cloned gene(s); may prove useful in *Erwinia* spp.	Nakamura and Inouye (1982)
pKEN045 (pIN-II-A1)	*Eco*R I, *Hind* III, *Bam* HI	Ap^r	5.0	An expression vehicle derived from pBR322; carries *lpp* promoter and *lac* promoter–operator region allowing controlled over-production of cloned gene(s); may prove useful in *Erwinia* spp.	Nakamura and Inouye (1982)
pSa727	*Bam* HI, *Pvu* II, *Sst* II *Kpn* I	Sp^r, Km^r Cm^r Cm^r (Km^s; Sp^s may also occur)	14.7	Broad host range derived from Inc W plasmid pSa; self-transmissible; however, frequency of transfer can be increased significantly by using pSa322 as the mobilizing plasmid	Tait *et al.* (1983)

with genes for extracellular polysaccharides (EPS), enzymes, toxins and hormones are discussed in the appropriate sections. Here we briefly present the observations with phytobacterial genes involved in virulence, avirulence and HR. Since we do not yet know the nature of the products of these genes or their specific (primary) functions, the cloned genes or DNA segments are identified by the composite phenotypes, i.e., pathogenicity, avirulence, HR, etc.

In *E. amylovora*, Bauer and Beer (1985) isolated Tn*5* insertion mutants that were nonpathogenic and concomitantly defective in HR production in tobacco leaves. In one mutant this pleiotrophic change appeared to result from a single Tn*5* insertion. The corresponding wild-type DNA segment, isolated by colony hybridization, partially restored both pathogenicity and HR inducibility. The partial (incomplete) restoration of the phenotypes was ascribed to the instability in *E. amylovora* of the ColE1-derived vectors under nonselective conditions, i.e., in plants. This explanation could be verified by transcloning (or subcloning) the insert DNA into broad host range cloning vectors (see Table XI) that may be more stable in *E. amylovora*. It should, however, be noted that selection pressure was required for the maintenance of the broad host range cosmid vector, pVK102 and the derived cosmid hybrids in *P. syringae* pv. *phaseolicola* (Ehrenshaft and Mills, 1985). Since the experiments in *E. amylovora* were conducted with *recA*$^+$ strains, genomic rearrangements also remain a possibility. Another possibility is that the large fragments (ranging from 17 to 29 kb) carry several pathogenicity (and HR) genes and an overproduction of some of the gene products due to a gene dosage effect (one copy of the chromosome and at least another on the plasmid) may affect the degree of complementation. Despite these uncertainties, the simultaneous cloning of pathogenicity and HR genes suggests a connection between these two host responses; similar observations have been made in several other host-bacterial systems.

Coplin *et al.* (1985) detected DNA segments of *E. stewartii* that complemented defects in watersoaking, in EPS production or in both. The genes for watersoaking (= virulence) were localized on a 9.3 kb and an adjacent 4.6 kb DNA segment, suggesting a clustering of virulence genes.

In *P. syringae* pv. *syringae*, Neipold *et al.* (1985b) isolated by colony hybridization a cosmid-hybrid (pOSU3101) with an insert of 30 kb that complemented a pleiotropic phenotype (nonpathogenicity, lack of HR induction and fluidal colony morphology) in a single site Tn*5* insertion mutant, PS 9021. Further subcloning and complementation indicated that about 8.5 kb of DNA was required in the restoration of these phenotypes. Upon subcloning of pOSU3101, Neipold *et al.* (1985b) detected abnormal chlorotic symptoms in tobacco but restoration of pathogenicity and colony morphology by mutants complemented with pOSU3103. This subclone lacks about 400

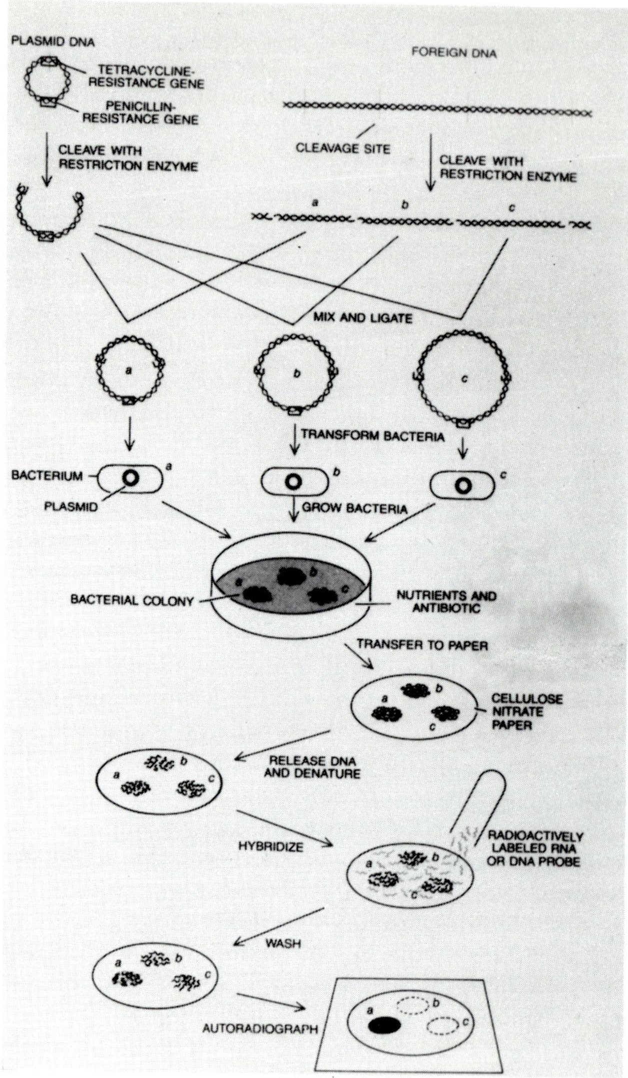
PLASMID DNA
TETRACYCLINE-RESISTANCE GENE
PENICILLIN-RESISTANCE GENE
FOREIGN DNA
CLEAVAGE SITE
CLEAVE WITH RESTRICTION ENZYME
CLEAVE WITH RESTRICTION ENZYME
a
b
c
MIX AND LIGATE
TRANSFORM BACTERIA
BACTERIUM
PLASMID
GROW BACTERIA
BACTERIAL COLONY
NUTRIENTS AND ANTIBIOTIC
TRANSFER TO PAPER
CELLULOSE NITRATE PAPER
RELEASE DNA AND DENATURE
HYBRIDIZE
RADIOACTIVELY LABELED RNA OR DNA PROBE
WASH
AUTORADIOGRAPH

nucleotide base pairs present on one end of the complementing clone, pOSU 3104, implicating the functions of separate gene products in the manifestation of these phenotypes. However, much additional work is needed to determine if separate genes, organized in an operon, specify all three phenotypes, as would be suggested by the phenotype of the Tn*5* insertion mutant and the amount of DNA required for its complementation. It should be noted that the findings of Panopoulos and associates (Panopoulos *et al.*, 1985; Lindgren *et al.*, 1985; Panopoulos and Peet, 1985) with *P. syringae* pv. *phaseolicola* are generally similar to those with *P. syringae* pv. *syringae*. Certain single site Tn*5* insertions caused avirulence as well as the inability to induce HR, and cosmid-hybrid DNAs restored both of these phenotypes. Similar studies have been done with *P. syringae* pv. *tomato* gene(s) specifying HR on soybean (Kobayashi and Keen, 1985). An additional significant observation in this regard is the DNA homology in pathogenicity/HR genes in *P. syringae* pathovars (Lindgren *et al.*, 1985; Neipold *et al.*, 1985b).

A clustering of pathogenicity genes on about a 10-kb DNA segment of the *X. campestris* pv. *campestris* genome was found by Turner *et al.* (1985). This conclusion was made after (1) isolating nitrosoguanidine induced mutants nonpathogenic on turnip seedlings (Daniels *et al.*, 1984a); (2) constructing a library of the wild-type genome and identifying a cosmid-hybrid (pIJ3000) which complemented a nonpathogenic mutant (8288) (Daniels *et al.*, 1984b); (3) inactivating gene functions in pIJ3000 by Tn*5* insertions and introducing the inactivated segment into the wild-type genome by marker exchange; and (4) testing the pathogenicity of the Tn*5* carrying strains. As with other bacterial species discussed above, the nature or function of the gene pro-

Fig. 5. Cloning of foreign DNA, insertional inactivation of vector phenotype, and detection of cloned DNA by colony hybridization. The plasmid and the foreign DNA are cleaved with a restriction endonuclease (see Fig. 3). The enzyme cuts the vector DNA at a unique site within a particular base sequence in a way that disrupts the gene sequence for tetracycline resistance. Incubation of the fragments with the plasmids leads to the annealing of a fragment with a plasmid by base pairing at the ends. After ligation the recombined plasmids are introduced into bacteria. Colonies containing a plasmid are identified by their resistance to antibiotics; in the case illustrated here they will be resistant to penicillin but sensitive to tetracycline. Colonies containing a particular foreign DNA fragment can then be identified by hybridization. The colonies are transferred to a filter paper, the cells are broken open, and their DNA is denatured (separated into single strands) and fixed in place. A probe is added: an RNA or DNA complementary to the desired fragment (in this case *a*) and labeled with a radioactive isotope. The probe anneals to any complementary sequences, forming a hybrid molecule and thus becoming fixed to the paper. The rest of the probe is washed away. When a photographic emulsion is exposed to the paper, a dark spot on the emulsion identifies clone *a*, which contains desired fragment *a*. (Reproduced from Chambon, 1981.) For methodological details see Grunstein and Wallis (1979).

duct(s) specified by the 10-kb DNA segment is not yet known. However, Daniels *et al.* (1984b) have also found that a mutant (8237) of *X. campestris* pv. *campestris*, defective in the export of pectate lyase and protease, was nonpathogenic and that pathogenicity along with extracellular enzyme production was restored by a cosmid hybrid (pIJ3020). Aside from implications in enzyme export (see Section IV.C.2.b.ii.), these findings suggest a role of extracellular enzymes in pathogenicity of this bacterium. Alternatively, a gene product(s) involved in pathogenicity, along with pectate lyase and protease, is not exported or the nitrosoguanidine induced mutant has mutations in two closely linked genes: one specifying pathogenicity and the other the export of extracellular enzymes.

Despite the preliminary nature of the data on pathogenicity (host interaction) genes, several general, albeit tentative, observations can be made. (1) The pathogenicity genes appear to be clustered and, in some instances, the data are consistent with an operon type of organization (i.e., a single transcriptional unit). (2) An apparent association is indicated between HR induction and the production of disease symptoms, implicating a common factor in these two responses – pathogenicity being an outcome of a susceptible (compatible) reaction whereas HR is believed to be a manifestation of a resistant (incompatible) reaction. (3) The genes for pathogenicity/HR, at least in *P. syringae* pathovars, appear to have nucleotide sequence homology suggesting that these traits may have been conserved in these bacteria.

The cloning of avirulence genes of *P. syringae* pv. *glycinea* (Staskawicz *et al.*, 1984, 1985) and *X. campestris* pv. *malvacearum* (Gabriel, 1985) as well as the occurrence of a similar gene(s) in *P. syringae* pv. *pisi* (Taylor *et al.*, 1984) and on a conjugative plasmid in *X. campestris* pv. *vesicatoria* (Stall *et al.*, 1984; Stall, 1985) constitute highly significant developments in understanding the molecular basis for disease resistance and host specificity. The occurrence of avirulence genes was predicted in the "gene for gene" hypothesis (Flor, 1955; see also Ellingboe, 1976, 1982) wherein the interactions between the products of dominant avirulence gene(s) of pathogens with dominant resistant gene(s) of hosts were believed to result in a resistant (= hypersensitive) reaction. While this hypothesis has been amply corroborated by genetic data in fungi, the supporting evidence in bacterial systems was lacking till the work of Staskawicz and associates (Staskawicz *et al.*, 1984). They took advantage of the occurrence of well-defined races in *P. syringae* pv. *glycinea* that elicit compatible (= susceptible) or incompatible (= resistant = hypersensitive) reactions on soybean cultivars. For example, the races 1, 4, 5 and 6 elicit different reactions on four soybean cultivars. While race 6 gives an incompatible reaction on all these cultivars, compatible reactions of races 1, 4 and 5 occur on "Acme" and "Peking", on all four cultivars, and on "Harosoy" and

"Peking", respectively. Staskawicz *et al.* (1984) isolated, by cosmid cloning, a DNA segment of the race 6 genome (pPG6L3) that, when introduced into strains of race 1, race 4 and race 5, caused an incompatible reaction similar to that seen in race 6. By restriction mapping, subcloning and Tn*5* mutagenesis, the avirulence (= race specificity) gene(s) was localized within 2.5 kb on a 3.4 kb *AccI* fragment (pPG6AC1) (Staskawicz *et al.*, 1984, 1985). The plasmid, pPG6AC1, when introduced into the strains of race 0 as well as into *X. campestris* pv. *glycines*, converted them to *P. syringae* pv. *glycinea* race 6 host-range-incompatibility, thus confirming the occurrence of a dominant avirulence gene(s) on the plasmid. A protein product specified by the race-specific incompatibility A (rSIA) gene has been isolated on SDS-PAGE gels (Napoli *et al.*, 1985). Further information on the gene product(s) and its function and localization should significantly contribute to our understanding of the molecular events in recognition and disease resistance.

IV. KNOWN AND PROSPECTIVE DETERMINANTS OF BACTERIAL PLANT PATHOGENICITY

Bacterial pathogenicity involves interactions between (macro)molecules of the host and pathogen. In this section we will examine properties of some of the bacterial factors that affect host associations. Where appropriate we will discuss recent advances with animal pathogenic and nonpathogenic (free-living) bacteria.

A. Cell Envelope Components

Bacteria generally fall into three groups with respect to the architecture of their cell envelopes. A Gram-negative bacterial cell envelope is a complex three-layered structure consisting of the cytoplasmic (inner) membrane, a wall of the peptidoglycan (mucopeptide; murein) layer, and an outer membrane (lipopolysaccharide layer) (Fig. 6). Gram-positive bacteria possess the cytoplasmic membrane and the peptidoglycan layer (cell wall) but lack the outer unit membrane. Bacterial groups such as mycoplasmas and spiroplasmas apparently lack the peptidoglycan layer and the outer membrane; the cell contents are bound only by a cytoplasmic membrane. Nevertheless, spiroplasmas stain Gram-positive, presumably due to cell wall remnants. In addition, many Gram-positive and Gram-negative bacteria form a polysaccharide-containing structure(s) or a glycocalyx external to the outermost envelope component (Costerton *et al.*, 1981a,b); this is not considered part of the envelope (Sokatch, 1979).

The bacterial envelope controls many physiological activities and affects

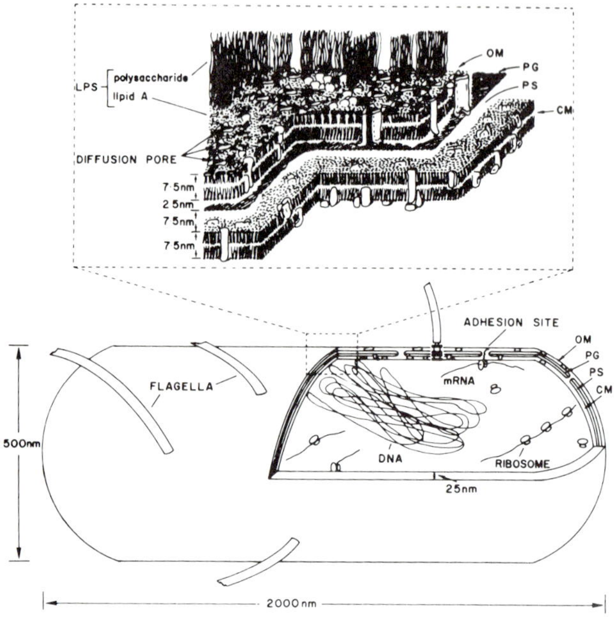

Fig. 6. Molecular architecture of the *Escherichia coli* envelope. OM, outer membrane; PG, peptidoglycan; PS, periplasmic space; CM, cytoplasmic membrane. The structure of the basal end of the flagellum is shown. (Reproduced from Inouye, 1979a.)

ecological behavior. Such cell components as the glycocalyx or lipopolysaccharide layer seem to affect plant pathogenicity by interacting directly with host cell components. On the other hand, the cytoplasmic membrane, which carries out a variety of metabolic activities, including the synthesis of cell envelope components (peptidoglycan, teichoic acids, or outer membrane), capsular material and extracellular proteins, solute transport, energy metabolism, genome replication, and cell division, would affect bacterial pathogenicity both directly and indirectly. Thus, it is appropriate to discuss here the nature of these cell surface components and examine their roles in bacterial phytopathogenicity.

1. Capsule

(a) *Introduction.* Capsular polymers (CP) are one of two main types of the glycocalyx; they are the outermost components of the cell periphery. Table XII illustrates the production of homopolysaccharides and heteropolysaccharides by various phytopathogenic bacteria. The structures of other polysaccharides are detailed in Sutherland (1977a), while a capsular polypeptide of *Bacillus licheniformis* is described by Troy (1979). The second type of glycocalyx, or S layer of globular glycoprotein units (Costerton *et al.*, 1981a,b), is not often seen. Some workers prefer to differentiate between the polymeric material adhering to the cell and that secreted into the growth medium. The latter substance, often referred to as slime or a flexible, peripheral capsule (Costerton *et al.*, 1981a,b), probably reflects hyperproduction and secretion or sloughing off of the material adhering to the cell, i.e., capsule. As will be seen later, various factors affect the biosynthesis of capsular polymers. The accumulation of polydisperse material in the culture supernatant, often interpreted as synthesis of diverse polymers, could also result from enzymatic (endodepolymerizing) or acid hydrolysis of the capsular polymer (Troy, 1979). Production of different polymeric species by a bacterial strain may occur under different cultural conditions. For example, Bennett and Billing (1980) found that the T strain of *E. amylovora* produced a fructose-rich polymer when grown in the presence of sucrose but produced a galactose-rich polymer when cultured in the presence of sorbitol or other carbon sources. This variation appears to be strain specific (Bennett and Billing, 1980; Thurn, Wyman, and Chatterjee, unpublished data). *P. syringae* pv. *phaseolicola* also appears to produce diverse polysaccharides: levan and alginate (predominantly composed of poly-D-mannuronic acid); in addition, a heteropolysaccharide containing rhamnose, fructose and galactose, a likely lipopolysaccharide component, was also detected in the EPS fraction (Gross, M., 1985). While both alginate and levan were present in heavily infected plants, only alginate was detected in early stages of infection (Gross, M., 1985). The production of alginate was also detected in soybean leaves inoculated with compatible strains of *P. syringae* pv. *glycinea* (Osman and Fett, 1985). The *in vitro* and *in planta* studies indicate that alginate production is a trait common to several pathovars (*glycinea*, *lacrymans*, *morsprunorum*, *papulans*, *phaseolicola*, *pisi*, *tabaci*, *tomato*) of *P. syringae* (Gross, M, 1985; Osman and Fett, 1985) and that alginate and other exopolysaccharides affected bacterial growth in plants (Rudolph *et al.*, 1985).

There are now considerable data supporting an essential role of capsule in phytopathogenicity of such bacteria as *E. amylovora*, *E. stewartii*, *P. solanacearum* and other *Pseudomonas* spp., *X. campestris* pv. *campestris*, *X. campestris* pv. *oryzae*, and other xanthomonads. The two principal observations for such a role are altered virulence of capsular mutants and pathological responses induced by fractionated capsular polymers.

In *E. amylovora*, noncapsular strains obtained as laboratory mutants or as naturally occurring strains (Billing, 1960; Bennett and Billing, 1978; Ayers *et al.*, 1979; Bennett, 1980; Bennett and Billing, 1980), are generally nonpathogenic or less virulent. Nonfluidal (noncapsular) mutants of *P. solanacearum* are known to be avirulent (Kelman, 1954; Husain and Kehman, 1958; Sequeira, 1979). An association between virulence and mucoid colony form in *X.*

Table XII. Composition of extracellular polysaccharides produced by phytopathogenic bacteria.

Class	Species	Components	Structure	References
Homopoly-saccharides	*A. radiobacter* *A. rhizogenes* *A. rubi* *A. tumefaciens*	(Glucose)$_n$ =glucan	$\beta(1\rightarrow2)$	Putman *et al.* (1950); Gorin *et al.* (1961)
	A. radiobacter *A. tumefaciens*	Cellulose	$\beta(1\rightarrow4)$	Matthysse *et al.* (1981)
	E. amylovora	(Fructose)$_n$=levans or polyfructose	?	Bennett and Billing (1980)
	Pseudomonas and *Xanthomonas* spp.	(Fructose)$_n$=levans	$\beta(2\rightarrow6)$?	Sutherland (1977a)
Heteropoly-saccharides	*A. tumefaciens*	Glucose, galactose, pyruvate, O-acetyl groups	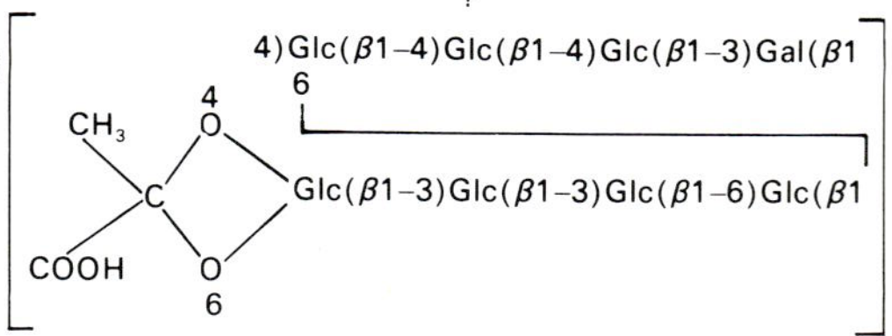	Zevenhuizen (1973); Jansson *et al* (1977); Hisamatsu *et al.* (1980)
	C. fascians	Glucose, mannose, galactose (trace)	?	Gorin and Spencer (1961a)
	C. flaccumfaciens subsp. *flaccumfaciens*	Glucose, mannose, fucose, rhamnose, aldobiouronic acid[a]	?	Gorin and Spencer (1961a)
	C. michiganense subsp. *insidiosum*	Glucose, galactose, fucose, pyruvate	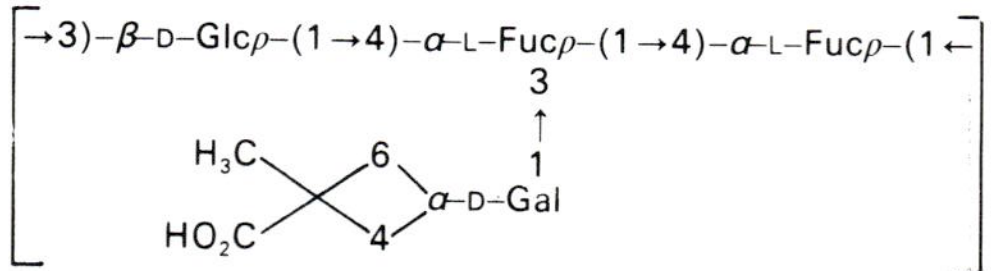	Gorin and Spencer (1964); Gorin *et al.* (1980)
	C. michiganense subsp. *michiganense*	Galactose, glucose, fucose, mannose	?	Gorin and Spencer (1961a)
	C. michiganense subsp. *sepedonicum*	Galactose, glucose, fucose	?	Gorin and Spencer (1961a)
	C. rathayi	Glucose, mannose, aldobiouronic acid, unknown	?	Gorin and Spencer (1961a)

C. tritici	Glucose, mannose, rhamnose, aldobiouronic acid	?	Gorin and Spencer (1961a)
E. amylovora	Galactose, glucose, mannose, uronic acid	?	Eden-Green and Knee (1974); Bennett and Billing (1980)
E. stewartii	Glucose, galactose, aldobiouronic acid (glucuronic acid)	?	Gorin and Spencer (1961b); Huang (1980)
P. solanacearum	Galactosamine	?	Dudman (1959); Whatley *et al.* (1980)
X. campestris pv. *campestris*	Glucose, mannose, glucuronic acid, pyruvate, acetate	...→4–β–D–Glc–1→4–β–D–Glc–1→... 3 ↑ 1 α–D–Man–6–0.CO.CH_3 2 ↑ 1 β–D–GlcA 4 ↑ 1 β–D–Man 4 6 ‖ Pyr	Jansson *et al.* (1975)
X. campestris pv. *carotae*	Aldobiouronic acid, glucose, mannose	?	Gorin and Spencer (1961b)
X. campestris pv. *hyacinthi*	Aldobiouronic acid, glucose, mannose	?	Gorin and Spencer (1961b)
X. campestris pv. *maculifolii-gardeniae*	Aldobiouronic acid, glucose, mannose	?	Gorin and Spencer (1961b)
X. campestris pv. *malvacearum*	Aldobiouronic acid, glucose, mannose	?	Gorin and Spencer (1961b)
X. campestris pv. *phaseoli*	Aldobiouronic acid, glucose, mannose	?	Gorin and Spencer (1961b)
X. campestris pv. *translucens*	Aldobiouronic acid, glucose, mannose	?	Gorin and Spencer (1961b)
X. campestris pv. *vignicola*	Aldobiouronic acid, glucose, mannose	?	Gorin and Spencer (1961b)

[a] Aldobiouronic acids are defined as dissaccharides in which one of the sugar components is a uronic acid linked in glycosidic union to a hexose or pentose unit. Thus, the precise composition of these disaccharides is not known.

campestris pv. *campestris* has been reported (Corey and Starr, 1957a), and, subsequently, also for other xanthomonads (Goto, 1972; Rao and Devdath, 1977). Reduced virulence and non-fluidal growth in *C. michiganense* subspecies occurs upon acquisition of plasmid pDG101 (Vidaver *et al.*, 1985).

In various instances, capsular materials have been fractionated from phytopathogenic bacteria and the pathological response of the introduced material has been followed. The findings with CP of *E. amylovora* and *Pseudomonas* and *Xanthomonas* spp. merit attention.

Amylovorin or CP production was first reported only in association with the host, as a wilt-inducing toxin (Goodman *et al.*, 1974; Hsu and Goodman, 1978). Fractionated polysaccharide was alleged to be host specific, causing both wilt and browning necrosis symptoms (Goodman *et al.*,1974; Ayers *et al.*, 1977; Goodman *et al.*, 1978). The biologically active fraction appears to be principally composed of galactose, glucuronic acid and glucose, with small amounts of mannose (Eden-Green and Knee, 1974; Bennett and Billing, 1980). However, Sjulin and Beer (1978) found that the CP produced by the pathogen under *in vitro* conditions and "amylovorin" produced in infected plants both caused wilting. Moreover, Sjulin and Beer (1978) showed that *E. amylovora* infection caused an alteration of cell permeability and that amylovorin caused wilting of the host plants by a nonspecific restriction of water movement. Thus, the alleged host-specific phytotoxicity of amylovorin has not been substantiated. Nevertheless, studies of mutants support an essential role of CP in pathogenicity of *E. amylovora*, perhaps in the establishment of the pathogen within host tissue. For example, ultrastructural studies show multiplication in plants and translocation of the capsulated strain but apparently restricted multiplication and no translocation of the noncapsulated strain (Huang *et al.*, 1975). However, studies of Billing and associates (Bennett, 1980; Bennett and Billing, 1980; Billing, 1984) revealed that in addition to CP, another factor, a cell leakage-inducing agent, determined virulence of *E. amylovora*.

Studies with extracellular (= capsular) polysaccharide (EPS) of pseudomonads and xanthomonads also show specificity of interactions between plants and pathogens (El-Banoby and Rudolph, 1978, 1979a,b, 1980). Partially purified EPS material caused water soaking in susceptible but not in resistant plants. Indeed, specificity was detected in exopolysaccharide obtained from bacterial races (strains specific for certain cultivars). These observations need to be extended with homogeneous EPS preparations and also with EPS from genetically modified strains.

Pathological conditions are accompanied by the introduction of high concentrations of EPS into plant tissues, at least with EPS from *Pseudomonas* and *Xanthomonas* species (El-Banoby and Rudolph, 1979a,b) and *E. amylovora* (Sjulin and Beer, 1978). Perhaps there is a general requirement for high levels of EPS in the elicitation of a biological response. Many phytopathogenic bacteria appear to produce copious amounts of EPS in plant tissues, much of which might be released into the intercellular spaces or in the vascular system. For example, *E. amylovora* (Van der Zwet and Keil, 1979) and *E. tracheiphila* (Smith, 1911) are known to produce copious amounts of ooze or threads, composed primarily of the EPS material, in infected plants. Similar, but less dramatic, production occurs with other phytopathogenic bacteria (see for example, Sutton and Williams, 1970; Braun, 1982; Smith and Mansfield, 1982; Gross, M., 1985; Osman and Fett, 1985).

Exopolysaccharide production in bacteria is influenced by a variety of physiological conditions and their genetic constitution (Sutherland, 1977a, 1982). For example, low incubation temperature, aeration (oxygen concentration), high C:N ratio, deficiency of P and S, and slower growth rate generally favor exopolysaccharide synthesis *in vitro*. During bacterial growth within plant tissues most of the conditions that are conducive to exopolysaccharide synthesis apparently are met. Technologies which should enable testing of physiological effects and accurate quantitation of bacterial EPS in plants are available.

Water soaking induced by EPS or live whole cells is a pathological condition resulting from movement of intracellular water into the intercellular space. It should be noted that the intercellular space within the mesophyll tissue is normally air filled, and thus does not normally permit physiological activities of bacteria. Conceivably, upon introduction into the mesophyll tissue, phytopathogenic bacteria synthesize (and excrete) EPS material which, by causing abnormal water movement from tissue, enables bacterial growth and multiplication. This would not exclude transient association of bacteria with host cell wall surfaces. A similar function has been attributed to lipomucopolysaccharides (Keen and Williams, 1971), lipopolysaccharides (LPS) (Epton *et al.*, 1977), and vesicles and wall membrane (Passmoor and Epton, 1978), although EPS was considered more effective and specific (El-Banoby and Rudolph, 1979a,b, 1980). Moreover, Rudolph *et al.* (1985) found a general stimulation of bacterial populations by several exopolysaccharides; this effect, however, was dependent upon the type of EPS and the physiological state of the host.

An additional function of capsules might be the masking of sites on the outer membrane complementary to plant components, thus preventing bacterial attachment to host cells. The findings with *P. solanacearum* and other phytopathogenic bacteria (Goodman, 1978; Sequeira, 1979; Smith and Mansfield, 1982; Fett and Jones, 1983) suggest that noncapsulated cells of the pathogen attach to host cells, elicit a host's defense reaction, and are ultimately inactivated or eliminated from the host tissue (see Section IV.A.2.a.vi.), while capsulated cells generally remain free in the intercellular space and multiply. Since capsules are known to mask cell surface antigens of many bacteria (Ørskov *et al.*, 1977), a similar function could be attributed to EPS in the interaction of LPS and other outer membrane components with plant cell wall proteins (for example, hydroxypoline-rich glycoprotein) or other cell wall components (see Section IV.A.2.b.vi. for HPRG-LPS interactions and a role of HPRG in host resistance). In at least *P. solanacearum*, this appears not to be the case, since EPS did not inhibit the LPS–HPRG interaction in plants (see Sequeira, 1984 for details). Additional investigations are needed to clarify the mechanisms by which capsules affect host interactions with phytopathogenic bacteria. Knowledge is required concerning monomeric composition, structure, immunological characteristics, genetics, and regulation of synthesis to facilitate such investigations. In contrast to phytopathogenic bacteria, much is known of capsules of certain animal pathogenic or free-living bacteria; the findings described below should serve as a model for future studies on phytopathogenic bacteria.

(b) *Demonstration of the Presence of Capsular Material.* Although light microscopy of cells treated with India ink, nigrosin, or other specific stains has generally proved useful for demonstrating capsular material, elegant electron microscopic procedures (Roth, 1977; Costerton, 1979) are quite effective in revealing the fine structure of the extracellular polymers. Electron microscopy of capsular polymers presents problems in that they are profoundly deranged upon dehydration and freezing. Moreover, they lack inherent electron density and resist heavy metal stains. In *E. coli*, Bayer and Thurow (1977) resolved these problems by treating cells with specific antibodies which both stabilized the exopolysaccharide and imparted sufficient electron density to render them visible. This simple yet elegant approach has revealed the fine structure of extracellular polymers in *E. amylovora and E. coli* as well as in other bacteria (Costerton, 1979; Politis and Goodman, 1980; Costerton *et al.*, 1981a,b) and also the structural changes occurring during adsorption by capsule-specific bacteriophages (Fig. 7). Stabilization with lectins and staining with cationic electron-dense stains, such as ruthenium red, have also been used in ultrastructural analyses of *E. amylovora* (Politis and Goodman, 1980), phytopathogenic and saprophytic *Pseudomonas* spp. (Smith and Mansfield, 1982), and other bacteria (Costerton *et al.*, 1981a,b).

Capsular polymers can be readily detected by a combination of immunological techniques such as agglutination, double diffusion in agar gels, and immunoelectrophoresis (Sutton and Williams, 1970; Ørskov *et al.*, 1977). Schaad (1979) has reviewed applications of immunological techniques in studies on phytopathogenic bacteria. Newer immunoelectrophoretic techniques such as rocket assay and counter immunoelectrophoresis, which have rarely been applied in studies on phytopathogenic bacteria (Ayers *et al.*, 1979),

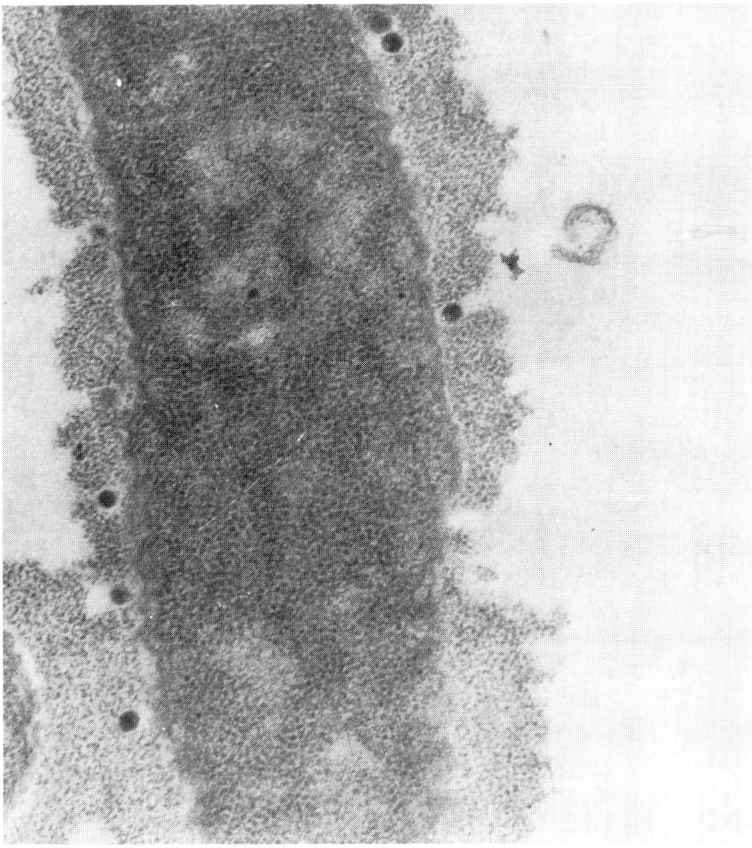

Fig. 7. *Escherichia coli* bacteriophage 29 after adsorption to encapsulated *E. coli* K29 cells. The bacteria were plasmolysed in 10% sucrose; the phages are seen at the cell wall, apparently preferentially at points of fusion between the wall and membrane after the endoglycosidase of the spikes at the phage tail has denuded the capsular layer. (Reproduced from Bayer *et al.*, 1979.)

should also prove valuable in quantitating the production of capsular polymers.

(c) *Isolation, Purification, and Characterization.* Polysaccharides excreted into the growth medium can be conveniently isolated by precipitating with ethanol (Bennett and Billing, 1980; El-Banoby and Rudolph, 1980), ammonium sulfate (Uchida *et al.*, 1973), acetone (Gahan *et al.*, 1967), and quaternary ammonium salts such as cetyltrimethylammonium bromide (cetavlon) (Zevenhuizen and Scholten-Koerselman, 1979).

Extraction of capsular material from cells can be accomplished by homogenizing in a blender or homogenizer (Adams and Young, 1965) and extracting with NaCl (Hungerer *et al.*, 1967) or aqueous phenol (Kamberger, 1979; Zevenhuizen and Scholten-Koerselman, 1979). Generally, drastic treatments such as boiling and alkali extraction should be avoided to prevent both contamination and degradation of the polymer.

Capsular polymers can be purified by applying a combination of such conventional techniques as fractional precipitation with cetavlon, solubilization in dimethyl sulfoxide, ultracentrifugation, molecular exclusion chromatography, ion exchange chromatography on Dowex 1, diethylaminoethyl (DEAE)-Sephadex, DEAE-cellulose, and gel electrophoresis (see Troy, 1979, for details). Nucleic acids and proteins can be removed by enzymatic (deoxyribonuclease, ribonuclease, pronase, trypsin) treatments. On a microscale, precipitation with specific antibody should produce polymer of high purity. Despite the knowledge that polysaccharides can specifically bind to plant lectins (Dazzo and Hubbell, 1975; Dazzo and Brill, 1977; Planque and Kijne, 1977; Kamberger, 1979), affinity purification of capsular polymers has not been reported.

Some of the criteria for purity of the preparations are elution of an apparent single molecular species from an ion exchange column, immunological homogeneity, NMR signals, and homogeneity by miniscus depletion sedimentation equilibrium (Troy, 1979).

A variety of chemical, immunological, and physical methods have been applied in structural investigations, some of which are highly sophisticated and require special skills and equipment. The details of these techniques can be found in Troy (1979).

(d) *Biosynthesis.* Based upon monosaccharide composition, exopolysaccharides can be placed under two major classes—homopolysaccharides and heteropolysaccharides. Both types of polymers occur in bacteria and some bacteria produce both simultaneously (Costerton *et al.*, 1981a,b); those produced by certain phytopathogenic bacteria are listed in Table XII. Extracellular homopolysaccharides include levans (*Pseudomonas* and *Xanthomonas* spp.) and glucans (*Agrobacterium* spp., *Rhizobium* spp.). Extracellular heteropolysaccharides, such as xanthan gum (from *X. campestris* pv. *campestris*), usually contain repeating units of 2–4 monosaccharides, uronic acid being the most common residue. Both homo- and heteropolysaccharides may possess some degree of branching and acyl and/or pyruvate substituents.

The biosynthesis of exopolysaccharides differs both in the type of glucosyl donor and the location of biosynthetic enzymes. In the biosynthesis of such homopolysaccharides as levans and dextrans, sucrose, but not nucleoside diphosphate sugar, is utilized as the glucosyl donor. The enzymes dextransucrase and levansucrase catalyze the following reactions:

$$n \text{ Sucrose} \xrightarrow{\text{(dextransucrase)}} (\text{dextran})_n + n \text{ fructose}$$
$$n \text{ Sucrose} \xrightarrow{\text{(levansucrase)}} (\text{levan})_n + n \text{ glucose}$$

Biosynthesis of these homopolysaccharides is predominatly extracellular, although cell-bound enzymes are often involved and a direct contact with the bacterial cell wall is required. In contrast, biosynthesis of heteropolysaccharides (see below) and some homopolysaccharides requires nucleoside diphosphate sugars and for the most part occurs on the bacterial cell membrane. Although certain phytopathogenic bacteria synthesize homopolysaccharides (Table XII), the biosynthetic steps are unknown.

Biosynthesis of heteropolysaccharides has been studied in various enteric and nonenteric bacteria and many reaction sequences have been elucidated. Although gaps remain in our overall knowledge, certain generalizations can now be made, some of which are outlined below. All are amenable to genetic analysis.

The mechanisms of synthesis of sugar nucleotides and their interconversions have been extensively reviewed (Glaser, 1973; Nikaido, 1973). The general reactions for Gram-negative bacteria are very similar; biosynthetic precursors for xanthan gum are shown in Fig. 8.

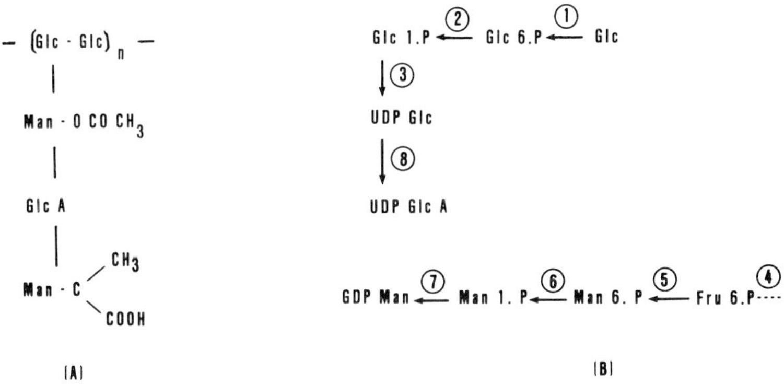

Fig. 8. Structure of the repeating unit of xanthan (A) and the synthesis of precursors (B). The enzymes of pathway (B) indicated by numbers are (1) hexokinase; (2) phosphoglucomutase; (3) UDP-glucose pyrophosphorylase; (4) phosphoglucose isomerase; (5) phosphomannose isomerase; (6) phosphomannomutase; (7) GDP mannose pyrophosphorylase; and (8) UDP-glucose dehydrogenase. Other enzymatic reactions mediated by pyruvate kinase and the enzymes for acetyl-CoA and isoprenoid lipid synthesis are not shown. Also note that reactions (1) through (6) participate in other metabolic systems (see text) whereas steps (7) and (8) are specifically required in the synthesis of the precursors of xanthan (see text and Sutherland 1977a,b for additional details). (Courtesy of I. Sutherland.) Glc, glucose; Glc A, glucuronic acid; Fru, fructose; Man, mannose; UDP, uridine diphosphate; GDP, guanosine diphosphate. Sugars are phosphorylated in either the 1 or 6 position.

Some precursors (UDP-galactose, UDP-glucose, UDP-*N*-acetyl glucosamine) may be common to more than one polymer. For example, UDP-galactose and UDP-glucose are used in the biosynthesis of both lipopolysaccharide and exopolysaccharide of many Gram-negative bacteria. This appears likely in at least *E. amylovora*, where both LPS and EPS appear to contain galactose (Eden-Green and Knee, 1974; Chatterjee *et al.*, 1977; Bennett and Billing, 1980). Moreover, certain noncapsular mutants were phenotypically reversed to capsular types in the presence of galactose (Billing, 1984). In such mutants, galactose inhibited growth, presumably due to a defect in UDP-galactose-4-epimerase and intracellular accumulation of galactose-1-phosphate (Billing, 1984). Assuming that UDP-galactose is a precursor of galactose in both LPS and EPS in *E. amylovora*, one would predict defects in both of these molecules in *GalE* mutants grown in the presence of glucose. In *E. stewartii*, galactose-sensitive mutants and the complementing wild-type DNA segments have been isolated by Dolph and Coplin (1985).

The second step, i.e., the synthesis of repeating units of the polymer, occurs by the sequential transfer of appropriate monosaccharides from sugar nucleotides to the carrier phospholipid undecaprenyl phosphate (P-C_{55} or lipid-P). The reactions shown in Fig. 9 were established for xanthan gum (Sutherland, 1977a; personal communication).

As stated above, alginic acid (i.e., the polymers of D-mannuronic acid) is produced by several pathovars of *P. syringae* (Gross, M., 1985; Osman and Fett, 1985). Although the presence of acetylated mannuronic acid and L-guluronic acid has not been shown in *P. syringae* alginic acid, it is reasonable to assume their occurrence based upon the evidence in *P. aeruginosa* and *Azotobacter vinelandii* (Sutherland, 1977a; Piggott *et al.*, 1981). This polysaccharide differs from other heteropolysaccharides (Table XII) in the absence of repeating units. Based upon the evidence in *P. aeruginosa* (Piggott *et al.*, 1981; Pugashetti *et al.*, 1983; Banerjee *et al.*, 1985; Darzino *et al.*, 1985), it seems that the biosynthesis of alginate in *P. syringae* pathovars may also occur via the following steps:

$$\text{fructose-6-phosphate} \xrightarrow[\text{isomerase}]{\text{phosphomannose}} \text{mannose-6-phospate} \xrightarrow[\text{mutase}]{\text{phosphomannose}} \text{mannose-1-}$$

$$\text{phosphate} \xrightarrow[\text{pyrophosphorylase}]{\text{GDP-mannose}} \text{GDP-mannose} \xrightarrow[\text{dehydrogenase}]{\text{GDP-mannose}} \text{GDP-mannuronic acid.}$$

Mannuronic acid is then incorporated into polyuronide; a glycosyl carrier lipid probably is involved in this step. The epimerization of polymannuronic acid, i.e., after the polymer synthesis, as shown in *P. aeruginosa* (Piggott *et al.*, 1981), may convert some of the mannuronic acid to L-glucuronic acid.

Further stages in the formation of capsular polysaccharide by polymerization of oligosaccharide repeating units or monosaccharides (viz. mannuronic acid) on the isoprenyl phosphate remain less clear. *In vitro* studies have shown polymerization of the lipid-linked tetrasaccharide into a macromolecular product identical to the native capsular polysaccharide of *K. aerogenes*. Nevertheless, several questions concerning final steps in the biosynthesis of the capsular polysaccharide await clarification. First the degree of polymerization of the oligosaccharide accumulated on the isoprenoid lipid is unknown. Second, transport of the oligosaccharide from its site of synthesis (presumably cytoplasmic membrane) to the cell exterior is unknown. Third, the mechanism of chain elongation and the enzymes involved are unknown.

Many exopolysaccharides, including that produced by *Xanthomonas* spp. (Table XII; Fig. 9), possess acyl and ketal substituents and longer side chains

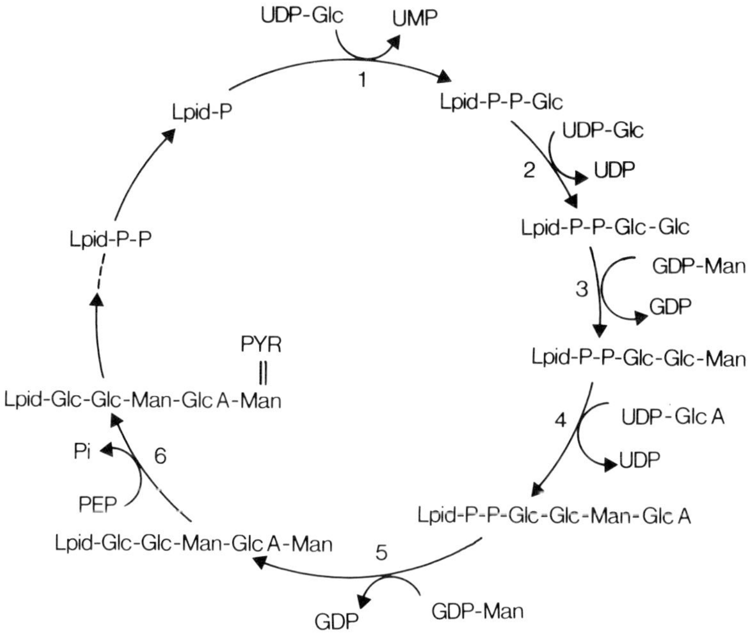

Fig. 9. Biosynthesis of the repeating unit of the xanthan polysaccharide in *Xanthomonas campestris* pv. *campestris*. The enzymes of the pathway indicated by numbers are (1) glucose-1-phosphate transferase; (2) glucosyl transferase; (3) mannosyl transferase I; (4) glucuronosyl transferase; (5) mannosyl transferase II; and (6) ketal transferase. The acetylation of mannose by mannose acetylase (not shown in the figure) occurs prior to the addition of the glucuronic acid residue (see Fig. 8 for the structure of the repeating unit). Lipid-P, carrier phospholipid; Lipid-P-P, carrier lipid pyrophosphate; PEP, phosphoenol pyruvate; P_i, inorganic phosphate; PYR, pyruvate. Other abbreviations are as in Fig. 8. For additional details see text and Sutherland (1977a,b). The precursors of carrier lipid and its interconversion are shown in Fig. 10. (Courtesy of I. Sutherland.)

than the exopolysaccharide of *K. aerogenes*. These findings raise the following questions: (1) are the acyl and ketal substituents added while the repeating units remain attached to the lipid carrier or at some later stage; and (2) are main and side chains constructed on separate carrier lipids prior to assembly?

The carrier lipid, (C_{55}) polyisoprenyl monophosphate, participates in the biosynthesis of repeating unit-containing glycan polymers located externally to the cell membrane, i.e., peptidoglycan, exopolysaccharide, and lipopolysaccharide or teichoic acid (Fig. 10). There is now some indirect evidence (Sutherland, 1977a,b) to suggest that the availability of isoprenoid lipid phosphate is critical in exopolysaccharide synthesis. In considering a regulatory role of isoprenoid lipid phosphate concentration, it is important to remember that this lipid carrier is required in the biosynthesis of both physiologically important structures such as peptidoglycan and the outer membrane and nonessential structures such as capsular polysaccharide.

Genetic evidence (Sutherland, 1977b), although indirect, suggests that the cell establishes a distinct set of priorities for isoprenoid lipid utilization. As might be expected, biosynthetic requirements in decreasing order of importance were peptidoglycan, lipopolysaccharide, and lastly exopolysacccharide (Sutherland, 1977a,b). This model is consistent with findings of low levels of exopolysaccharide synthesis by exponentially growing cells but high levels of exopolysaccharide synthesis at the end of log phase, during stationary phase, and under low incubation temperatures. However, several questions pertaining to this model require clarification: (1) can bacteria regulate the concentration of isoprenoid lipid phosphate and, if so, how; and (2) how does the cell establish biosynthetic priorities?

The biosynthesis of amylovorin (capsular heteropolysaccharide of *E. amylo-*

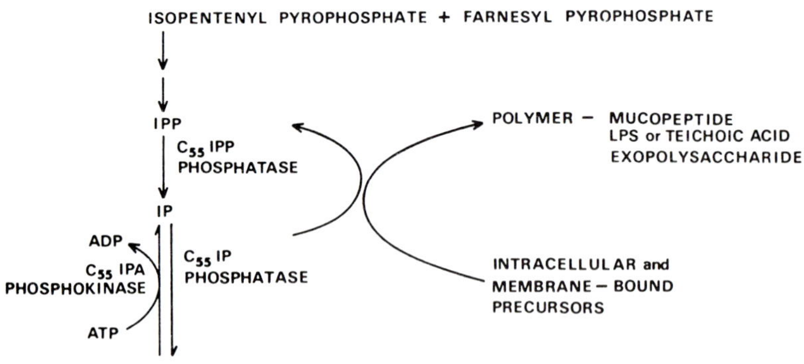

Fig. 10. Precursors of carrier lipid and its interconversion during polymer synthesis. Abbreviations: IPP, C_{55}—isoprenyl pyrophosphate; IP, C_{55}—isoprenyl phosphate; IPA, C_{55}—isoprenoid alcohol. (Reproduced from Sutherland, 1977b.)

vora) may follow a different pathway from that found in *K. aerogenes*. Transferase activity was detected with such acceptors as amylovorin (= EPS), *N*-acetyl glucosamine, and glucose; the latter acceptor role was remarkably stimulated in the presence of β-lactalbumin (Huang, 1979). However, the finding that amylovorin primer served as galactosyl acceptor does not necessarily prove that the enzyme is involved in the biosynthesis of capsular polysaccharide of *E. amylovora*. A similar galactosyl transferase was found in *E. stewartii* and implicated in the biosynthesis of EPS (Huang, 1980). Genetic approaches should prove useful in establishing a role of the galactosyl transferases in amylovorin and EPS synthesis.

(e) *Genetics*. Although capsule-defective mutants have been obtained in a few species, the genetics of capsular polysaccharides of phytopathogenic bacteria generally remains unexplored. As stated above, capsules seem to affect host associations of several phytopathogenic bacteria. Genetic approaches can be effective in defining the capsular role in pathogen establishment and host specificity. We briefly consider here some genetic approaches that might prove fruitful in understanding the biosynthesis, regulation, and function of capsular polysaccharides of phytopathogenic bacteria.

(i) Mutant phenotypes. Based upon the current knowledge of the biosynthesis of capsular polysaccharides, we predict that several gene functions are required in (1) synthesis of sugar components, (2) assembly of the oligosaccharide on the isoprenoid lipid phosphate, (3) polymerization of the oligosaccharide, and (4) transport of the polymer to the cell exterior followed by attachment onto the existing basal structure. Additional or alternate gene functions probably are also required in the biosynthesis of those exopolysaccharides that possess acetyl and pyruvyl groups as in *Xanthomonas* spp. (Table XII) or for Gram-positive bacteria.

Mutations in either structural or regulator genes could alter the capsular phenotype to yield the following phenotypic classes: (1) cells are noncapsulated, i.e., underproduce capsular material; (2) cells are heavily capsulated, i.e., overproduce capsular material; (3) cells synthesize capsular material different from that produced by the parent strain; (4) cells produce capsular material only under certain cultural conditions; and (5) cells excrete much of the polymer (or oligomer) into the growth medium. In the light of the findings for *E. coli* and *K. aerogenes* (see Markovitz, 1977), the occurrence of these phenotypic classes could be ascribed to the quantity or quality of the precursors, defects in polymerization steps, quantity of the isoprenoid lipid phosphate, or synthesis of depolymerizing enzymes.

Many bacteria, including some phytopathogens, are known to excrete capsular polysaccharide or slime material (= EPS) into the growth medium (El-Banoby and Rudolph, 1978; Ayers *et al.*, 1979; El-Banoby and Rudolph,

1980; Politis and Goodman, 1980). Obviously, in these instances all of the polymerized material either is not assembled onto the preexisting basal structure, or, if assembled, is not retained as a component of adherent capsular material. There is also the possibility that the steps of polymerization and assembly are loosely coupled. Whether or not excretion of capsular polysaccharide is required in bacterial pathogenicity remains to be established.

Structural gene mutations could mimic phenotypes generally ascribed to regulatory gene mutations by creating a deficiency of either the sugar precursors common to other envelope components (such as UDP-glucose, UDP-galactose) or isoprenoid lipid phosphate. During exponential growth, limiting quantities of sugar nucleotides and isoprenoid lipid phosphate might be preferentially utilized in the synthesis of lipopolysaccharide, but not the capsule, resulting in noncapsular cells. On the other hand, during stationary phase, limited synthesis of lipopolysaccharide would spare biosynthetic components, thereby enabling synthesis of the capsular polysaccharide so that cells in stationary phase would be capsulated. Elucidation of the mechanisms underlying such phenotypic repression and depression would necessitate both genetic and biochemical investigations.

(ii) Isolation and characterization of capsular mutants. Defects in capsular biosynthesis that alter colony morphology can be recognized on appropriate growth media. Thus, mucoid bacteria may give rise to nonmucoid colony types due to a lack of capsular synthesis or to watery, spreading (fluidal) colonies due to the overproduction of capsular material. Such mutant classes have been obtained in several phytopathogenic bacteria such as *E. amylovora* (Ayers *et al.*, 1979; Wyman and Chatterjee, unpublished data), *P. solanacearum* (Kelman, 1954; Message *et al.*, 1978), and *Xanthomonas* spp. (Corey and Starr, 1957a; Goto, 1972; Cadmus *et al.*, 1976; Otta, 1976; Rao and Devadath, 1977; Whitfield *et al.*, 1981). In *X. campestris* pv. *campestris*, Whitfield *et al.* (1981) recognized a class of mutants of unusual colony morphology and designated them as crenated mutants. The crenated mutants produced exopolysaccharide of an unusual composition containing sugars normally found in the lipopolysaccharide.

Conditions can be selected to enrich capsular mutants. The enrichment of presumptive noncapsular mutants of *P. solanacearum* in static cultures was attributed to motility of such mutants (Kelman and Hruschka, 1973). In static culture, the oxygen concentration is higher at the surface. Noncapsular mutants, apparently responding to an aerotactic stimulus, swim to the surface, while capsulated cells, which exhibit limited flagellar activity due to the viscous capsular material or lack flagella, remain under partially anaerobic conditions. Since *P. solanacearum* is an aerobic organism, noncapsular cells present in an oxygen-enriched environment multiply more rapidly than the capsulated cells localized in a partially anaerobic environment. These events might explain the accumulation of noncapsular mutants of *P. solanacearum* in stationary phase cultures or under storage. It should prove useful to determine if a similar enrichment is effective in the isolation of capsular mutants of other phytopathogenic bacteria, including such facultative anaerobes as *Erwinia*.

Certain bacteriophages are specific for capsular polysaccharides (Lindberg, 1977; Ayers *et al.*, 1979; Ritchie and Klos, 1979). Mutants resistant to such bacteriophages generally lack capsule or synthesize capsular material different from that produced by the bacteriophage-susceptible parent strain. Bacteriophage-resistant mutants associated with differences in colony morphology have been obtained in *E. amylovora* (Ayers *et al.*, 1979), *X. campestris* pv. *oryzae* (Goto, 1972), and *P. solanacearum* (Hendrick and Sequeira, 1984). Certain bacteriophage-resistant mutants of *E. amylovora* (Ayers *et al.*, 1979) and *P. solanacearum* (Hendrick and Sequeira, 1984) possessed little or no exopolysaccharide. Use of capsule-specific bacteriophage should prove highly effective in isolating spontaneous capsular mutants in other phytopathogenic bacteria.

Capsule-specific bacteriophages may be recognized by their plaque morphology since they may produce plaques surrounded by halos (Lindberg, 1977; Ayers *et al.*, 1979; Ritchie and Klos, 1979). The halos apparently result from the action of bacteriophage-induced, capsule-depolymerizing enzymes. In *E. amylovora* (Ayers *et al.*, 1979), the capsule-depolymerizing enzyme was associated with the capsule-specific bacteriophage [PEal (h)] particles. A similar situation may prevail in other phytopathogenic bacteria. However, it should be noted that not all capsule-specific bacteriophages produce plaques with halos (see Lindberg, 1977).

Another potentially efficient approach that remains to be exploited is the application of capsule-specific antibody in the enrichment of capsular mutants. The binding of antibody to capsulated cells should result in agglutination or altered density. Thus, cells that lack capsule or possess antigenically distinct exopolysaccharide could be recovered by differential centrifugation.

Separation of capsular mutants by sequential elution from DEAE-cellulose offers promise. Hall and Jafri (1980) reported that capsulated cells of *K. aerogenes* eluted with 200 m*M* NaCl, whereas noncapsulated cells eluted with 90 m*M* NaCl. This sequential elution, seemingly controlled by electrical charge, might be applicable in separation of capsular mutants as well as mutants that are defective in other cell envelope components. Indeed, using a similar approach, Hall *et al.* (1978) resolved fimbriate and nonfimbriate cells of *S. typhimurium* (see below for hydrophobic interaction chromatography).

Presumptive capsular mutants obtained by using any of the above procedures must be subjected to rigorous biochemical, immunological, and genetic characterization to assess the quality and quantity of exopolysaccharide produced. Genetic characterization is required to locate the mutational site and investigate the expression of the mutated (regulatory) gene (i.e., dominant or recessive). Moreover, isogenic lines, constructed by using recombinational methods, should be employed in physiological and pathological investigations.

It is also desirable, and often essential, to know that a capsular mutation has not produced pleiotropic change(s) in the bacterial cell envelope components. In certain bacteria the same precursors are utilized in the synthesis of both the cell envelope and capsules. Genetic changes causing a deficiency of the precursor would produce defects in both components. Erroneous conclusions regarding the role of cell envelope components in bacterial phytopathogenicity could be drawn in the absence of knowledge of such a pleiotrophic change. In previous studies on presumptive noncapsular mutants of *E. amylovora* (Ayers *et al.*, 1979) or *Xanthomonas* spp. (Corey and Starr, 1957a; Goto, 1972), characterization of cell envelope components was not done. Although organized capsules visible under the microscope have not been detected, it is noteworthy that certain spontaneous (nonfluidal) mutants of *P. solanacearum* were also defective in the lipopolysaccharide component (Whatley *et al.*, 1980). However, Hendrick and Sequeira (1984) found that the LPS moiety could vary from normal (wild-type) to mutant type; only if EPS was produced were mutants of *P. solanacearum* virulent. Coplin *et al.* (1982) found that *E. stewartii* noncapsulated mutants lacked the enzyme UDP-galactose epimerase, and suggested that avirulence might have resulted from a lipopolysaccharide defect. Also, the sugar composition of the lipopolysaccharides of nonmucoid (= noncapsular) and crenated mutants, such as in *X. campestris* pv. *campestris*, can be altered (Whitfield *et al.*, 1981).

Expression of bacterial mutational phenotypes in plants deserves attention. The possibility of phenotypic reversions in plants of certain regulatory gene mutations must be considered unless proved otherwise. In *E. amylovora*, some mutants that apparently lack capsular polysaccharide *in vitro* produced necrosis and copious ooze when pear seedlings were inoculated (Wyman and Chatterjee, unpublished). The ooze is probably composed of bacterial exopolysaccharide. Bacteria isolated from infected plants were nonmucoid on isolation media. This suggests that bacterial production of capsular polymers can be affected by host factors and that certain genetic defects of the pathogen may be phenotypically reversed by the host. Thus, gene expression should be investigated both under *in vivo* and *in vitro* conditions.

(iii) Location of capsular genes. Genetic determinants of capsular polysaccharides have not been mapped in any phytopathogenic bacterium. The spontaneous appearance of noncapsular mutants at an apparent high frequency in

certain phytopathogenic species (such as *P. solanacearum*) suggests either transposon or plasmid involvement, but there are no data to support this hypothesis. Similarly, transformation of colony type and modifications in colony morphology following lysogeny have been reported in *X. campestris* pv. *citri* (Wu, 1972a,b), but the genetic basis is unknown.

In *E. coli*, many gene products participate in capsule (cholanic acid) synthesis, thus indicating the complexity of its biosynthesis and regulation (Markovitz, 1977; Bachmann, 1983). As in *E. coli*, biosynthesis of capsular polysaccharides in phytopathogenic bacteria (*E. amylovora*, *E. stewartii*, *P. solanacearum*, X. *campestris* pv. *campestris*) probably requires expression of several structural and regulatory genes. The recent work of Coplin and asociates in *E. stewartii* supports this contention. By examining the complementation of capsular (Cps^-) mutants by a 27-kb cloned DNA (pES2144) and the phenotype of its derivatives constructed by Tn*5 lac* mutagenesis, Dolph and Coplin (1985) identified five separate regions of pES2144 that contained *cps*. They also constructed *cps–lac* fusions (see Section III.C.3. for gene fusions) which should facilitate investigations on the regulation of EPS production in *E. stewartii*.

Although transformation of type-specific capsular genes in pneumococci (Griffith, 1928; Avery *et al.*, 1944) was the first demonstration that gene transfer was mediated by DNA, the genetics of capsule synthesis in pneumococci has not been intensively studied. However, the phenomenon of so-called *binary capsulation* observed in pneumococcal transformants should be noted. In transformation of pneumococcal capsular types (type I, donor; type III, recipient), a majority of transformants was of the donor type (type I) while a minority was of a binary type, i.e., synthesized both donor and recipient types (type I and type III) of polysaccharides (Austrian *et al.*, 1959; Bernheimer *et al.*, 1967). Binary type transformants seem to have resulted from additive recombination and correction of the biochemical defect responsible for the noncapsular phenotype of the recipient. Since the possibility of plasmid-mediated capsule production has not been ruled out, explanations other than additive recombination are plausible. Despite the uncertainty of the genetic mechanism, the implications of binary capsulation in host associations of phytopathogenic bacteria are apparent. As an example, exopolysaccharides of pseudomonads and xanthomonads were found to elicit a pathological response (water soaking) in compatible but not in incompatible hosts. If the specificity of such a reaction (and host compatibility) is determined by exopolysaccharides, it could be predicted that recombinants which produce binary capsules but are otherwise genetically indistinguishable should exhibit a broader host range. Such investigations, now technically feasible (see Sections III. E. and F.), might provide insight into the genetic bases for races in phytopathogenic bacteria.

2. *Outer Membrane*

Electron microscopy of the envelope of *Escherichia coli*, the best studied bacterium, reveals five electron-dense layers: the innermost double track represents the cytoplasmic membrane (CM), the central track is the peptidoglycan layer, and the outer double track is the outer membrane (OM). Although such a tripartite envelope occurs in most Gram-negative bacteria, variations in specific components (OM or peptidoglycan) have been detected (see Costerton, 1979; Costerton *et al.*, 1981a,b, for details). The presence of an OM in the envelope of several Gram-negative bacteria has also been confirmed by biochemical criteria such as differential enrichment of OM and CM markers in fractionated membranes. Since the outer membrane is usually denser than the CM, it can be separated from the CM upon isopycnic sucrose gradient centrifugation of the total membrane (Osborn *et al.*, 1972); this technique has been used in membrane studies of some plant pathogens.

The unit outer membrane is composed of proteins, phospholipids, and lipopolysaccharide (Lugtenberg and Van Alphen, 1983). LPS is present exclusively in the outer leaflet of the OM bilayer, and the fatty acids of the lipid A portion cover 30–40% of the layer. OM, as opposed to the CM, contains fewer proteins. The two membranes also differ in phospholipid composition—OM is usually enriched in phosphatidyl ethanolamine and deficient in phosphatidyl glycerol and cardiolipin (Osborn *et al.*, 1972; Lugtenberg and Peters, 1976).

Despite current uncertainty over the distribution of lipids in the OM bilayer (Bayer *et al.*, 1975; Irvin *et al.*, 1975; Smit *et al.*, 1975; Kamio and Nikaido, 1976; Van Alphen and Lugtenberg, 1977; Wilkinson, 1977; Cronan, 1979), it is clear that proteins and lipopolysaccharide present in the outer leaflet are exposed to the external environment. The exposed segments of such molecules might well affect bacterial associations with plant hosts.

The OM's of *A. tumefaciens*, *E. amylovora*, *E. chrysanthemi*, *P. cepacia*, *P. solanacearum*, *P. syringae* pv. *syringae* and various pathovars of *X. campestris* have been fractionated substantially free from CM (Sprang and Lindow, 1981; Thurn *et al.*, 1981; Dianese and Schaad, 1982; Thurn and Chatterjee, 1982b; Anwar *et al.*, 1983; Hurlbert and Gross, 1983; Minsavage and Schaad, 1983; dos Santos and Dianese, 1985) (Fig. 11). Additional studies are required, however, to adequately characterize membrane components with respect to their composition and function.

Functions generally attributed to the OM are (1) to allow passive diffusion of small hydrophilic molecules; (2) to act as a selective barrier to certain chemicals, enzymes, and antibiotics; (3) to exclude hydrophobic compounds; (4) to interact with natural environments such as plant and animal tissues; (5) to harbor components toxic against animal and plant cells; (6) to provide

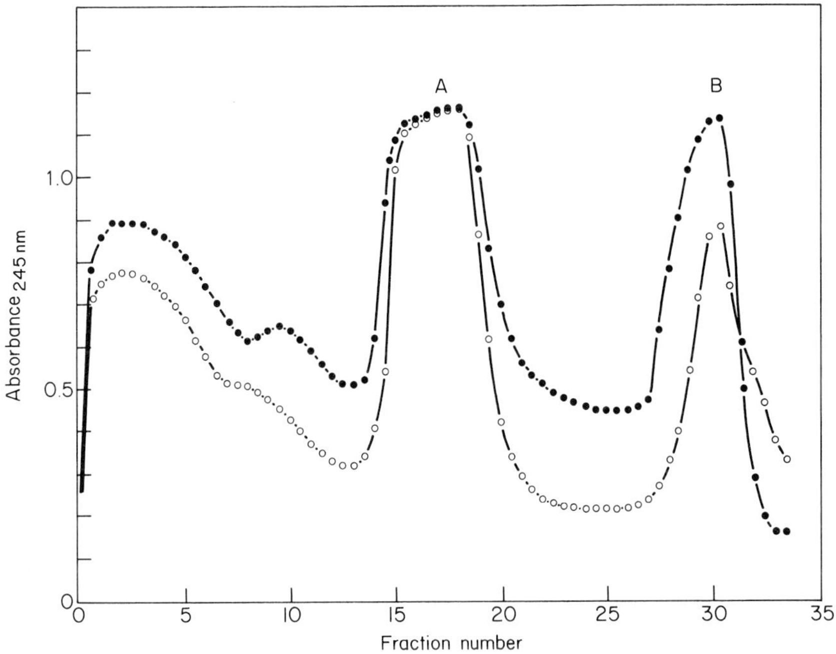

Fig. 11. Cytoplasmic inner (peak A) and outer (peak B) membranes of *Erwinia chrysanthemi* strains originating on corn (○) and chrysanthemum (●) as separated by isopycnic sucrose gradient centrifugation. Procedures described by Thurn and Chatterjee (1982b) were followed in the preparation of cell lysates, isolation of total membranes, and fractionation of membranes on sucrose gradients. Density of the gradients increases from the left to the right. Besides density, the peaks (A and B) differ with respect to the relative contents of lipopolysaccharide (ketodeoxyoctonate) and NADH oxidase activity.

receptors for bacteriophages, bacteriocins, and nutrients; and (7) to mediate cellular interactions during conjugation.

An additional function (8) of the OM is to confine degradative enzymes and various proteins into the periplasmic space, a compartment between the CM and the OM (Fig. 6). That space, which can include 20–40% of the total cell mass (Stock *et al.*, 1977), contains various degradative enzymes and transport (binding) proteins (Heppel, 1971; Costerton *et al.*, 1974). Certain mutations in OM components such as LPS cause release of enzymes normally found in the periplasm (Chatterjee *et al.*, 1976).

Ice nucleation is the ability of certain bacteria such as *E. herbicola* and *P. syringae* pv. *syringae* to cause ice crystal formation at relatively high temperatures (ca. −4°C) (Maki *et al.*, 1974; Hirano *et al.*, 1978; Lindow *et al.*, 1978;

Paulin and Luisetti, 1978). Sprang and Lindow (1981) associated ice nucleation activity of *P. syringae* pv. *syringae* with a wall fraction, predominantly composed of outer membrane. Recombinant DNA technology enabled isolation and characterization of a gene(s) that conferred upon *E. coli* the ability to become ice-nucleation positive (Orser *et al.*, 1983, 1985; Gies *et al.*, 1985; Govindarajan and Lindow, 1985). Genetic and physiological evidence (Kozloff *et al.*, 1984) indicates that phosphotidyl inositol, present in the outer surface of *P. syringae* and *E. herbicola*, is a major component of the ice-nucleating activity.

Individual components, such as LPS and proteins, are responsible for outer membrane functions outlined above. In the following sections we briefly consider the composition, biosynthesis, genetics, and functions of these membrane components.

(a) *Lipopolysaccharide.* Lipopolysaccharide (LPS) is an integral component of the outer membrane. It is an amphipathic molecule, i.e., it contains a polysaccharide (hydrophilic) tail and a lipid (hydrophobic) head. The molecule consists of three segments: lipid A, core polysaccharide, and the O-side chain (Fig. 12); the core polysaccharide is further subdivided into an "inner" core and an "outer" core. As seen later, the three segments differ in their composition, biosynthesis and biological functions. Although this model of LPS was originally proposed for *Salmonella* spp., the basic features appear valid for most, if not all, Gram-negative bacteria. In this section we present a brief account of the location, composition, isolation and characterization, biosynthesis, genetics, and functions of lipopolysaccharides, with emphasis on *Salmonella* spp. in which lipopolysaccharides have been most thoroughly investigated.

(i) Location. LPS is present exclusively in the outer leaflet of the outer membrane with the lipid A portion of the molecule anchored in the membrane and the polysaccharide (core and the O-side chain) extending onto the cell surface. LPS appears to be stabilized in the outer membrane by hydrophobic and ionic interactions. Treatment with chelating agents such as ethylenediamine-tetraacetic acid releases LPS from the membrane (Leive, 1965; Leive *et al.*, 1968; Rogers *et al.*, 1969; Mazzucchi and Pupillo, 1976). Within the membrane, LPS appears to be firmly associated with lipids and proteins, and mild extraction procedures generally yield a lipid–protein–LPS complex usually referred to as "complete" O-antigens. Although certain bacteria excrete LPS–protein complexes into the growth medium (Knox *et al.*, 1967; Marsh and Crutchley, 1967; Rothfield and Pearlman-Kothencz, 1969; Cadieux *et al.*, 1983), the bulk of the LPS remains attached to the outer membrane.

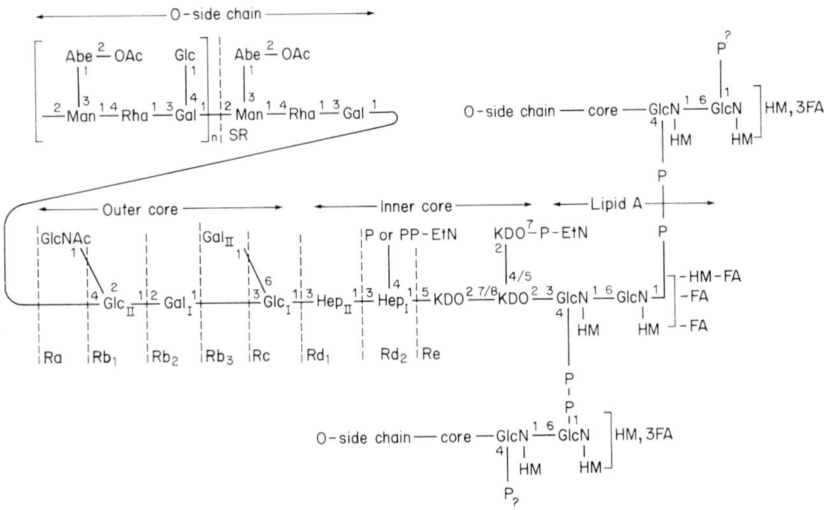

Fig. 12. Tentative structure of the lipopolysaccharide of *Salmonella typhimurium*. The number of chains cross-linked by pyrophosphate bridges between the diglucosamine residues was assumed to be three to suggest the polymeric structure. The fatty acids (HM, FA) are bound to the free hydroxyl groups of the diglucosamine residues. The vertical dotted lines indicate the core structures of the different mutants *SR*, *Ra*, *Re*. The components to the left of the line are missing in such mutants. Abbreviations: Abe, abequose; Ac, acetic acid; EtN, ethanolamine; FA, fatty acid ($C_{12:0}$, $C_{14:0}$, $C_{16:0}$); Gal, galactose; Glc, glucose; GlcN, glucosamine; Hep, 1-glycero-D-mannoheptose; HM, β-hydroxymyristic acid; KDO, 3-deoxy-D-mannooctulosonic acid; Man, mannose; Rha, rhamnose. (Reproduced from Braun and Hantke, 1974.)

(ii) Composition. The structure of LPS as it occurs in *S. typhimurium* is complex (Fig. 12). The molecule is composed of common hexoses (D-glucose, D-galactose, D-mannose), 6-deoxyhexoses (L-rhamnose), hexosamines (glucosamine), fatty acids [3-hydroxyalkanoic acids (3-OH 14:0) and nonhydroxy acids (12:0, 14:0, 16:0)], phosphate, acetate, ethanolamine, and such distinctive compounds as 3-deoxy-D-mannooctulosonic acid ("2-keto-3-deoxyoctonic acid"; KDO) and aldoheptose (L-glycero-D-mannoheptose).

The lipid A unit of the LPS component consists of a β-1,6-linked disaccharide of D-glucosamine with fatty acids attached to glucosamine by both ester and amide linkages. The disaccharide units are phosphorylated in positions 1 and 4′ and the 3′ position is the site of attachment of core polysaccharide via KDO. The findings of Muhlradt *et al.* (1977) and Rosner *et al.* (1979a,b) indicate that pyrophosphoryl ethanolamine is present at the reducing terminus of the lipid A of *Salmonella* and the 4′ phosphate is linked to the 1-position of 4-aminoarabinose, thus suggesting a non-cross-linked monomeric structure of lipid A. In the light of these findings it is unlikely that LPS occurs as a covalently cross-linked polymeric structure as depicted in Fig. 12.

We stress that variations in both structure and composition of lipid A have been noted (for details see Wilkinson, 1977, and Weckesser *et al.*, 1979), including those of plant pathogens.

Significant differences from *S. typhimurium*, the most studied bacterium, are already apparent in core polysaccharides among certain phytopathogenic bacteria. For example, LPS of *Agrobacterium* spp. lacks heptose but contains KDO and 4-*O*-methyl-D-glucuronic acid, glucosamine, quinovosamine, galactose, mannose, rhamnose, and xylose (Manasse and Corpe, 1967; Santaolalla, 1971, 1974; Salkinoja-Salonen and Boeck, 1978; Helander *et al.*, 1980). Sandulache and Prehm (1985) proposed the following structure for the core oligosaccharide of *E. carotovora* LPS:

$$\begin{array}{c} \text{Hep} \\ \beta \qquad \alpha \;\; \vdots 7\alpha \;\; \alpha \qquad \alpha \\ \text{Glc1---6Glc1---3Hep1---3Hep1---(KDO)}_n\text{---lipid A} \end{array}$$

This core structure is similar to that of *E. coli* B (Prehm *et al.*, 1975) except for the linkage of the terminal glucose residues: *E. coli* B contained a Glcα1–3Glcα1 unit, while *E. carotovora* had Glcβ1–6Glcα1 residues (Sandulache and Prehm, 1985). LPS of *X. campestris* pv. *campestris*, upon phenol–water extraction, resolved into phenol-soluble and water-soluble components (Hickman and Ashwell, 1966; Volk, 1968; Schlabach, 1970) which differed markedly in composition. The phenol-soluble LPS contained as major components D-rhamnose, 3-acetamido-3,6-dideoxy-D-galactose, D-glucuronic acid, and D-glucose, but not heptose or KDO. The water-soluble LPS of this and other *Xanthomonas* spp. possessed D-glucuronic acid (as 1-phosphate), glucose, mannose, rhamnose, KDO, and either fucose or xylose. These two lipopolysaccharides also differed in the configuration of rhamnose residues. The findings of Volk *et al.* (1972) with water-soluble LPS also suggest that the inner core region probably contains a single KDO residue substituted by mannose phosphate.

In pseudomonads, a pectolytic strain of *P. cepacia* contains heptose in LPS, but not KDO (Manniello *et al.*, 1979). The lipid A in the LPS of *P. solanacearum* contains fatty acids and glucosamine (five fatty acids for two glucosamine residues) (Drigues *et al.*, 1982); the polysaccharide is made of KDO, heptose, rhamnose, xylose, glucose, glucosamine, and 2-amino-2-deoxyglucose (Whatley *et al.*, 1980; Drigues *et al.*, 1982; Baker *et al.*, 1984). Bacteriophage-resistant LPS mutants of *P. solanacearum* differed in O-antigen and core sugars. Some had only core sugars (rhamnose, glucose, heptose, and KDO), some had no core rhamnose, and some had apparently only glucosamine and KDO (Hendrick and Sequeira, 1984). Heterogeneity was also apparent in the composition of neutral sugars in LPS from *P. syringae* pathovars (Anderson, 1984): *P. syringae* pv. *pisi* and *P. syringae* pv. *syringae* contained rhamnose and glucose, but no fucose, as the major components. All three sugars were, however, detected in LPS from *P. syringae* pvs. *phaseolicola* and *tabaci*. The predominance of rhamnose residues in lipopolysaccharides of *P. solanacearum*, *P. syringae* pathovars, *X. campestris* pv. *campestris*, and other *Xanthomonas* pathovars is rather striking.

O-side chains, also referred to as O-antigen chains, are polymers of oligosaccharide repeating units attached to the R core of the lipopolysaccharide (Fig. 12). Some enterobacteria such as *E. coli* and *Salmonella* spp. are classified into hundreds of O-groups (serogroups or serovars) based upon the presence of immunologically distinct O-antigens. Serological (O)-specificity is conferred by the composition and structure of the O-side chain (Ørskov *et al.*, 1977; Wilkinson, 1977). In many instances, particularly in *Salmonella* spp., O-antigenic diversity results from multiple variations of the same basic structure, such as the presence of group-specific residues of didehexoses like paratose, abequose, or tyvelose, and types of linkages between the O-side chain and the R core. The structure of the O-side chain is genetically determined and bacteriophages and plasmids are known to cause various sorts of structural modifications (see below), thus contributing to O-antigen diversity. It is noteworthy that despite extensive heterogeneity in O-side chains of these enterobacteria, the composition of the core polysaccharides is relatively uniform. Since *Erwinia*, *Pseudomonas*, and *Xanthomonas* species have adapted to different plant hosts, it should prove useful to determine the extent of diversity in the O-side chain structure and immunological properties.

The O-antigen composition (and structure) in *P. solanacearum* (Akiyama *et al.*, 1984; Hendrick and Sequeira, 1984; Baker *et al.*, 1984) and in races of *P. syringae* pv. *glycinea* (Barton-Willis *et al.*, 1984) do indicate variability. For example, in the K60 strain the O-antigen consists of rhamnose, 2-amino-2-deoxyglucose, and xylose in a ratio of 4:1:1, with terminal xylose, and substituents at 2, 3 and 4 positions of rhamnose and 3 position of 2-amino-2-deoxyglucose (Hendrick and Sequeira, 1984; Baker *et al.*, 1984). In contrast, the repeating unit of the O-antigen of an avirulent *P. solanacearum* strain is a tetrasaccharide of the following structure (Akiyama *et al.*, 1984):

$$\rightarrow 3)\text{-}\alpha\text{-}_{\text{D}}\text{-Glc}p\text{Nac-}(1\rightarrow 2)\text{-}\alpha\text{-}_{\text{L}}\text{-Rha}p\text{-}(1\rightarrow 2)\text{-}\alpha\text{-}_{\text{L}}\text{-Rha}p\text{-}(1\rightarrow 3)\text{-}\alpha\text{-}_{\text{L}}\text{-Rha}p\text{-}(1\rightarrow .$$

In light of these differences it may be desirable to examine the O-antigen structure of the various *P. solanacearum* "races" (Buddenhagen and Kelman, 1964) or "biotypes" (Hayward, 1964). Although the observations of Barton-Willis *et al.* (1984) did not associate race-specificity with LPS, they did show O-antigen heterogeneity in *P. syringae* pv. *glycinea* races. For example, while the major O-antigen components in races 1, 4, 5 and 6 were rhamnose and *N*-

acetylglucosamine, a marked difference in the ratio of these residues was noted in race 4 compared to races 1, 5 and 6. The LPS sidechain of *P. syringae* pv. *morsprunorum* strain C58 is composed entirely of rhamnose; this can be split off by phage A7, which contains a rhamnanase, to yield tri- or hexa-saccharide products (Smith *et al.*, 1984).

Based upon migration in polyacrylamide gels, Sandulache and Prehm (1985) tentatively concluded that the *E. carotovora* LPS was similar to those of "rough" bacteria and probably contained core oligosaccharide only. If this is indeed confirmed in additional strains, the serological diversity of the *E. carotovora* subspecies (see, e.g., DeBoer, 1985) may result from differences in outer membrane components other than the O-antigen.

(iii) Purification. The most extensively used procedures to extract LPS from whole cells or envelope preparations, among those currently available (see Lüderitz *et al.*, 1971; Wilkinson, 1977), are phenol–water (Westphal and Jann, 1965) and aqueous phenol–chloroform–light petroleum (PCP; Galanos *et al.*, 1969). Hot phenol extraction appears to be a method of choice for "smooth" LPS, i.e., those that possess a full complement of the molecule. The PCP method is more efficient in extracting R-type LPS that have reduced water solubility due to deficiencies in the O-side chain and core polysaccharide. Since LPS has been recovered from the phenol phase of *X. campestris* pv. *campestris* (Hickman and Ashwell, 1966) and other bacteria (Moreno *et al.*, 1979), care should be exercised while extracting LPS from phytopathogenic bacteria, particularly if the recovery of LPS in the aqueous phase is poor. In such an event, the extraction protocol used with *Brucella abortus* (Moreno *et al.*, 1979) might prove useful.

Purification of extracted LPS can be achieved by using a combination of such methods as sequential enzymatic treatments (DNAse and RNAse followed by Pronase), solubilization in dimethyl sulfoxide in the presence of sodium iodide and methanol precipitation of the solubilized LPS (Moreno *et al.*, 1979), chromatography on Sephadex G-200 (Moreno *et al.*, 1979; Hendrick, 1982), ultracentrifugation (Galanos *et al.*, 1969; Perry *et al.*, 1975), and precipitation with cations and cetavlon (Gormus and Wheat, 1971; Kropiniski *et al.*, 1979).

The extent of EPS (capsular polysaccharide) contamination can be minimized by growing bacteria under conditions that reduce synthesis of the EPS, washing cells thoroughly to remove much of the bound polysaccharide, and by using some of the procedures (such as precipitation with cetavlon) outlined above. Complete removal of EPS on a microscale might be accomplished by using capsule-specific antibody.

The absence of protein, phospholipid, and RNA are generally accepted as reasonable criteria of purity of LPS. Homogeneity in purified preparations is difficult to establish for the following reasons: (1) aggregation of LPS in aqueous solution and during phenol extraction; (2) microheterogeneity and phenotypic variation in LPS; and (3) partial degradation during isolation. Numerous methods are now available for cytological, physicochemical, structural, and biological characterizations of LPS, and the details can be found in Lüderitz *et al.* (1971), Ørskov *et al.* (1977), and Wilkinson (1977).

(iv) Biosynthesis. The synthesis of core polysaccharides and O-side chains in *Salmonella* spp. has been intensively studied and the mechanism of the formation of oligosaccharides, polymerization of the repeating units of the O-side chain, and their assembly on the core polysaccharide are well understood. Although some of the precursors in the biosynthesis of lipid A have been identified, many details remain unclear. Since numerous reviews and book chapters (Osborn, 1969; Heath, 1971; Osborn, 1971; Osborn and Rothfield, 1971; Glaser, 1973; Nikaido, 1973; Galanos *et al.*, 1977; Osborn, 1979) deal with the biosynthesis of LPS, we present here only a very brief overview of the subject.

Some of the final steps in the biosynthesis of lipid A, mostly derived from studies of *Salmonella*, are now recognized (see Osborn, 1979, for details). These developments include (1) elucidation of the structure of lipid A (Rietschel *et al.*, 1977; Rosner *et al.*, 1979a,b); (2) knowledge of the biosynthesis of KDO and the formation of the nucleotide (CMP)-KDO derivative (Levin and Racker, 1959; Ghalambor and Heath, 1966; Ghalambor *et al.*, 1966; Lim and Cohen, 1966); (3) isolation of mutants conditionally defective in KDO synthesis (Lehmann, 1977; Rick and Osborn, 1977; Rick *et al.*, 1977); (4) characterization of precursors accumulating in conditionally KDO-defective mutants (Lehmann, 1977; Lehmann and Rupprecht, 1977; Rick *et al.*, 1977), and (5) properties of CMP-KDO: lipid A-KDO transferase systems (Lehmann *et al.*, 1977, 1978; Munson *et al.*, 1978). The acetylated lipid A containing KDO residues serves as the acceptor of sugar residues which are added sequentially by specific sugar nucleotide transferases (Osborn, 1971; Osborn and Rothfield, 1971; Nikaido, 1973).

Biosynthesis of repeating oligosaccharide units of the O-side chain follows the general pattern discussed above for the synthesis of capsular polysaccharide in *K. aerogenes* (Section IV.A.2.a.iv.) Monomeric residues are sequentially transferred from nucleotide–sugar precursors to isoprenoid lipid phosphate (Osborn, 1969; Robbins and Wright, 1971) located on the inner face of the cytoplasmic membrane; the oligosaccharide lipid carried then moves to the outer face of the membrane where subunits are assembled onto the growing chain. The polymerized O-side chain is then transferred to the core polysaccharide, thus completing LPS biosynthesis.

Biosynthesis of LPS occurs on the cytoplasmic membrane (Osborn *et al.*,

1972), and the molecule is transported and assembled into the outer face of the outer membrane. The details of the mode(s) of transport and assembly of LPS into the outer membrane are not known. Elegant ultrastructural studies (Bayer, 1979) have revealed areas of adhesion between inner and outer membranes (Fig. 13). There is speculation that these zones serve as specialized sites of intermembrane transport of phospholipids, proteins, and LPS. Various models for the transport of LPS through the areas of adhesion have been postulated (Osborn, 1979).

(v) Genetics. Extensive genetic studies on the *Salmonella* LPS have contributed much to the knowledge of the structure and biosynthesis of this molecule. Bacteriophages with receptor sites in the LPS have proved extremely useful in the isolation of LPS-defective mutants and their characteriza-

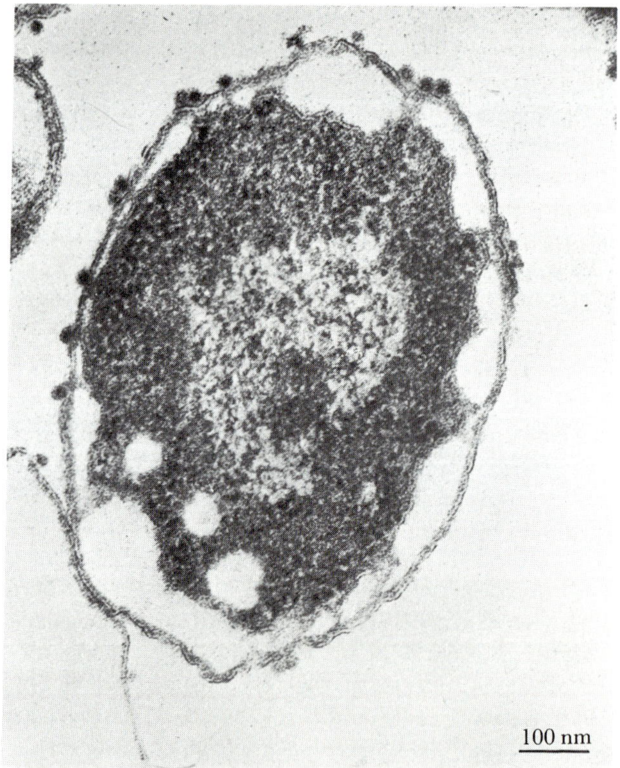

Fig. 13. Electron micrograph of the cross-section of plasmolyzed *Escherichia coli* C1 to which φ X had been adsorbed for 3 min with a multiplicity of infection of 200. Note that most of the virus particles are positioned over areas of wall membrane contact. × 120,000. (Reproduced from Bayer and Starkey, 1972.)

tion. In fact, modifications in LPS structure can be accurately predicted from the reactions of the bacterium to a limited number of bacteriophages. Many gene loci that control LPS biosynthesis have been mapped and the genetic mechanisms of LPS modifications have been clarified. Isolation of mutants conditionally defective in KDO biosynthesis and characterization of precursors accumulating under nonpermissive conditions have provided significant insights into the synthesis of lipid A and the inner core region of the LPS. The *Salmonella* system clearly serves as a model for genetic investigations on LPS of other bacteria. We present here some of the technologies and findings that may apply to studies on phytopathogenic bacteria; further details can be found in several reviews (Mäkelä and Stocker, 1969; Stocker and Mäkelä, 1971; Osborn, 1979).

Isolation and characterization of LPS-defective mutants. The LPS determines such bacterial properties as bacteriophage sensitivity (Lindberg, 1973, 1977), drug resistance (Leive, 1974; Roantree *et al.*, 1977), bile salt (deoxycholate) sensitivity (Sanderson *et al.*, 1974), periplasmic leakiness (Chatterjee *et al.*, 1976), and immunological characteristics (Ørskov *et al.*, 1977). LPS-defective strains may differ from normal LPS-containing strains in such cultural properties as smooth or rough colony morphology and growth characteristics (uniform turbidity or sediment) in liquid cultures (Stocker and Mäkelä, 1971). Only some of these properties are useful in the selection of LPS mutants. Resistance of bacteria to LPS-specific bacteriophages is the most effective tool in the isolation of spontaneous LPS-defective mutants. In *S. typhimurium*, there are available bacteriophages (P22, P22h, and 9NA) that attach to the O-side chain or those (GSR, Br2, Ffm, Br60, P221, C21) that attach to the R core chemotypes (Lindberg, 1973). (Chemotype = classification of LPS based upon chemical composition; see Fig. 12.)

By employing successive selections in *S. typhimurium* for bacteriophage resistance (Lindberg, 1973) it is feasible to obtain mutants defective in different regions of the LPS molecule. Although such a sophisticated approach is currently unavailable for phytopathogenic bacteria, the potential does exist. First, we must acquire a knowledge of the structure of the LPS and the sites of bacteriophage attachment on the molecule. Bacteriophages active against most phytopathogenic bacteria can be obtained readily (see Section III.E.2.). The receptor site for some of these phages of Gram-negative bacteria probably resides in the LPS, such as in *P. solanacearum* (Whatley *et al.*, 1980; Hendrick, 1982), and *P. syringae* pv. *morsprunorum* (Quirk *et al.*, 1976; Hignett and Quirk, 1979). In *E. amylovora*, bacteriophage-resistant mutants presumably defective in LPS were also obtained (Billing, 1984). Bacteriophage-resistant mutants of *P. syringae* pv. *morsprunorum* (cherry strain) were reduced in virulence (Garrett *et al.*, 1974). Bacteriophage Mu-resistant mutants of *E. carotovora* were defective in the core oligosaccharide of the LPS (Sandulache and Prehm, 1985). Similar studies with other phytopathogenic bacteria might prove useful in the isolation of LPS-defective strains. In characterizing receptors, either purified LPS should be employed or wall fraction(s) with bacteriophage still attached should be obtained from infected whole cells. Techniques which allow such determinations are available (Adams, 1959; Lindberg, 1973, 1977).

Drug resistance is sometimes associated with changes in the LPS molecule (Monner *et al.*, 1971; Brown and Wood, 1972; Gilleland and Murray, 1976). Liao and Heberlein (1979) obtained spontaneous polymyxin-resistant mutants of *A. tumefaciens* which generally exhibited reduced tumorigenicity compared to the parent. LPS alteration in these mutants seems likely in view of the findings with other polymyxin-resistant bacteria and the evidence for a role of LPS agrobacterial attachment to plant cells (Whatley *et al.*, 1976).

In certain bacteria, bacteriocin resistance apparently results from alterations in LPS structure. For example, receptor sites of R-type pyocins (R_1, R_2, and R_5) of *P. aeruginosa* are located in the LPS, and pyocin resistance in mutants was attributed to alterations in the LPS (Fig. 14). (R pyocins are resistant to trypsin; Konisky, 1978.) Pyocin resistance was also used to isolate LPS-defective mutants of *Neisseria gonorrhoeae* (Guymon *et al.*, 1982). In other

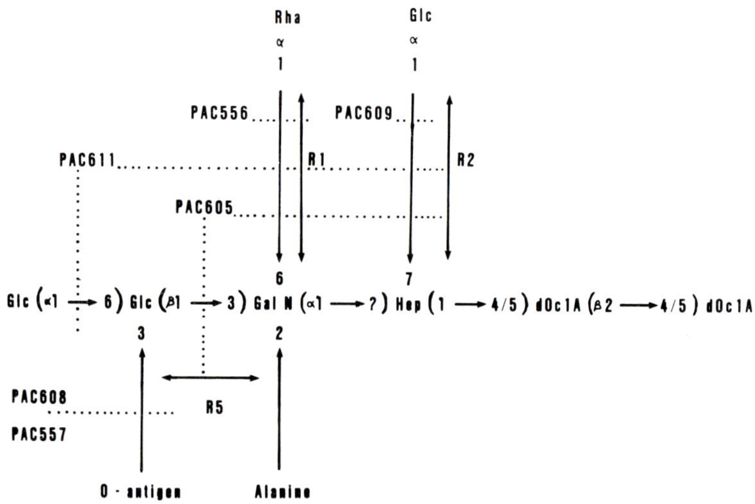

Fig. 14. Partial structure of the core part of the lipopolysaccharide of *Pseudomonas aeruginosa* PAC. Receptor sites for the R-type pyocins (R_1, R_2, R_5) are shown. PAC numbers represent mutants of *P. aeruginosa*, the dotted lines show the absence of the residues in mutants, and the arrows indicate the linkage between sugar residues. Abbreviations: dOclA, deoxyoctulsonic acid; Hep, heptose; GalN, galactosamine; Glc, glucose; Rha, rhamnose. (Courtesy of P. M. Meadow.)

bacteria, such as *E. coli*, outer membrane proteins are receptors of bacteriocins (colicins) and resistant mutants either possess an altered outer membrane protein or lack the receptor protein. It is likely that receptor sites of some of the bacteriocins produced by phytopathogenic bacteria (Vidaver, 1976) also reside in such outer membrane components as LPS and proteins.

To determine the role of LPS structure in host associations of phytopathogenic bacteria, it is essential to fully characterize putative LPS mutants isolated by using any of the approaches outlined above. Knowledge is required of the composition of the LPS, immunological properties, location of the gene(s), and pleiotropic changes in the outer membrane.

Although changes in LPS structure in putative mutants can be inferred by immunological criteria (passive hemagglutination inhibition; Stocker and Mäkelä, 1971) and rates of migration of the fractionated LPS upon SDS–polyacrylamide gel electrophoresis (Jann *et al.*, 1975; Whatley *et al.*, 1980; Bradbury *et al.*, 1984; Hendrick and Sequeira, 1984), detailed structural investigations are necessary to ascertain modifications in the LPS of the mutant strains. Technologies, such as NMR spectroscopy and gas–liquid chromatography combined with mass spectrometry (Perry *et al.*, 1975; Wilkinson and Taylor, 1978), have greatly facilitated structural investigations of the LPS.

Genetic characterization of LPS-defective mutants is useful in probing biosynthetic mechanisms, locating mutations, and constructing near-isogenic lines for studies on LPS effects on host associations. Certain alterations in LPS also induce loss of outer membrane proteins and profound changes in outer membrane properties (Ames *et al.*, 1974; Koplow and Goldfine, 1974). Thus it is desirable to characterize other outer membrane components, such as phospholipids and proteins, particularly in LPS mutations which seem to affect host associations.

Genetics of LPS biosynthesis. Although much is now known about the genetics of synthesis of the R-core and O-side chain of the LPS of *S. typhimurium*, genetic and biochemical aspects of lipid A synthesis remain obscure. Mutants defective in lipid A synthesis have not been obtained. This has led to the hypothesis (1) that this LPS component probably performs a vital cellular function, or (2) that the synthesis of lipid A is coupled with the synthesis of other essential cellular component(s) and thus a defect in lipid A synthesis is lethal to the cell. However, the inability to isolate conditional mutants cannot be reconciled on either of these grounds.

The biosynthesis of lipid A appears to be intimately associated with the synthesis of KDO and its incorporation into the lipid A precursors. Characterization of the precursors accumulating under nonpermissive conditions in mutants conditionally defective in KDO synthesis (see Obsorn, 1979) has revealed that incorporation of KDO residues precedes attachment of nonhydroxylated fatty acids to lipid A. Thus, terminal steps in lipid A biosynthesis do not occur in the absence of KDO synthesis and the attachment of KDO residues to the lipid A precursor. Isolation of conditional KDO-synthetase mutants in *P. solanacearum* should prove extremely useful in further probing the role of lipid A in the induction of host resistance to bacterial infection, as suggested by the findings of Sequeira and associates (Graham *et al.*, 1977). Such mutants may facilitate testing biological activities of various acidic and neutral forms of lipid A precursors. Moreover, phenotypic properties of KDO-defective mutants of phytopathogenic bacteria would reveal if the findings with *S. typhimurium* are generally applicable to other bacterial systems.

Mutants defective in the incorporation of sugar residues beyond KDO (i.e., R-core and O-side chain) have been obtained and the gene loci have been located on the chromosome of *S. typhimurium* (Sanderson and Roth, 1983). In addition to these, certain catabolic genes (*galE*, *galU*, *pmi*) affect the synthesis of nucleotide sugars, the precursors of the R-core and O-side chain oligosaccharides.

Despite the evidence for a role of LPS in host associations, the genetics of biosynthesis of this molecule has not been investigated in phytopathogenic bacteria. The few mutants that have been obtained remain to be characterized biochemically, immunologically, and genetically. Biochemical characterization has been done, however, for several *P. solanacearum* mutants (Hendrick and Sequeira, 1982; Baker *et al.*, 1984). Since homology is apparent in the locations of certain chromosomal genes on *Erwinia* and *S. typhimurium* chromosomes (Section III.E.1.) it would be of interest to determine if LPS genes of enterobacterial phytopathogens are functionally equivalent to *rfa* and *rfb* and are clustered as in *E. coli* or *S. typhimurium*.

LPS modification. Modification of the composition or structure of *Salmonella* LPS can occur upon bacteriophage infection or plasmid introduction. Variations in O-antigens arise upon *lysogenic conversion*, i.e., infection of the parental strain by temperate bacteriophages capable of modifying the basic O-antigen structure. Four types of structural changes have generally been detected (Osborn, 1969): (1) loss of substituents such as O-acetyl by phage ε^{15}; (2) change in configuration such as α-galactosyl to β-galactosyl by ε^{15}; (3) addition of a substituent such as glucosyl by P22 or ε^{34}; and (4) change in linkage such as Gal-1,4-Man to Gal-1,6-Man by ψ 27.

Bacteriophage-mediated modification can occur soon after infection (within 7–10 min) and irrespective of lysogeny or lysis of the host cell. Converting phages either carry genes for modifying activities or repress the function of bacterial genes. While modification of newly synthesized LPS has been documented (Robbins and Uchida, 1962), whether or not the alteration of old (preformed) LPS occurs upon phage infection remains to be established.

The biochemical and genetic bases for plasmid-mediated modification of LPS await elucidation. Thus far there is no evidence to suggest that R or Col plasmids, reported to cause LPS modification (Derylo *et al.*, 1975), carry genes for modifying activities (enzymes). Repression of host function(s) by plasmids remains a possibility.

Since many phytopathogenic bacteria are lysogenic and harbor one to several plasmids (Section III), the effects of these genetic elements merit elucidation. For example, introduction of various plasmids into *E. amylovora* resulted in enhanced susceptibility of transconjugants to a variety of antimicrobial agents (Chatterjee *et al.*, 1977); whether or not LPS modification also occurs is unknown.

Growth conditions may also affect LPS composition. For example, temperature-dependent variation in LPS structure has been reported in *Y. enterocolitica* (Acker *et al.*, 1980; Kawaoka *et al.*, 1983). The bacterium synthesized smooth LPS at a low temperature (22–25°C) but a partial smooth–rough transition (i.e., the loss of O-antigen) occurred upon growth at 37°C or higher. Whether or not temperature-dependent alteration in LPS structure also occurs in phytopathogenic bacteria is not known. However, in *P. solanacearum*, LPS composition was not affected by growth either in the presence of different carbon sources or for various incubation periods (Baker *et al.*, 1984).

(vi) LPS in plant host associations. An important role of LPS in bacterial associations with plants is suggested for *A. tumefaciens*, *E. chrysanthemi*, and *P. solanacearum*. For example, induction of the hypersensitive reaction in plant cells appeared to be related to LPS structure of *P. solanacearum* (Whatley *et al.*, 1980) but later was shown to be independent of LPS structure (Hendrick and Sequeira, 1984). The lipid A·R core segment of LPS seems to induce resistance in tobacco to *P. solanacearum* (Graham *et al.*, 1977). Infiltration of tobacco leaves with a protein–LPS complex of *E. chrysanthemi* prevents the hypersensitive necrosis reaction of tobacco to *P. syringae* pv. *syringae* (Mazzuchi and Pupillo, 1976). On the other hand, the LPS–protein complexes of *P. syringae* pv. *morsprunorum* and *P. syringae* pv. *syringae* elicit a toxic response in cherry leaves (Hignett and Quirk, 1979). Development of crown gall disease requires bacterial attachment to the wound site (Lippincott and Lippincott, 1976, 1980; Douglas *et al.*, 1982), and the competition experiments of Whatley *et al.* (1976) indicated LPS as the site of attachment. However, data of Gurlitz and Matthysse (1982) suggested that a surface (outer membrane or periplasmic) protein of *A. tumefaciens* was required in its attachment to callus-cultured carrot cells. The surfaces of such cultures may well differ from the surfaces of intact plants. Most of these findings need to be extended to rigorously establish the role of LPS (see Addendum).

Hydrophobic interactions may also play a role in bacterial attachment to plant cells. Cell surface hydrophobicity is an ecologically important factor in various microorganisms including certain human and animal pathogens (viz. *E. coli*, *Mycoplasma* spp., *Neisseria* spp., *Streptococcus* spp.), since it allows microbial adherence and subsequent proliferation on solid surfaces and interfaces (Marshall, 1976; Gibbons, 1980; Dahlback *et al.*, 1981; Rosenberg *et al.*, 1981; Gaastra and de Graff, 1982). In animal or human pathogenic *E. coli*, cell surface structures (e.g., K88, K99, or colonization factor antigens), collectively called *adhesins*, contribute to bacterial colonization of epithelial cells (Gaastra and de Graff, 1982). Such structures also contribute to cell surface properties such as hydrophobicity and hemagglutination (Smyth *et al.*, 1978; Wadstrom *et al.*, 1978; Faris *et al.*, 1981). Cell surface hydrophobicity can be measured by hydrophobic interaction chromatography (HIC) (Smyth *et al.*, 1978) and by adherence to hydrocarbons (Rosenberg *et al.*, 1980) or polystyrene (Rosenberg, 1981).

Preliminary results suggest an apparent correlation between cell surface hydrophobicity of certain phytopathogenic bacteria and their ability to attach to plant cells. For example, comparing *P. solanacearum* strain K60 (mucoid; virulent) with strain B1 (spontaneous nonmucoid and avirulent variant of K60, Sequeira *et al.*, 1977), revealed that the B1 cells bound to the hydrophobic gel phenyl-sepharose whereas the mucoid parent (K60) cells did not bind (McEvoy and Chatterjee, 1983). Moreover, a majority of B1 cells partitioned into the xylene (hydrocarbon)–water interface whereas the bulk of K60 cells remained in the aqueous phase; partition of cells into the xylene (hydrocarbon)–water interface also indicates cell surface hydrophobicity (Rosenberg *et al.*, 1982). Thus, the nonmucoid B1 strain, known to attach to plant (tobacco) cells (Sequeira *et al.*, 1977; Duvick and Sequeira, 1984b), exhibited hydrophobicity. In contrast, the parent K60 strain, which generally remains free in the intercellular spaces in plant (tobacco) tissue (Sequeira *et al.*, 1977), was not hydrophobic. Additional work is needed to identify bacterial components conferring cell surface hydrophobicity in these and other phytopathogenic bacteria and to assess the role of hydrophobicity in host–pathogen interactions.

In examining the role of LPS and other cell surface components in host–pathogen interactions, it is essential to consider the phenomenon of the hypersensitive reaction (HR). Investigations with many phytopathogenic bacteria, principally pseudomonads, have shown that upon introduction of high numbers (ca. 10^6–10^8) of cells of an incompatible pathogen, host cells rapidly collapse and the bacterial population rapidly declines; this response is generally referred to as the *hypersensitive reaction* (HR) (Klement and Goodman, 1967). When inoculated with fewer (i.e., about 10^4) cells, although there is a decline in pathogen population, no visible host reaction is discernible. Nonphytopathogenic, as well as some phytopathogenic bacteria, generally do not elicit HR.

Based upon various ultrastructural and physiological findings, the following model for the HR can be proposed. Upon introduction into plant tissue, incompatible bacteria attach to host cells. This event triggers a set of host responses which usually results in bacterial engulfment by material or a pellicle of host origin. Physiological activities of immobilized bacteria are generally suppressed, although certain metabolic activities do occur. Bacterial cells eventually disintegrate ("lyse") and are thus eliminated from the host tissue. Plant cells, physiologically altered by a diffusible product (HR factor) of the attached bacterium, also ultimately die. Whether or not a visible necrotic reaction occurs is simply a matter of the population of host cells that are physiologically altered by the pathogen (Turner and Novacky, 1974). This model predicts that attachment and the production of an "HR factor" are two discrete events.

Sequeira and associates (Sequeira *et al.*, 1977) have presented evidence for a lack of specificity in bacterial attachment to plant cells. For example, such saprophytic bacteria as *B. subtilis* and *E. coli* attached to the tobacco cell wall but did not elicit the HR. This lack of specificity can be explained by invoking the presence of complementary molecules (such as lectins and carbohydrate residues) on cell surfaces. Thus, once inside the tissue most bacteria, including saprophytes, avirulent mutants, and certain incompatible pathogens, bind to plant cell walls and are subsequently immobilized and inactivated. Hence, surface composition of plant cells, particularly those components that promote attachment of invading bacteria, can be viewed as constitutive traits contributing to the overall defense mechanism of the host. It then follows that a pathogenic bacterium, by producing a shielding (protective) device, prevents this interaction which is inimical to its activities.

Electron micrographs of thin sections clearly reveal ultrastructural changes in host cells following inoculation with compatible and incompatible bacteria (Fig. 15a and b). Attachment to plant cells occurs with incompatible bacteria (live or dead) (Fig. 15b), saprophytic bacteria, or purified LPS from wild-type cells. Accompanying changes are seen in the separation of plasmalemma from the cell wall and accumulation of vesicular and granular material in the space between the cell wall and plasmalemma. These structural derangements do not necessarily culminate into the HR. This view is supported by the finding that purified LPS and dead (heat-killed) bacteria attach to the plant cell wall and induce initial ultrastructural changes, but do not elicit the HR reaction. Support for the view that an additional bacterial factor(s) (HR factor) is required is that streptomycin or proteolytic enzymes prevented the HR (Sasser, 1978), suggesting a need for newly synthesized material, perhaps a protein.

Ultrastructural findings of Hildebrand *et al.* (1980) raise questions concerning the mechanism underlying bacterial attachment and immobilization to

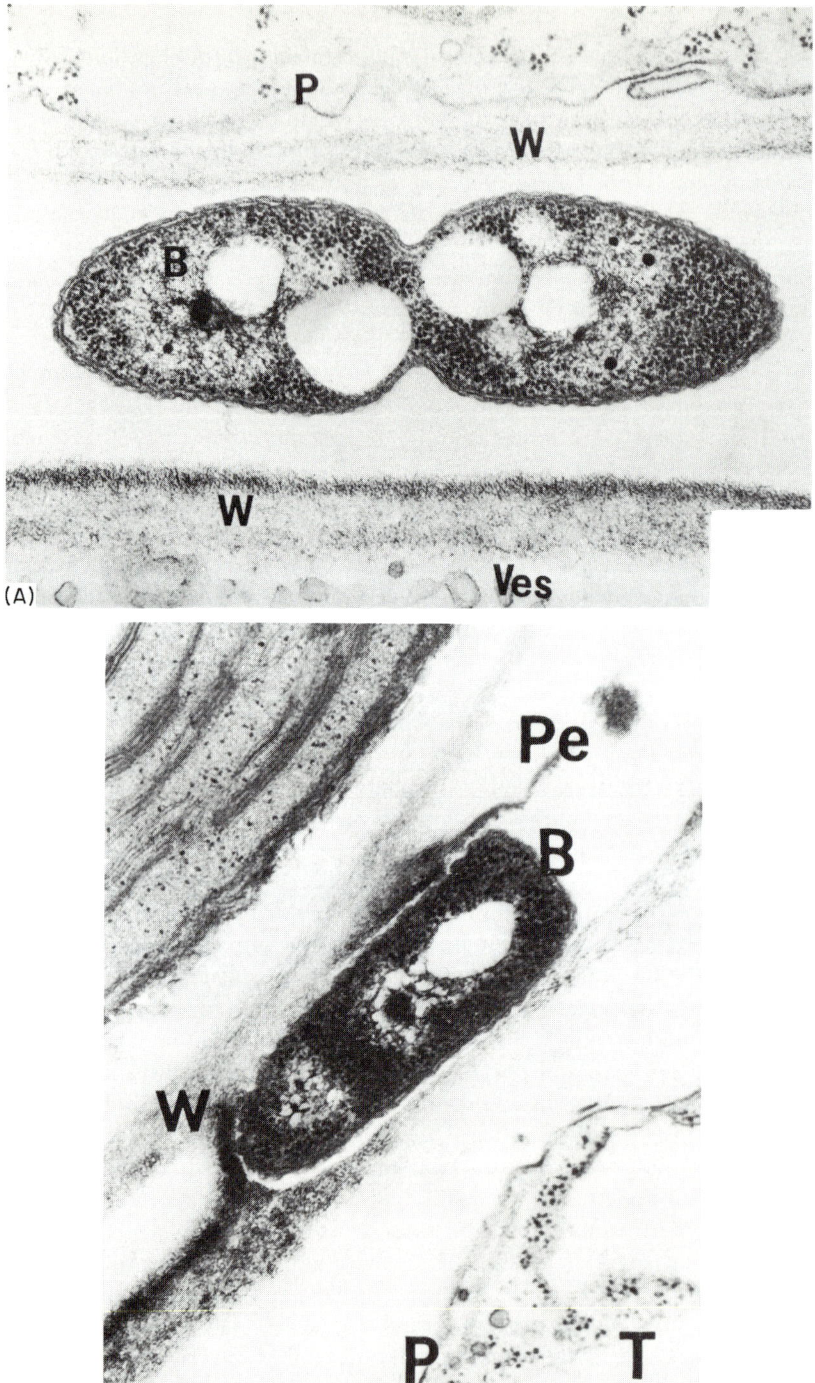
P
W
B
W
Ves
(A)
Pe
B
W
P
T
(B)

plant cell walls. These authors postulated that the fibrillar material implicated in pathogen entrapment was not formed in response to bacteria but resulted from physical forces generated during evaporation of water from intercellular spaces. Their data suggested that attachment of *P. syringae* pv. *syringae* and *P. syringae* pv. *phaseolicola* to the bean plant cell wall was not an active process. Evidence supporting an active process is cited by Duvick (1982) that Sequeira and Gaard found that latex spheres showed no association with mesophyll cell walls, thus nonspecific settling of particles onto cell wall surfaces may not occur. However, starch granules do attach (cited in Duvick, 1982) so that charge, size, and/or composition may affect attachment. Although the possibility of artifacts cannot be entirely ruled out, several findings (Graham *et al.*, 1977; Sequeira *et al.*, 1977; Sequeira and Graham, 1977; Al-Issa and Sigee, 1982; Leach *et al.*, 1982a,b; Tejerina *et al.*, 1982; Duvick and Sequeira, 1984a,b) support a physicochemical interaction leading to bacterial attachment and HR development. While the nature of the bacterial cell surface component(s) necessary for attachment is not yet known (Hendrick and Sequeira, 1982), there is some evidence to suggest that LPS is not essential. Nevertheless, the possibility remains that differences in charged substituent groups may be important.

Recent evidence strongly implicates hydroxyproline-rich glycoproteins (HPRG's) present in plant cell walls in early stages of bacterial attachment. Leach *et al.* (1982a) purified a protein from Katahdin potato tubers which strongly agglutinated avirulent strains of *P. solanacearum*, but only weakly agglutinated virulent strains. This glycoprotein, of an apparent molecular weight of 91,000 and a pI of > 11 was different from potato lectin (Allen and Neuberger, 1973; Sequeira and Graham, 1977). This glycoprotein was composed of 61% carbohydrates [predominantly arabinose (91%) along with galactose (5%), glucose (3%), and glucosamine (1%)]. The peptide moiety was rich in hydroxyproline (42%), lysine (16%), serine (9%), and proline (9%), hence the protein was called a hydroxyproline-rich glycoprotein. Immunofluorescent localization studies of Leach *et al.* (1982b) revealed the presence of HPRG or cross-reacting "HPRG-like" molecules in walls of tobacco and potato parenchyma cells as well as in the cell walls of xylem vessels, epidermis, and collenchyma. Moreover, the immunological evidence revealed the presence of HPRG-like molecules in cell walls of such nonsolana-

Fig. 15. Electron micrographs showing the interaction of host (tobacco) cell wall with compatible *Pseudomonas solanacearum* (strain K60) (A) and incompatible *P. solanacearum* (B1) (B). Abbreviations used: B, bacterium; Ch, chloroplast; Er, endoplasmic reticulum; M, mitochondrion; Mb, microbody; P, plasmalemma; Pe, pellicle; T, tonoplast; V, vacuole; Ves, vesicle; W, cell wall. (Reproduced from Sequeira *et al.*, 1977.) (A) At 4 hr after infiltration bacteria are dividing in the intercellular space. Note the separation of the plasmalemma and the accumulation of vesicles next to the cell wall. × 26,000. (B) Granular and fibrillar material progressively surround the incompatible strain B1 as it rests on the host cell wall 3 hr after infiltration. Note containment of the granular material by the host cell wall pellicle. × 32,000.

ceous plants as soybean, corn, and begonia. These observations led to the suggestion (1) that HPRGs are common components of plant cell walls and (2) that they may play a role in bacterial attachment (immobilization).

We wish to point out that several lines of evidence in fungal–plant interactions strongly implicate HPRG (=HRGP) in host resistance (Esquerré-Tugayé *et al.*, 1979; Esquerré-Tugayé and Lamport, 1979; Hammerschmidt *et al.*, 1984), although the underlying mechanism is not yet known. It also appears that HPRG accumulation is stimulated by ethylene or its biosynthetic precursor, 1-aminocyclopropane-1-carboxylic acid (Toppan *et al.*, 1982). Recent studies of Roby *et al.* (1985) showed that phytoalexin elicitors stimulated HPRG accumulation probably by enhancing ethylene synthesis. These findings implicate a common signal in the synthesis of molecules that could contribute to a generalized host resistance. In light of these observations and those with *P. solanacearum* (see above) it is desirable to examine the regulation of HPRG accumulation and the role of the cell wall protein in relation to bacterial infections of plants.

Genetic approaches might prove useful in unraveling the nature of a hypothetical HR factor(s). It is conceivable that the production of the HR factor by bacteria is induced by the host. Since macroscopic manifestation of HR depends upon a high inoculum concentration of bacterial cells, it could be argued that synthesis of an HR factor(s) either is induced in only a minority of input cells or a small quantity is produced per cell, which collectively is manifested in the HR. If the former hypothesis is correct, it should be possible to isolate by chemical or transpositional mutagenesis derepressed mutants that constitutively produce high levels of HR factor. Such mutants should elicit a typical HR at low (10^3–10^4) inoculum concentrations. It may also be feasible to obtain conditional mutants of the pathogen which would elicit HR under one set of circumstances but not in another. If mutants that hyperproduce the HR factor can be isolated, it should be feasible to purify and analyze the molecule(s). Tn*5* insertion mutants that no longer elicit the HR response have been obtained in *E. amylovora* (Bauer and Beer, 1985) and in pathovars of *P. syringae* (Panopoulos *et al.*, 1984; Lindgren *et al.*, 1985; Neipold *et al.*, 1985; Deasey *et al.*, 1985). This development has facilitated the isolation of HR genes (see Section III.G. for further details).

Alterations in membrane permeability have been observed consistently in plant cells that subsequently show the HR. Such an alteration could be attributed to sequestering divalent cations (such as Ca^{2+}) by LPS or to the activity of the HR factor. LPS-mediated change in cellular permeability is plausible for the following reasons: (1) LPS molecules avidly bind such divalent cations as Ca^{2+} and Mg^{2+}; (2) divalent cations such as Ca^{2+} affect the permeability of plant cell membranes; and (3) LPS binds to plant cell walls and produces structural changes in the wall and plasmalemma. The possibility remains that bacterial cells within plant tissue continue to synthesize and excrete LPS which is transported to the cytoplasmic membrane of the host. By sequestering Ca^{2+}, LPS then would enhance membrane permeability. This hypothesis can be tested both by examining LPS effects on the physiology and structure of plant cells and protoplasts and mutants showing various defects in the LPS.

Induction of plant resistance to bacterial infection is another function attributed to LPS. There are several instances where preinoculation with saprophytic, avirulent, or heat-killed Gram-negative or Gram-positive bacterial cells (which are devoid of LPS) protected plants against pathogenic bacteria (Carroll and Lukezic, 1972; Goodman, 1978; Sequeira, 1979, 1983). Much effort has been directed to the identification of the putative protective factor produced by various bacteria. Of those cellular fractions tested, LPS (from *P. solanacearum*) and LPS–protein complex (from *E. chrysanthemi*) prevented the HR in tobacco caused by *P. solanacearum* and *P. syringae* pv. *syringae*, respectively. The findings were interpreted as the induction of host resistance. Using chemically modified LPS, Graham *et al.* (1977) associated inducer activity with the core polysaccharide linked to the lipid A; free lipid A, R-core, or O-side chain did not confer a protective effect. Since deacylation of lipid A destroyed inducer activity, fatty acids substituted to diglucosamine residues are apparently required in the protective effect. LPS derived from such nonphytopathogenic bacteria as *E. coli* and *S. marcescens* also possessed inducer activity. In a *P. syringae* pv. *glycinea*-soybean system, it appears that restriction of bacterial multiplication in tissue undergoing HR is at least partly due to the accumulation of glyceollin, a phytoalexin (Keen and Holliday, 1982; Long *et al.*, 1985; see also Section IV.C.3.). In this pathovar (Barton-Willis *et al.*, 1984) and other *P. syringae* pathovars (Anderson, 1984), LPS did not seem to stimulate phytoalexin production.

Goodman and associates (Romeiro *et al.*, 1981a,b) detected the presence in apple seed, leaf, and stem tissue of a protein that agglutinated a noncapsulated avirulent strain of *E. amylovora*; the capsulated virulent parent was also agglutinated but at a much reduced efficiency. The protein (agglutinin; malin) was inferred not to be a lectin, based upon its lack of interaction with monosaccharides. The receptor for this agglutinin appeared to reside in the R-core of the LPS of *E. amylovora*. Production of proteins such as HPRGs (discussed above) may contribute to a generalized resistance of plants to invading bacteria. It remains to be determined if LPS or other bacterial components induce the synthesis in plant tissue of such biologically active proteins. There is some evidence that LPS induces the synthesis in plant tissue of an unidentified new protein which, along with two others, is correlated with resistance to bacterial multiplication (Leach *et al.*, 1983).

The induction of resistance by amphipathic molecules raises questions concerning the contributions of hydrophilic and hydrophobic components. It is possible that the acetylated lipid A facilitates inducer transport and subsequent interactions with the host cell membrane. The oligosaccharide (R-core) is probably the inducer, and the loss of activity in the absence of hydrophobic lipid A could be due to the lack of transport and/or destruction of oligosaccharides by host enzymes. Genetically modified and radioactively labeled LPS or chemically synthesized model compounds facilitate testing of these hypotheses.

In examining an inducer role for LPS it has been tacitly assumed that prevention of the HR is similar operationally to the prevention of disease development. Whether this assumption has any validity needs to be rigorously established.

(b) Outer Membrane (OM) Proteins. Compared to lipopolysaccharide, OM proteins of phytopathogenic bacteria have received little attention and only recently. In nonphytopathogenic bacteria, such as *E. coli*, *S. typhimurium*, *P. aeruginosa*, and *Neisseria gonorrhoeae*, important physiological functions and properties have been associated with OM proteins (Braun, 1978; DiRienzo *et al.*, 1978; Inouye, 1979a,b; Inouye and Halegoua, 1979; Osborn and Wu, 1980; Hall and Silhavy, 1981; Lugtenberg and Van Alphen, 1983). These include (1) structural formation of passive diffusion pores, (2) F-pilus-mediated conjugation, (3) maintenance of integrity of outer membrane structure, (4) uptake of nutrients, and (5) attachment sites for bacteriophages and bacteriocins. Lambden *et al.* (1979) suggest that certain (OM) proteins are important virulence factors of *N. gonorrhoeae*. In *Y. enterocolitica* several OM polypeptides were present in virulent but not in avirulent strains (Chang and Doyle, 1984). It seems reasonable to postulate that OM proteins of phytopathogenic bacteria might well affect bacterial attachment to plant cells, induction of disease resistance, phytotoxicity, transport of solutes, and structural integrity of the bacterial membrane. To associate such functions with OM proteins of phytopathogenic bacteria, both genetic and biochemical studies are essential. The means to do so now are available, and some are presented here to stimulate investigations with phytopathogenic bacteria.

(i) Fractionation. Investigations on OM proteins have been facilitated by development of such technologies as (1) differential separation of CM from OM by equilibrium centrifugation (see Section IV.A.2.), (2) differential solubilization of CM proteins in detergents (Schnaitman, 1971; Moldow *et al.*, 1972; Filip et al., 1973; Rosenbusch, 1974), (3) gel electrophoresis and electrofocusing (Laemmli, 1970; Lugtenberg *et al.*, 1975; O'Farrell, 1975; Ames and Nikaido, 1976; O'Farrell *et al.*, 1977; Jones *et al.*, 1980; Kamio and Takahashi, 1980), (4) immunoelectrophoretic techniques (Kroll, 1973; Owens and Salton, 1975; Smyth *et al.*, 1976, 1978; Collins *et al.*, 1979), (5) reconstitu-

tion of artificial membranes (Nakae, 1975; Nixdorff *et al.*, 1977; Hancock and Nikaido, 1978; Nikaido, 1979) and (6) differential radiolabeling of OM proteins, including those oriented on the outer face of the OM (Kamio and Nikaido, 1977; Halegoua and Inouye, 1979).

Separation of CM and OM by isopycnic sucrose gradient centrifugation (see Section IV.A.2. for details) provides a powerful tool for the enrichment of OM proteins. With the *E. coli* envelope, anionic (Triton X-100) and mild ionic (sodium lauryl sarcosinate) detergents solubilized CM proteins but not the OM proteins (Schnaitman, 1971; Filip *et al.*, 1973); thus the use of these detergents yielded preparations enriched in the OM proteins and largely free from CM proteins. Likewise, sodium lauryl sarcosinate selectively solubilized the CM of *E. amylovora*, enabling fractionation of OM proteins (Thurn and Chatterjee, 1982b). As in *E. coli* (Schnaitman, 1971), a preferential solubilization by 0.4% (w/v) SDS of the CM of *P. syringae* pv. *syringae* has been indicated (Hurlbert and Gross, 1983). In contrast, the detergent did not appear to effectively solubilize the CM of *X. campestris* pv. *campestris* (Minsavage and Schaad, 1983). Thus, in investigations on phytopathogenic bacteria, differential solubility of CM and OM by various agents (such as guanidinium compounds, sodium thiocyanate, cetyltrimethylammonium bromide, sodium deoxycholate, sodium lauryl sarcosinate, Triton X-100, or lithium chloride) should be tested in developing a standard protocol for isolation of OM proteins.

Lithium chloride or lithium acetate has proven useful in extracting the OM from *N. gonorrhoeae* (Johnston *et al.*, 1976; Heckels, 1977; Jones *et al.*, 1980). Yakrus and Schaad (1979) isolated a "membrane protein complex" (MPC) from *E. chrysanthemi* by extracting whole cells with LiCl but the extent of contaminating CM or other proteins was not reported. In *E. amylovora* (strain E9), Thurn and Chatterjee (unpublished) were unable to efficiently extract membranes using various concentrations of LiCl and different temperature conditions.

Hofstra *et al.* (1980) reported the isolation of membrane fractions enriched in OM components following sonication of various enterobacteria (in the presence of EDTA) and ultracentrifugation. The efficiency of this simple procedure in studies on phytopathogenic bacteria remains to be tested.

OM obtained by any of these methods contains, in addition to proteins, such envelope components as peptidoglycan and LPS. Proteins are associated with these components either by covalent linkage or by hydrophobic interactions. Boiling of the outer membrane in the presence of sodium dodecyl sulfate (SDS) solubilizes most of the OM proteins (Rosenbusch, 1974). Lipoprotein, covalently attached to peptidoglycan, is released by lysozyme treatment of the peptidoglycan–protein complex (Suzuki *et al.*, 1976; Braun, 1978; Thurn and Chatterjee, 1982b). Most of the OM proteins of *E. coli* have been extensively purified by using standard protein purification procedures (Rosebusch, 1974; Hindennach and Henning, 1975; Uemura and Mizushima, 1975; Inouye *et al.*, 1976; Lee *et al.*, 1979). Such procedures also should be applicable to proteins of phytopathogenic bacteria.

(ii) Properties. Studies with both enteric and nonenteric bacteria have shown the presence of relatively few major proteins in the OM compared to the CM (Inouye, 1979b; Osborn and Wu, 1980). For example, in *E. coli* K12, two to four dominant protein species are detected, depending upon cultural conditions and genetic mutations. Likewise, in various phytopathogenic bacteria, three to six dominant proteins are detected in the OM of the wild-type strains (Thurn and Chatterjee, 1982b; Minsavage and Schaad, 1983).

Studies with *E. coli*, *S. typhimurium*, *P. aeruginosa*, *Rhodopseudomonas sphaeroides*, and *N. gonorrhoeae* have shown that OM proteins are peptides, glycoproteins, or lipoproteins. There are at least three classes of dominant proteins in the OM of *E. coli* K12: the matrix proteins (or porins), OmpA (TolG) protein, and lipoprotein (Osborn and Wu, 1980). So far as is known at present, these protein classes differ in their structure and location (organization) as well as in function.

Several types of evidence suggest that porins affect cellular structure and physiology. For example, in mutant strains that lack normal porins there is a strong tendency to produce new porins. Moreover, *E. coli* cells lysogenized with the bacteriophage PA2 synthesize a new peptidoglycan-associated protein which seems to behave as a porin (Schnaitman *et al.*, 1975; Pugsley and Schnaitman, 1978). There is also evidence for regulatory control of cultural conditions (Schnaitman, 1974; Schnaitman *et al.*, 1975; Nakamura and Mizushima, 1976; Bassford *et al.*, 1977; Van Alphen and Lugtenberg, 1977; Smit and Nikaido, 1978; Kawazi *et al.*, 1979; Overbeeke and Lugtenberg, 1980) on the production of porin proteins. Such findings implicate porins functioning in bacteria in their natural habitats.

There is a strong rationale for the production of multiple types of porin proteins, each of which produces channels of apparently indistinguishable exclusion limit. The production of multiple porin species may ensure porin activity if one or more are lost in strains resistant to bacteriophages or bacteriocins. Or a channel produced by each of the porin species might have slightly different permeability properties enabling discrimination between the transfer of various solutes. Such subtle differences might well affect bacterial activities in nature.

Based upon the results with nonphytopathogenic bacteria, we postulate that all Gram-negative phytopathogenic bacteria also possess OM matrix protein(s) that function as porins. Whether porin-like molecules exist in Gram-positive bacteria is not known. Confirmation of their existence would require isolation of peptidoglycan-associated proteins and establishment of their porin functions in reconstituted vesicles. By using mutants deficient in porin proteins, it should also be feasible to determine if porin functions are required in bacterial multiplication in plants. In such studies two factors that affect interpretation of results are: (1) the effect of the host (metabolites) on the differential expression of genes for matrix protein(s); and (2) selection of strains with new porin proteins due to mutations in normally silent gene(s). A regulatory effect of the host is more likely in view of the findings that cultural conditions and physiological factors affect expression of matrix–protein genes in *E. coli*. The technical limitations in studying OM proteins of bacteria growing within or on the plant might be overcome by using model systems such as plant cell and tissue cultures.

Certain OM *proteins* are *heat modifiable* (HM), i.e., there is a shift in the molecular weight of the polypeptide as a function of incubation temperature in the presence of SDS (DiRienzo *et al.*, 1978; Hancock and Carey, 1979). As an example, the OmpA protein of *E. coli* undergoes a characteristic conformational change in the protein upon heating to 100°C in the presence of SDS; this has the anomalous effect of raising the apparent molecular weight from 28 to 33.5k (Beher *et al.*, 1980). Heat-modifiable OM proteins of *E. coli* K12 possess β-structured sequences which are stable in SDS solution at low temperatures (ca. 0–60°C) but which are destroyed at higher temperatures (ca. 70–100°C). Heat-modifiable proteins occur in the OM of all bacteria examined to date, including such phytopathogens as *E. amylovora* (Thurn and Chatterjee, 1982b), *E. carotovora* subsp. *carotovora* (Beher *et al.*, 1980) and X. *campestris* pvs. *cassavae* and *manihotis* (dos Santos and Dianese, 1985), although their physiological function is unclear in most bacteria. In *E. coli*, the HM (OmpA or TolG) protein is required in F-pilus-mediated conjugation (Manning and Achtman, 1979). Mutants of *E. coli* K12 deficient in OmpA protein can be isolated by using bacteriophages (K3 or TuII*), colicin L, or by loss of recipient function (Reeves, 1979). The HM proteins can be detected in a variety of bacteria and are immunologically related in enterobacteria (Beher *et al.*, 1980). The HM protein of *E. carotovora* subsp. *carotovora* cross-reacted with anti-HM protein of *E. coli*, although peptide maps of HM proteins of these bacteria were quite different (Beher *et al.*, 1980). Such findings show that HM proteins from bacteria distantly related to *E. coli* shared at least some properties of the *E. coli* protein. The hypothesis that the structure (and, presumably, the function) of HM protein(s) of Gram-negative bacteria is conserved during evolution (Beher *et al.*, 1980) in phytopathogenic bacteria awaits confirmation.

Lipoprotein (LP), of about 700 Da, is another major OM protein of *E. coli*. It exists covalently linked to the peptidoglycan or in a free form in the membrane (Halegoua and Inouye, 1979). According to Inouye (1979b) this protein is the most abundant in *E. coli* cells; the total (free and bound) LP acounts for 7.2 × 10^5 molecules per cell. The complete chemical structure (Braun, 1975, 1978; Halegoua and Inouye, 1979) and the biosynthesis of LP in *E. coli* have been extensively studied, and further details can be found in Inouye (1979a,b).

Based upon both direct and indirect evidence, it is likely that most, if not all, Gram-negative bacteria possess LP in their OM. Inouye and associates (Nakamura and Inouye, 1979) made interesting observations on the degrees of homology in lipoprotein gene(s) (*lpp*) of various enteric and nonenteric bacteria. From the results of hybridization with ^{32}P-labeled purified LP–*m*RNA from *E. coli*, bacteria were segregated into three groups on the basis of different degrees of homology: group I contained *E. coli*, *Shigella*, *Salmonella*, *Citrobacter*; group II contained *Enterobacter*, *Klebsiella*, *Serratia*, *Erwinia*; and group III contained *Proteus*. Nonenterobacterial DNA (i.e., from *Pseudomonas aeruginosa*, *Acinetobacter*, *C. crescentus*, *M. xanthus*) did not hybridize with *E. coli* LP–*m*RNA. Subsequently, Yamagata *et al.* (1981) cloned the *E. amylovora lpp* gene into *E. coli* using a labeled *lpp* DNA probe from *S. marcescens*. The DNA sequence data strongly suggested that the structure of the LP was highly conserved in various enterobacteria (*E. amylovora*, *E. coli*, and *S. marcescens*).

Despite intensive studies, the precise function of LP remains obscure. Studies of mutants (Suzuki *et al.*, 1976; Torti and Park, 1976; Yem and Wu, 1977, 1978) reveal that LP plays an important role in the integrity of the OM structure in *E. coli*. Whether this is true for other bacteria remains to be elucidated. Indeed, in *P. mirabilis*, Gmeiner (1979) failed to detect covalently linked LP in exponentially growing cells, although cells in stationary phase possessed covalently linked LP in amounts similar to those found in *E. coli* or *S. typhimurium* during all growth phases.

In concluding this section we would like to emphasize that more attention should be given to OM proteins of phytopathogenic bacteria. Research on OM proteins is both feasible and attractive for the following reasons.

1. There are fewer proteins in relatively large quantities in the OM, thus facilitating their purification and subsequent investigations.
2. Mutants defective in OM protein can be readily obtained using bacteriophages, bacteriocins, drugs, and certain toxic elements such as Cu^{2+} and Ag^{+}.
3. Biochemical, biophysical, and immunological techniques are available which allow fractionation of OM proteins and investigations of their properties.

Future research involving OM proteins of phytopathogenic bacteria should help in resolving questions such as: (1) whether homology in the OM proteins could be used as a criterion of taxonomic relatedness, in diagnosis, and in epidemiological investigations (see, for example, dos Santos and Dianese, 1985); (2) the role of OM proteins in bacterial–host associations (see below), particularly in the context of compatible and incompatible reactions, including elicitation of host defense reactions; (3) the role of OM proteins in bacterial resistance to Cu^{2+}-based biocides, as has been noted in *X. campestris* pv. *vesicatoria* (Stall *et al.*, 1984), and ice-nucleating activity; and (4) OM protein function(s) in infectivity of *A. tumefaciens*, specifically in bacterial attachment and the transfer of Ti or Ri DNA.

In the context of a role of OM in host–pathogen interactions, the findings of Expert and Toussaint (1985) are of considerable significance. They isolated bacteriocin-resistant mutants of *E. chrysanthemi* (strain 3937 JRH) that were nonpathogenic on *Saintpaulia*. The mutants were deficient in one to three OM polypeptides, the production of which, in the parent, was stimulated under iron limited growth conditions. The following two hypotheses were invoked (Expert and Toussaint, 1985): (1) the OM proteins may be normally involved in surface interactions between the host and the pathogen and the mutations, by altering the OM proteins, caused changes in recognition; or (2) the mutants were unable to procure iron from the host tissue due to the loss of receptors for iron siderophores—in this case the missing OM protein(s) were assumed to function as the receptors since they were induced under conditions of iron deficiency. (*Siderophores* are iron chelating molecules produced by various microorganisms including many phytopathogenic bacteria; see, for example, Leong and Neiland, 1982; Neiland, 1982.) Despite the current uncertainty of the mechanism underlying the loss of pathogenicity, the findings are highly significant since they demonstrate a role of cell surface components, in addition to pectolytic enzymes (see below), in plant pathogenicity of *E. chrysanthemi*. Additional work is needed to assess a generalized role of OM proteins in host associations of other *E. chrysanthemi* strains originating on different hosts (Dickey, 1981).

3. *Peptidoglycan and Other Wall Components*

Pathogenicity is not known to be correlated with the structure or composition of the peptidoglycan in either Gram-negative or Gram-positive bacteria. However, the peptidoglycan and lipoteichoic acid (see below) of Gram-positive bacteria associated with human infections can apparently possess some endotoxin-like properties and produce inflammatory skin reactions (Heymer and Rietschel, 1977; Peterson and Quie, 1981). Thus, it is notable that of all species studied to date (Tipper and Wright, 1979), the related phytopathogenic corynebacteria *C. flaccumfaciens* subsp. *poinsettiae*, *C. michiganense* subsp. *insidiosum*, and *C. michiganense* subsp. *sepedonicum* and a few other saprophytic and pathogenic bacteria share a distinctive, so-called type B2, peptidoglycan (Tipper and Wright, 1979). Effects of these purified and isolated peptidoglycans in plants would be of interest. The peptidoglycan of virulent and avirulent *P. solanacearum* strains lacked biological activity (Wacek and Sequeira, 1973). The peptidoglycans showed no significant compositional differences between the strains and were similar to those found in other Gram-negative bacteria.

Teichoic acids are common in Gram-positive cell walls, but these polymers containing alditol phosphates have not been detected in *C. flaccumfaciens* subsp. *poinsettiae* and *C. flaccumfaciens* subsp. *betae* (Diaz-Maurino and Perkins, 1974) or other corynebacteria (unspecified; Fiedler *et al.*, 1981). However, it is possible that the teichoic acids are in the membrane covalently linked to lipid. Tipper and Wright (1979) proposed that most Gram-positive bacteria contain such lipoteichoic acids. No analyses for such compounds have yet been reported for any phytopathogenic bacterium, so the significance of their presence or absence cannot be assessed. On the other hand, teichuronic acids, a group of acidic polysaccharides (see Rogers, 1983), are likely components of the cell walls of at least *C. flaccumfaciens* subsp. *poinsettiae* and *C. flaccumfaciens* subsp. *betae* (Diaz-Maurino and Perkins, 1974), since glucuronic acid was detected in hydrolysates of their cell walls. Other corynebacteria may also contain teichuronic acids, based on hydrolysis products (Gorin and Spencer, 1961a). Confirmation and extension of these reports would be of interest, despite lack of knowledge of the functions of such compounds (Rogers, 1983). Readers should be cautioned that drastic changes in cell wall structure, affecting synthesis of teichoic and teichuronic acids, can occur under different growth conditions (Tipper and Wright, 1979; Shockman and Barrett, 1983). Knowledge is rapidly increasing on the synthesis, regulation, and interrelationships of these compounds to other cell wall components in nonphytopathogens (Shockman and Barrett, 1983).

B. Flagella, Pili, and Fimbriae

Flagella are principally organs of motility and also function in chemotaxis. In plant pathogens, there is no apparent correlation with pathogenicity and motility in several pathogens; e.g., corynebacteria contain both motile and nonmotile species. In *P. solanacearum* K-60, however, nonmotile cells are pathogenic (Kelman and Hruschka, 1973), and this applies to most other strains with some exceptions (Kelman, personal communication). Likewise, nonmotile mutants of *A. tumefaciens* were able to induce tumors and attach to plant cells (Bradley *et al.*, 1984). Raymundo and Ries (1980, 1981a,b) studied the effects of cultural conditions on flagellar synthesis and factors affecting the chemotactic response of *E. amylovora*. However, motility appeared not to be necessary either for entry or in pathogenesis of this bacterium. Nor is motility required for the flagella-less pathogen *Spiroplasma* (Townsend *et al.*, 1977), an organism which is normally helical and motile by a yet undetermined mechanism. Motility may, however, be correlated with virulence, since motile *P. syringae* pv. *phaseolicola* caused more infection foci than paralyzed mutants (Panopoulous and Schroth, 1974).

Pili and fimbriae are much shorter and thinner than flagella and cannot be resolved without an electron microscope. Pili can serve as adsorption organs for bacteriophages, mediate conjugation (sex pili), and can be plasmid determined. Sex pili are undoubtedly involved in at least some of the reports of conjugation in plant pathogenic bacteria (see Sections III.D. and III.E.1.); their genetic and chemical characterization would be of interest. The term fimbriae is often used synonymously with pili, but it has been suggested that they are hairlike appendages not involved in transfer of bacterial or viral nucleic acids (Ottow, 1975).

Fimbriae are of different morphological types, and can have different distinguishable biological properties, which has resulted in some efforts to classify them (Ottow, 1975). They are found more commonly thus far on Gram-negative bacteria, especially on freshly isolated strains. While there is much uncertainty about the function of fimbriae, it is clear that they function in adherence to surfaces both *in vivo* and *in vitro* (Ottow, 1975; Sokatch, 1979). D-Mannose and certain derivatives inhibit bacterial adherence and hemagglutination mediated by some fimbriae but not by others (Ottow, 1975). Some cells may have both or neither type of fimbriae, but may still have adhesins or colonization factors that are not associated with visible fimbriae (Ottow, 1975).

Fimbriae are usually prepared by shearing liquid-grown cultures in a blender or by harvesting them from liquid suspensions of cells grown on solid media (Ottow, 1975; Tomoeda *et al.*, 1975; Date *et al.*, 1977; Eshdat *et al.*, 1981; Fader *et al.*, 1982). Differential centrifugation separates whole cells and flagella (Date *et al.*, 1977; Paranchych *et al.*, 1979). Purification is usually by ammonium sulfate or polyethylene glycol precipitation (Helmuth and Achtman, 1978; Paranchych *et al.*, 1979; Fader *et al.*, 1982) followed by one or more cycles of sucrose or cesium chloride gradient centrifugation (Ottow, 1975; Paranchych *et al.*, 1979). Detergents, sucrose, or guanidine hydrochloride can solubilize pili or fimbriae which subsequently can be reaggregated to obtain pure preparations (Date *et al.*, 1977; Helmuth and Achtman, 1978; Eshdat *et al.*, 1981; Fader *et al.*, 1982). Purity is assessed by criteria used in protein analyses along with electron microscopy.

The genetics of fimbriae are well studied only in *E. coli*, where *fim* genes, or the genes controlling fimbriation, have been reported (Ottow, 1975). Differences in colony morphology can differentiate Fim^+ and Fim^- cells at least in some human pathogens (Ottow, 1975; Stephens and McGee, 1983). Techniques for obtaining overproducing mutants and selection for self-aggregating clones are available (Helmuth and Achtman, 1978). While there is a correlation in some animal pathogens that fimbriated cells are pathogenic and nonfimbriate are not, no such correlation has been detected for others (see Ottow, 1975; Arbuthnott and Smyth, 1979). Thus, Ottow (1975) suggests that fimbriae are more likely to be associated with adhesive properties in general than pathogenicity in particular.

The adherence or colonization factors of animal and human pathogens are considered virulence factors (Elwell and Shipley, 1980; Jones and Isaacson, 1983); these are proteinaceous, pilus-like surface antigens, a number of which have been purified. At least some of these are plasmid determined and some of these (K88ab, K88c, K88ad, and K99) have been cloned by recombinant DNA techniques (Elwell and Shipley, 1980; van Embden *et al.*, 1980; Mooi *et al.*, 1981; Cerf and Desmettre, 1982; de Graff *et al.*, 1984).

Among plant pathogens, there are reports of the presence of fimbriae in *P. solanacearum* (Fuerst and Hayward, 1969; Stemmer and Sequeira, 1981, 1985), *P. cepacia* (Fuerst and Hayward, 1969), *P. syringae* pv. *savastanoi*, *P. syringae* pv. *syringae*, *A. tumefaciens* (Stemmer and Sequeira, 1981, 1985), and in certain members of the "carotovora" group of *Erwinia* (Christofi *et al.*, 1979). However, little is known about the generality of their occurrence, structure, composition, presence on cells in infected plants, or role, if any, in pathogenicity.

General reviews on isolation, properties, and genetics of pili and fimbriae (Brinton, 1965; Ottow, 1975; Tomoeda *et al.*, 1975; Sokatch, 1979; Pearce and Buchanan, 1980) should be of interest to phytobacteriologists.

Other surface-associated structures that may play a role in pathogenicity are not well characterized and are not yet reported for plant pathogens. These include the "fibrils" from a *Spiroplasma* isolated from honey bees (Townsend *et al.*, 1980), hairlike projections or "fuzz" of some streptococci (see Peterson and Quie, 1981) and adhesive factors of streptococci that may contribute to attachment of cells in culture and to tooth surfaces (Peterson and Quie, 1981; Fukushima *et al.*, 1982). These adhesive factors may be capsular components or structures.

C. Extracellular Enzymes

1. *Introduction*

Bacterial enzymes are located in different parts of the cell: they may be intracellular (cytoplasmic), membrane bound, periplasmic, or extracellular. Distinction between membrane-bound and extracellular enzymes may be difficult because the activity can either be loosely bound to the membrane or cell wall so that it is released by gentle extraction, such as by washing, or bound to the envelope components in exponential growth phase and released during cessation of growth in the stationary phase. Nevertheless, many enzymes are truly extracellular in that they are synthesized inside the cell, transported across the envelope, and released into the environment, i.e., they become completely dissociated from the producing cell.

Many Gram-positive and Gram-negative phytopathogenic bacteria produce a number of extracellular enzymes, many of which cause breakdown of polymeric substrates. Some of these enzymes are important in the interactions of certain phytopathogenic bacteria with hosts. For example, certain pectolytic enzymes destroy preformed structures in plant tissues and thus may be required for bacterial pathogenesis. The role of other degradative enzymes such as cellulases, proteases, or phospholipases is not known but they may have a purely catabolic function in generating monomeric residues which serve as sources of nutrients and energy.

Because of the apparent significance of certain extracellular pectolytic enzymes in host associations of some phytopathogenic bacteria, such enzymes have been investigated with respect to their distribution, structure, genetics, regulation, and pathological effects. In contrast, little is known of other extracellular enzymes such as cellulases, proteases, and phospholipases of phytopathogenic bacteria. Therefore, in the following sections we discuss what is currently known of pectolytic enzymes and indicate areas that should receive increased attention in the future.

2. *Pectic Substances and Pectic Enzymes*

During the past two decades or so pectic substances and the enzymes that act on these substrates have been intensively investigated and numerous reviews (Bateman and Millar, 1966; Codner, 1971; Rombouts, 1972; Fogarty and Ward, 1974; Rexova-Benkova and Markovic, 1976; Chesson, 1980; Whitaker, 1984) and book chapters (Aspinall, 1970; Bateman, 1976; Bateman and Basham, 1976; Darvill *et al.*, 1980) discuss the available information. Here we present a pertinent synopsis to provide the reader with background for later discussion on genetics and regulation of pectic enzymes.

(a) Pectic Substances

(i) Nomenclature. Various terms are used to describe pectic substances that can be distinguished by their chemical properties (see Fogarty and Ward, 1974). Since some of these terms are used in the following sections, to minimize confusion we provide here definitions and features that distinguish pectic substances:

1. Protopectin refers to water-insoluble native pectic substances of plants which, upon restricted hydrolysis, yield pectic substances.
2. Pectic substances, complex colloidal polysaccharides of plant origin, are

largely made up of anhydrogalacturonic acid units, the carboxyl groups of which may be partially esterified by methyl groups or partially or totally neutralized by bases.

3. Pectinic acid and pectins are differentiated by the degree of esterification of carboxyl groups with methanol; the quantity of methoxyl groups in pectinic acid, although variable, is generally less than in pectins, which have 75% or more of the carboxyls in uronic acid methylated. Both pectinic acid and pectin are capable of forming gels with sugars and acids under suitable conditions.

4. Pectic acid is a pectic substance primarily composed of poly-D-galacturonic acid, essentially free of methyl ester groups; it is also referred to as polygalacturonic acid.

(ii) Distribution and composition. Pectic substances, essential structural constituents of cell walls of higher plants, are abundant in nature. They are the primary constituents of the middle lamella (i.e., the intercellular cementing layer), which holds plant cells together into fairly rigid tissues. They occur in lesser amounts throughout the remainder of the plant cell walls. Young and actively growing plant tissues are generally high in pectin, while lignified tissues contain small amounts of pectic substances (Fogarty and Ward, 1974). The average pectin content ranges from a low of 0.7% for cotton fibers to a high of 32% for lemon rind (Fogarty and Ward, 1974). The type and amount of pectin affect the texture of vegetables and fruits. For example, protopectin, the dominant pectic substance of unripe fruits, is converted into pectin during ripening and maturation (Hulme, 1958).

The structure of pectic substances shows a primary chain composed of D-galacturonic acid residues linked by α-1,4 glycosidic bonds and some rhamnose residues. The occurrence of oligosaccharides rich in rhamnose and rhamnosyl–rhamnose linkages suggests an uneven distribution of rhamnose in the galacturonorhamnan chain, i.e., regions of uninterrupted galacturonic acid residues and regions where rhamnose residues predominate (Aspinall *et al.*, 1968a,b; Darvill *et al.*, 1980). Oligosaccharide side chains are usually linked to a galacturonorhamnan chain at C-3 of some galacturonic acid residues and at C-4 of some rhamnose residues. The structure and composition of the side chain oligosaccharides in pectic substances vary with the plant material. The anhydrouronic acid content of pectic substances ranges from 55 to more than 99%. Depending on the plant material the nonuronide fraction is generally made up of the neutral sugars arabinose, galactose, rhamnose, and xylose. Darvill *et al.* (1980) and McNeil *et al.* (1984) have discussed the structures of pectic and other plant cell wall polysaccharides.

The methoxyl content of pectic substances varies with the source: for example, 0.2% in strawberry and over 10% in apple (Fogarty and Ward, 1974). A methoxyl content of 16.32% is the theoretical limit (100%) of esterification of the COOH groups in pure polygalacturonic acid (Fogarty and Ward, 1974). In addition to methoxylation, some hydroxyl groups on C-2 and C-3 of D-galacturonic acid residues may be acetylated. The degree of acetylation varies with the source of pectin. For example, 0.2% acetylation was detected with apple and citrus pectins whereas the degree of acetylation in sugar beet pectin ranged from 2–4% (for details see Fogarty and Ward, 1974).

Pectic substances are generally soluble in water, dimethyl sulfoxide, and formamide, but insoluble in most organic solvents. The longer the chain length, the lower the solubility. Pectic substances, for example polygalacturonic acid and pectin, are viscous in aqueous solution; viscosity is affected by various factors such as concentration of pectin or polygalacturonic acid, temperature, pH, and ionic strength.

Pectic polysaccharides can be precipitated from aqueous solutions by various substances such as ethanol, acetone, water-soluble basic polymers, polyvalent cations, and quaternary detergents (Rombouts, 1972; Fogarty and Ward, 1974). Such precipitants have proved useful in the isolation and purification of pectic substances (Aspinall *et al.*, 1968a,b) and in scoring for microbial degradation of pectinic and pectic acids (Hildebrand, 1971; Chatterjee and Starr, 1977; Starr *et al.*, 1977).

Acids hydrolyze ester or glycosidic bonds in pectic substances; at low temperatures ester bonds are preferentially cleaved, whereas at higher temperatures depolymerization occurs (Doesburg, 1965). Thus, pectins of low methoxyl content can be prepared by acid hydrolysis under a low-temperature regimen. Acid treatment of pectic substances also lowers the content of nonuronide saccharides (such as araban) because of their greater acid sensitivity (Fogarty and Ward, 1974).

Alkali treatment of pectic substances results in either saponification of ester groups or depolymerization, depending upon the treatment temperature (Fogarty and Ward, 1974). Saponification is favored at low temperatures. Alkaline depolymerization occurs at high temperatures due to the cleavage of the α-1,4 bond by a *trans*-elimination (β-elimination) reaction (Neukom and Deuel, 1958). It should be noted that the *trans*-eliminative cleavage under alkaline conditions only occurs at α-glycosidic bonds adjacent to an esterified carboxyl group. Thus, pectates (salt form of pectic acid), which lack such groups, are far more resistant than pectinates (salt form of pectinic acid) to alkaline degradation at high temperatures (Albersheim *et al.*, 1960).

(*b*) *Pectic Enzymes*

(*i*) *Introduction and classification.* As might be expected from their ubiquitous occurrence in nature, pectic substances are substrates for an array of plant and microbial enzymes (Ashwell, 1962; Albersheim, 1966; Bateman and Millar, 1966; Moran *et al.*, 1968; Dehority, 1969; Gradel and Dehority, 1972; Fogarty and Ward, 1974; Sato and Kaji, 1975; Ulrich, 1975; Gardner and Kado, 1976; Lund and Brocklehurst, 1978; Chesson, 1980; Sato and Kaji, 1980; Wojciechowicz *et al.*, 1980; Collmer *et al.*, 1982b; Whitaker, 1984; Ofuya, 1984), collectively called pectic enzymes. These enzymes can be produced both by phytopathogenic and nonphytopathogenic microorganisms. Such enzymes act on polymethylgalacturonans and/or polygalacturonans (Fig. 16) and are of three types: (1) pectin methyl esterases (saponifying enzymes) that cause the hydrolysis of ester groups on the substrate pectin or pectinic acid, with the production of methanol and demethylated polygalacturonic (pectic) acid; (2) pectin or pectic lyases that cleave α-1,4 glycosidic bonds between adjacent uronic acid monomers by a *trans*-eliminative mechanism; and (3) pectin or pectic glycosidases that cleave α-1,4 glycosidic bonds between uronic acid residues by a hydrolytic mode. The *trans*-eliminative or hydrolytic cleavage is either exo- or endo-: the enzymes either remove the terminal monomer or dimer residues or produce a random intrachain splitting of the glycosidic bonds. The exocleavage can be differentiated from endocleavage by (1) the reaction products, (2) the rate of reduction in viscosity, and (3) metal ion activation (see below and Bateman and Millar, 1966; Hatanaka and Ozawa, 1973; Hatanaka and Imamura, 1974; Sato and Kaji, 1975, 1977a,b, 1980; Stack *et al.*, 1980; Collmer *et al.*, 1982b).

Pectic enzymes have been variously classified (Bateman and Millar, 1966; Fogarty and Ward, 1974; Hatanaka and Ozawa, 1971); one such classification (Chesson, 1980; Table XIII) takes into account the action pattern and the preferred substrates for the enzymes. Not included in this scheme is pectin esterase that transforms pectin into pectic acid by deesterification of methoxyl residues (Bateman and Millar, 1966; Fogarty and Ward, 1974). The likely existence of other enzymes can be predicted on the basis of the structure of pectic substances. Such putative enzymes might cause removal of oligosaccharide side chains or cleavage of the main chain between rhamnosyl residues. Thus, any classification scheme will probably require modification as additional enzymes that degrade pectic substances are detected and characterized. Further knowledge of the isozymes, those enzymes that act on the same substrate, should enable more detailed classification of pectate lyases and polygalacturonases.

The properties of pectic enzymes of bacteria, fungi, and plants have been admirably presented by Bateman and Millar (1966), Rombouts (1972), Fogarty and Ward (1974), Bateman and Basham (1976), and Chesson (1980); the reader should refer to these reviews for detailed information. Bacterial enzymes [e.g., polygalacturonic acid *trans*-eliminase (PATE) or pectate lyase (PL = PEL) and hydrolytic polygalacturonase (PG = PEH)] generally seem to

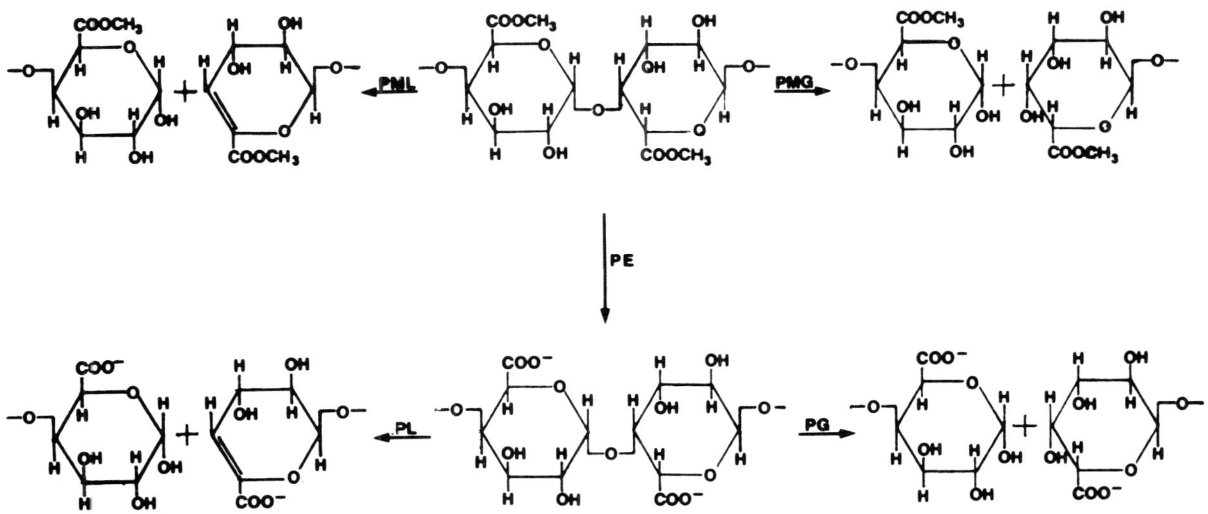

Fig. 16. Modes of action of pectinases. PMG, polymethylgalacturonase (pectin hydrolase); PML, polymethylgalacturonate lyase (pectin lyase); PE, pectin esterase (pectin methyl esterase); PG, polygalacturonase; PL, pectate lyase (polygalacturonate *trans*-eliminase, PATE). (Adapted from Fogarty and Ward, 1974.)

Table XIII. Classification of pectic depolymerases.[a]

Trivial name	Action pattern	Modified EC systematic name	EC number	Ability to macerate
Hydrolases preferentially active against pectic and low esterified pectinic acids				
Endo-polygalacturonase	Random	Poly-(1→4)-α-D-galactosiduronate glycanohydrolase	3.2.1.15	+
Subterminal polygalacturonase	Near chain end	—	—	−
Exo-polygalacturonase I	Terminal	Poly-(1→4)-α-D-galactosiduronate galacturonohydrolase	3.2.1.67	−
Exo-polygalacturonase II	Penultimate	Poly-(1→4)-α-D-galactosiduronate digalacturonohydrolase	3.2.1.82	−
Hydrolases preferentially active against oligouronides				
Oligogalacturonate hydrolase	Terminal	—	—	−
Lyases preferentially active against pectic and low esterified pectinic acids				
Endo-pectate lyase	Random	Poly-(1→4)-α-D-galactosiduronate lyase	4.2.2.2	+
Exo-pectate lyase	Penultimate	Poly-(1→4)-α-D-galactosiduronate exolyase	4.2.2.9	−
Lyases preferentially active against highly esterfied pectinic acids				
Endo-pectin lyase	Random	Poly(methyl-D-galactosiduronate) lyase	4.2.2.10	+
Lyases preferentially active against oligouronides				
Oligogalacturonate lyase	Terminal	Oligo-D-galactosiduronate lyase	4.2.2.6	−

[a] Reproduced from Chesson (1980).

preferentially cleave demethylated polygalacturonic acid (Codner, 1971; Starr and Chatterjee, 1972). However, the enzymes from *P. marginalis* (Ohuchi and Tominaga, 1973) and *Streptomyces* spp. (Sato and Kaji, 1979, 1980) showed higher activity with pectins than with polygalacturonic acid. In bacteriocinogenic strains of *E. carotovora* subsp. *carotovora*, pectin lyases were detected following nalidixic acid, mitomycin C, or bleomycin treatment or UV irradiation of cells (Tomizawa and Takahashi, 1971; Kamimiya *et al.*, 1974; Itoh *et al.*, 1980, 1982). These enzymes possessed good activity against 91% esterified polygalacturonate, reduced activity against 67% esterified polygalacturonate, and no activity against nonesterified polygalacturonate (Itoh *et al.*, 1982). Moreover, pectin and pectate lyases of *E. carotovora* subsp. *carotovora* were immunologically and biochemically distinct (Itoh *et al.*, 1982). It is now evident that mitomycin C- or UV-induced pectin lyase production occurs in most soft rot *Erwinia* spp., including *E. carotovora* subsp. *atroseptica*, *E. carotovora* subsp. *carotovora*, and *E. chrysanthemi* (Tsuyumu and Chatterjee, 1983a,b, 1984). Whether or not esterified polygalacturonate is also the preferred substrate for hydrolytic enzymes of certain phytopathogenic bacteria remains to be determined.

The reaction products of pectolytic enzymes differ depending upon the lyase or hydrolase type of activity and upon the exo- or endo-mode of cleavage. For example, in *E. carotovora* subsp. *carotovora* (strain EC 153,) the primary products of an endo-PL activity were unsaturated digalacturonic acid [*O*-(4-deoxy-β-L-5-*threo*-hexopyranos-4-enyluronic acid)-(1→4)-D-galacturonic acid] and lesser amounts of saturated mono- and digalacturonic acid (Moran *et al.*, 1968). The production of exopectate lyase by this bacterial strain has not been shown. In another isolate (EC14) both endo- and exo-types of PL were detected (Stack *et al.*, 1980). The endo-PL yielded different unsaturated oligouronides as reaction products of polygalacturonate, ranging in size from trimers to hexamers, whereas exo-PL produced unsaturated dimers or unsaturated dimers and trimers. In another strain of *E. carotovora* subsp. *carotovora* (= *E. aroideae*, isolated from Japanese radish; Okamoto *et al.*, 1964), exo-PL also produced unsaturated digalacturonic acid as the primary reaction product from polygalacturonate. In contrast to the exo-PLs of *Erwinia*, those of *Streptomyces* spp. produced different reaction products depending upon the species (Sato and Kaji, 1977a,b, 1980). For example, the enzyme from *S. nitrosporeus* produced unsaturated trigalacturonate whereas that from *S. massasporeus* produced unsaturated digalacturonate.

Endo- and exo-types of PL also differ with respect to metal ion activation. While the endoenzymes, irrespective of bacterial origin, are usually activated by Ca^{2+} (Moran *et al.*, 1968; Stack *et al.*, 1980), no consistent pattern exists for exo-PL. As an example, of the two exolyases of *E. carotovora* subsp. *carotovora* (strain EC14), one was strongly stimulated by Mn^{2+} and to a lesser extent by

Mg^{2+} and Ca^{2+}, while the other was equally stimulated by Ca^{2+}, Mn^{2+}, or Mg^{2+} (Stack *et al.*, 1980). In contrast, exo-PL from an unspecified *Erwinia* sp. (isolated from irrigation water of a paddy field), was activated strongly by Na^{+} and to a much lesser extent by K^{+} or by the divalent cations Ca^{2+}, Mn^{2+}, or Co^{2+} (Hatanaka and Ozawa, 1973).

As with PL, exo- and endo-types of polygalacturonases also produce different reaction products. The primary products of endo-PG activity were D-galacturonate and digalacturonate in *E. carotovora* subsp. *carotovora* strain EC153 (Nasuno and Starr, 1966), or many different products ranging in size from saturated dimers to hexamers in the strain EC14 (Stack *et al.*, 1980). In contrast, exo-PG of *E. carotovora* subsp. *carotovora* (= *E. aroideae*; Hatanaka and Ozawa, 1969), *E. chrysanthemi* (Collmer *et al.*, 1982b), and *Pseudomonas* sp. (= *P. marginalis*) (Hatanaka and Imamura, 1974) released only saturated digalacturonate from polygalacturonic acid.

Aside from their reaction mechanisms and metal ion requirements, PL and PG activities also differ markedly in their pH optima; the former is most active in an alkaline pH range of 8 to 9 whereas maximal hydrolytic activity of the latter is detected in the acidic pH range of pH 6 or below. This differential pH optimum has been exploited in the isolation of mutants defective in PL activity (Chatterjee and Starr, 1977). Hildebrand (1971) also used pectin-containing media adjusted to various pH levels to screen *Pseudomonas* and other spp. for the production of various pectolytic enzymes.

In certain bacterial species there may be an enzyme complex or a single polypeptide possessing both PL and PG activities, as reported for *E. carotovora* subsp. *carotovora* strain EC14 (Stack *et al.*, 1980). However, this appears not to be true in the strain Ecc71 (De Boer strain 71), since the genes for PG and PL were resolved upon cosmid cloning and transpositional mutagenesis (Zink and Chatterjee, 1984b, 1985; also see the section on genetics of pectolytic enzymes).

There are two enzymes reported that degrade unsaturated and saturated diuronides, the primary products of PL and PG activities, respectively. An oligogalacturonate lyase (OGL) of *E. carotovora* subsp. *carotovora* cleaves both the diuronides; unsaturated digalacturonate is cleaved into two molecules of 4-deoxy-L-threo-5-hexoseulose uronic acid (DTH) and saturated digalacturonate into one molecule of DTH and one molecule of D-galacturonic acid (Fig. 17), indicating a *trans*-eliminative reaction (Moran *et al.*, 1968). Hatanaka and Ozawa (1970) also have characterized an OGL produced by *Erwinia aroideae* (= *E. carotovora* subsp. *carotovora*; Young *et al.*, 1978). An oligogalacturonate hydrolase that preferentially hydrolyzed short-chain uronides was isolated from a *Bacillus* sp. (Hasagawa and Nagel, 1968). The enzyme was most active toward saturated trigalacturonate, and did not attack unsaturated uronides. It is reasonable to predict the presence in *Erwinia* spp. and other pectolytic phytopathogens of an α-galacturonidase activity with a lower K_m and a higher V_{max} for di(oligo)-galacturonate than noted with the PG activity. In addition to their role in the utilization of pectic substrates (see Fig. 17), these activities are probably required in the expression of genes that specify PL or PG (see below). It should therefore prove useful to further investigate these oligouronide-degrading enzymes, both genetically and biochemically.

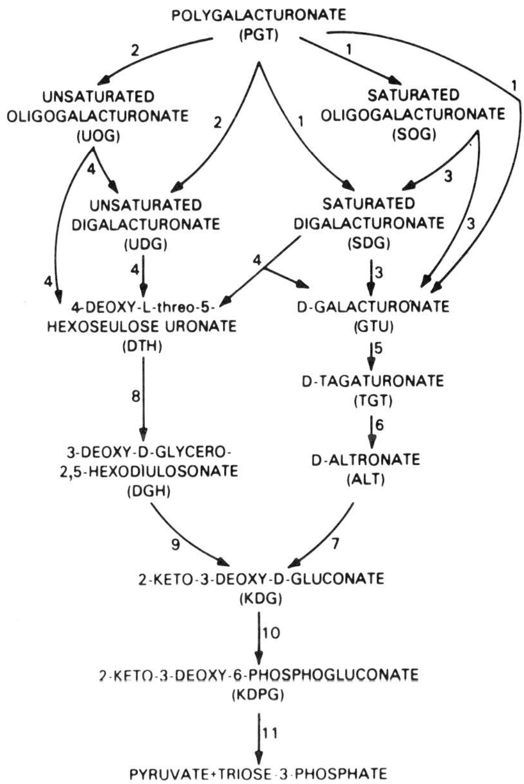

Fig. 17. Main pathways of polygalacturonate catabolism in bacteria. Enzymes for the catabolic steps are (1) polygalacturonase, PG (*peh*); (2) pectate lyase, PL (*pel*); (3) α-galacturonidase or oligogalacturonate hydrolase, OGH (*ogh*); (4) oligogalacturonate lyase, OGL (*ogl*); (5) D-glucuronate-ketol-isomerase (*uxaC*); (6) D-altronate oxidoreductase (*uxaB*); (7) D-altronate hydrolyase (*uxaA*); (8) 4-deoxy-L-*threo*-5-hexoseulose uronic acid isomerase (*kduI*); (9) 3-deoxy-D-glycero-2,5-hexodiulosonate dehydrogenase (*kduD*); (10) ketodeoxy-gluconate kinase (*kdgK*); (11) ketodeoxyphosphogluconate aldolase (*kdgA*). D-galacturonate (hexuronate; EXU) pathway consists of steps 5, 6, 7; 4-deoxy-L-*threo*-5-hexoseulose uronate (ketodeoxyuronate; KDU) pathway follows the sequence 8, 9; 2-keto-3-deoxy-D-gluconate (KDG) pathway consists of steps 10, 11. See text for the occurrence of depolymerizing enzymes and their catalytic properties, and appropriate references.

Surprisingly, little is known of pectin methyl esterases (PME) of phytopathogenic bacteria. These enzymes cause deesterification of native pectic substances along the pectin chain in a linear fashion (see Fogarty and Ward, 1974). Some indirect evidence suggests a role of PME in host associations of certain phytopathogenic bacteria. For example, *in vitro* studies on bacterial PL and PG have generally shown a much reduced activity of the enzymes toward methylated substrates but good activity with the demethylated polymer, polygalacturonic (pectic) acid. Since native pectic substances are known to possess various degrees of methylation, it could be argued that either PME and PL or PME and PG probably act in a concerted manner. Since PME causes deesterification of pectic substances, it facilitates subsequent depolymerization by PG or PL. We might note that the purified PLs of *E. carotovora* subsp. *carotovora* and *E. chrysanthemi* were found to macerate certain plant tissues (Mount *et al.*, 1970; Garibaldi and Bateman, 1971; Alberghina *et al.*, 1973; Basham and Bateman, 1975; Stephens and Wood, 1975), suggesting the activation of a host PME (Bateman and Millar, 1966).

Further support for the requirement of PME in maceration is the report of Bagley and Starr (1979), who detected plant tissue maceration by *Klebsiella*-PL in the presence of exogenously supplied PME and not in its absence; control phytopathogenic *Erwinia* spp. included in this study macerated plant tissues in the absence of exogenous PME.

Of the various pectolytic enzymes produced by phytopathogenic bacteria, the PL of *E. carotovora* subsp. *carotovora* and *E. chrysanthemi* and the pectin lyases of *E. carotovora* subsp. *carotovora* have been investigated to some extent, both genetically and biochemically. There now is strong evidence for an essential role of these enzymes in the pathogenicity of *Erwinia* spp. PL of *Erwinia* spp. might be considered as a prototype of extracellular proteinaceous virulence factors of phytopathogenic bacteria, and the knowledge acquired with this model system might very well apply to similar virulence factors of other phytopathogenic bacteria. It is therefore appropriate to analyze briefly some of the recent findings on PL of *Erwinia* spp. and to indicate directions of future research.

(ii) Location and excretion of pectate lyase (PL=PEL). Both theoretical considerations and experimental findings underscore the significance of extracellular PL in bacterial tissue maceration and pathogenicity. In those bacteria where PL is totally intracellular, the enzyme, in the absence of cellular lysis, would not be able to reach pectic substrates present in the middle lamella and plant cell wall. Therefore, tissue maceration would not occur. Experimental findings are consistent with this hypothesis. For example, differences were detected in the location of PL in phytopathogenic and nonphytopathogenic enterobacteria (Chatterjee *et al.*, 1979). It appears that

phytopathogenic species, particularly those that cause tissue-macerating diseases, excreted most or all of the PL, whereas nonphytopathogenic species secreted little or none of the total PL. Moreover, mutants of *E. chrysanthemi* that no longer secrete PL and accumulate the enzyme within the cell, invariably fail to macerate plant tissues (Chatterjee and Starr, 1977; Chatterjee and Thurn, unpublished data). Extracellular pectate lyases may also be important in pathogenicity of *Xanthomonas campestris* pv. *campestris* (Daniels *et al.*, 1984b) and *Pseudomonas fluorescens* (Zucker and Hankin, 1970, 1971).

Since secretion of PL is important in bacterial pathogenicity, it would be useful to learn how it is transported across the cell envelope. Various models (loop model, Halegoua and Inouye, 1979; linear model, Blobel and Dobberstein, 1975a,b; membrane trigger hypothesis, Wickner, 1979) have been proposed to explain the movement of polypeptides across the membrane. For additional information the reader should consult reviews on the subject (Halegoua and Inouye, 1979; Ramaley, 1979; Silhavy *et al.*, 1979, 1983; Wickner, 1979; Davis and Tai, 1980; Kreil, 1981; Michaelis and Beckwith, 1982; Benson *et al.*, 1985).

None of these models adequately explains the export of those enzymes that are synthesized in the bacterial cytoplasm or on the cytoplasmic membrane, actively transported across the membrane and cell wall, and finally released into the exterior. Location of such enzymes is clearly distinct from those that usually reside in the periplasmic space. In both *E. carotovora* subsp. *carotovora* and *E. chrysanthemi*, active secretion by exponentially growing cells of pectate lyase occurs but not of acid phosphatase or cyclic phosphodiesterase; the latter enzymes are known to reside in the periplasmic space of these bacterial species (Chatterjee *et al.*, 1979). Such findings are inconsistent with the hypothesis that exoenzyme secretion is simply mediated by diffusion of a periplasmic pool of proteins.

The physiological and genetic findings in various Gram-negative bacteria strongly suggest the occurrence of at least two different pathways in the export of exoproteins. For example, Lory *et al.* (1983) postulated that the peptide precursor of exotoxin A in *P. aeruginosa* was held to the inner face of the cytoplasmic membrane by the signal peptide, and that polypeptide chain elongation occurred in this state. Following completion of elongation, the product was transported laterally through the Bayer junction to the outer face of the outer membrane. During this transport process the signal peptide was cleaved by a signal peptidase releasing the peptide into the growth medium. According to this model no exotoxin or its precursor should be recovered from the periplasm, and this indeed was the case. In this model the functions of gene products, other than those required in the interaction of the nascent polypeptide with the cytoplasmic membrane (for example, the Sec system; Silhavy *et al.*, 1983; Benson *et al.*, 1985), may not be necessary in the translocation of the polypeptide from the cytoplasmic membrane to the cell exterior.

In contrast, in other Gram-negative bacteria (Springer and Goebel, 1980; Howard and Buckley, 1983, 1985; Wagner *et al.*, 1983; Andro *et al.*, 1984; Thurn and Chatterjee, 1985a), both genetic and physiological evidence supports a two-step model (Springer and Goebel, 1980; Wagner *et al.*, 1983) of protein export. In this model, step I involves the synthesis and translocation of polypeptides across the cytoplasmic membrane. In step II, the emerging polypeptides may interact with components that direct the translocation of the molecules across the outer membrane and their subsequent release into the growth medium. This model predicts (i) that mutational changes in the "export-components" of step II would have little effect on the synthesis of exoprotein polypeptides and their secretion into the periplasm; and (ii) that mutations in the step I components would lead to the accumulation of polypeptide precursors in the cytoplasmic membrane or the cessation of polypeptide synthesis; some such mutations may be lethal.

The phenotypes of Out$^-$ (export-defective) mutants of *E. carotovora* subsp. *carotovora* strains Ecc71 and Ecc193 (Chatterjee *et al.*, 1985a; McEvoy, Thurn and Chatterjee, unpublished) and *E. chrysanthemi* strain EC16 (Thurn and Chatterjee, 1985a) and strains B374, 3665 and 3937j (Andro *et al.*, 1984) are consistent with the two-step model. The mutants were pleiotropically defective in the export of PL, PG and CL. In the single-site Tn*5* insertion Out$^-$ strain of EC16, the enzymes were synthesized normally but accumulated in the periplasm. In addition, the mutations caused the loss of a 35 KD periplasmic protein, but did not alter the outer membrane protein composition or the export of protease (Thurn and Chatterjee, 1985a; see below for the protease export system). Whether or not the 35 KD protein is a component of the step II export system is not known. Despite that, genetic and physical evidence in *E. chrysanthemi* (EC16) revealed that single gene mutations in at least three different chromosomal loci produced such a pleiotropic phenotype, implicating the cooperative function of several gene products in the step II export system. Studies currently in progress in several laboratories should indicate if functions of additional gene products may also be required.

Pleiotropic effects on protein export resulting from single gene mutations have been noted in various bacteria including *Aeromonas hydrophila* (Howard and Buckley, 1983), *E. carotovora* subsp. *carotovora* (Chatterjee *et al.*, 1985a), *E. chrysanthemi* (Andro *et al.*, 1984; Thurn and Chatterjee, 1985a), *E. coli* (Dassa and Boquet, 1981; Oliver and Beckwith, 1981; Kumamoto and Beckwith, 1983; Oliver, 1985), *P. aeruginosa* (Wretlind *et al.*, 1977; Stapleton *et al.*, 1984), and *V. cholerae* (Neill *et al.*, 1983). In *A. hydrophila*, two classes of export-defective mutants were obtained (Howard and Buckley, 1983). In class I mutants, only the export of protease, hemolysin, and glycerophospholipid:

cholesterol acyl transferase was impaired and the composition of the major outer membrane proteins was unaffected. In class II mutants, the export of those enzymes as well as the composition of the outer membrane proteins were affected. The *E. chrysanthemi* mutants phenotypically resemble class I *A. hydrophila* mutants, except that the pleiotropic effect did not extend to protease or phospholipase C.

Such pleiotropic effects implicate a common pathway in the export of exoproteins. Strong evidence for this comes from the elegant studies in *V. cholerae* (Neill *et al.*, 1983). Neill *et al.* (1983) constructed a mutant of *V. cholerae* which was impaired in the export of cholera toxin (CT). A plasmid carrying the *E. coli* heat-labile enterotoxin (LT) gene was then introduced into the mutant and its parent and the localization of both CT and LT examined. (LT is normally localized in the periplasm of *E. coli*; Hofstra and Witholt, 1984.) While both CT and LT were exported by the parent *V. cholerae*, both toxins largely remained cell associated in the mutant. The diversity of the pleiotropic effect in various export mutants may reflect alterations in the "recognition specificity" of step II "export components", depending upon a mutation.

The two-step export process is also supported by the localization in *E. coli* of various exoprotein gene products. *E. coli* strains carrying *Erwinia* genes for pectate lyases and polygalacturonase synthesize these molecules and secrete them into the periplasm (Keen *et al.*, 1984; Collmer *et al.*, 1985a; Lei *et al.*, 1985; Zink and Chatterjee, 1985). Similar localization of *Cellulomonas* cellulase and *V. cholerae* cholera toxin was reported (Gilkes *et al.*, 1984a; Neill *et al.*, 1983). These observations suggest that step I but not step II occurs in *E. coli*. An implication of these findings is that genes for the enzymes cloned into *E. coli* contain the information for their secretion into the periplasmic space, but they lack the information that specifies step II "export components".

As opposed to PL, PG and CL, the export of protease, at least in *E. chrysanthemi*, appears to follow a different pathway. As stated above, the Out::Tn*5* mutants of *E. chrysanthemi* (EC16) produce normal levels of protease, all (>98%) of which is exported (Thurn and Chatterjee, 1985a). While Tn*5* insertion mutants have been obtained that are deficient in both intra- and extracellular protease and produce normal levels of extracellular PL and PG, those defective in protease export were not recovered. Moreover, *E. coli* strain HB101 carrying the *E. chrysanthemi* protease gene exported all of the enzyme to the cell exterior, i.e., secreted into the growth medium (Barras, Thurn and Chatterjee, in prep.). This sharply contrasts with the localization of pectolytic enzymes in *E. coli* (HB101) carrying the appropriate *Erwinia* genes. Likewise, the strain carrying the *E. carotovora* subsp. *carotovora* protease gene retained all of the activity in the periplasmic space (Willis and Chatterjee, 1985). This differential localization of *Erwinia* proteases in *E. coli* and the phenotype of *E. chrysanthemi* mutants may be explained by invoking the occurrence of a protease-specific export system. It may be that in *E. chrysanthemi* protease export and structural genes are tightly linked and hence were simultaneously cloned, thereby allowing enzyme export in *E. coli*. These genes may be organized in an operon, with export gene(s) being proximal. If so, this may explain the lack of recovery of Out (Prt)::Tn*5* mutants that would produce intracellular protease, since Tn*5* causes polar mutations. In contrast to *E. chrysanthemi*, in *E. carotovora* subsp. *carotovora* strain Ecc71 the structural and export genes for protease may be unlinked and hence were not simultaneously cloned resulting in a periplasmic localization of the enzyme in *E. coli*. Another alternative could be that while *E. chysanthemi* protease follows a pathway similar to that of exotoxin A export in *P. aeruginosa* (see above), *E. carotovora* subsp. *carotovora* protease is exported via a two-step process. Research in progress should enable us to distinguish between these alternatives.

As is evident, much additional research, both at the genetic and biochemical levels, is necessary to gain an insight into the mechanisms underlying the secretion of PL and protease as well as other extracellular enzymes in *Erwinia* spp. Such studies should also be extended to other phytopathogenic bacteria known to produce extracellular peptide or glycopeptide toxins. The presence in toxin molecules of hydrophobic and basic amino acids and the general lack of acidic amino acids suggest a role for these residues in the transport of toxins (see Section IV.D. for additional information on peptide toxins).

(iii) Isozymes. Multiple forms of PL or PG active on polygalacturonate, or isozymes, that differ in isoelectric points, tissue-macerating ability and cellular location have been reported (Garibaldi and Bateman, 1971; Mazzucchi *et al.*, 1974; Basham and Bateman, 1975; Pupillo *et al.*, 1976; Stack *et al.*, 1980). Garibaldi and Bateman (1971) found two to four isozymes of PL in strains of *E. chrysanthemi*, stimulating work on the immunological relationships (Mazzucchi *et al.*, 1974), plant tissue macerating ability (Garibaldi and Bateman, 1971) and distribution of the isozymes in wild-type strains under *in vivo* and *in vitro* conditions (Basham and Bateman, 1975; Pupillo *et al.*, 1976). Pupillo *et al.* (1976) have classified PL-isozymes of *E. chrysanthemi* into alkaline, central and acidic types. More recently, a better fractionation of pectate lyases has been achieved through isoelectric focusing in ultrathin gels and *in situ* activity assays (Bertheau *et al.*, 1984; Collmer *et al.*, 1985a,b). Although the total number of pectate lyases seem to vary (up to 24 detected by Bertheau *et al.*, 1984), four to five major extracellular pectate lyases have been consistently detected in several strains of *E. chrysanthemi*. For these pectate lyases the following gene symbols have been used: acidic enzyme: *pelA*, pI 4.7; neutral (= central) enzymes: *pelB*, pI 7.7, *pelC*, pI 8.3; basic enzymes: pI >9.5, *pelD* and *pelE*, with PelE being the most basic. While this general pattern holds true for several *E. chrysanthemi* strains (Belgium strain, B374; Paris strain, 3937j; Cornell strain,

CUCPB1237), strain EC16 (ATCC 11662) appears to produce PelA, PelB, PelC, and either PelD or PelE, but not both. Despite the subtle differences in Pel profile in *E. chrysanthemi* strains, the gene and the corresponding protein designations seem useful for the present.

While all of these isozymes occur predominantly in the extracellular state in *E. chrysanthemi*, in *E. carotovora* subsp. *carotovora* strain EC14, only one isozyme (PDI, pI 9.4) was detected extracellularly (Stack *et al.*, 1980); the other isozymes remained cell bound. This observation is unusual since in strain Ecc71 of *E. carotovora* subsp. *carotovora* (Willis *et al.*, 1985) and in other strains of *E. carotovora* subsp. *carotovora* as well as *E. carotovora* subsp. *atroseptica* (Reid and Collmer, 1985) at least three extracellular pectate lyases are readily detected on isoelectric focusing (IEF) gels; their pIs ranging from about 8.3 to > 10.0. It should, however, be noted that until recently pI values were derived from sucrose gradient IEF, a procedure which appears far less sensitive compared to ultrathin-layer IEF (see, e.g., Bertheau *et al.*, 1984; Collmer *et al.*, 1985a). Recent evidence (Collmer *et al.*, 1985a; Thurn *et al.*, 1985) also indicates that homogeneity in SDS-PAGE is, by itself, not a reliable criterion for the absence of other PL species. For example, some of the "purified" pectate lyases of EC16, homogeneous on SDS-PAGE, revealed the presence of other pectate lyases upon IEF on ultrathin gels and *in situ* activity assays (Thurn *et al.*, 1985). Likewise, Collmer *et al.* (1985a) found that a 38 KD PL band on SDS-PAGE resolved into three different pectate lyases upon IEF. In light of these observations it would seem desirable to re-examine the pectate lyase profiles of soft-rot *Erwinia* spp., in particular those that are being studied genetically.

Genetic aspects of isozyme production are being studied in several laboratories. While minor differences in the pI (such as 9.3 as opposed to 9.5) of the isozymes could be due to variability between laboratories or artifacts of some sort (Righetti and Drysdale, 1979), the vast differences in pIs of certain isozymes (viz. pI 9.4, 7.8 or 4.7) are most likely due to the dominance of acidic or basic amino acid residues in the polypeptide. The latter hypothesis predicts that these isozymes are either coded by different structural genes or result from the activity of a modifier gene(s). The evidence presented below clearly shows that the major pectate lyases are indeed coded for by different genes in *E. chrysanthemi* as well as in *E. carotovora* subsp. *carotovora*.

The physiological function of the PL isozymes is not yet known. Previous studies of Garibaldi and Bateman (1971) revealed that some but not all isozymes of *E. chrysanthemi* macerated plant tissues. In contrast, Bagley and Starr (1979) detected tissue maceration with all PL isozymes of *E. chrysanthemi* strain EC16. Likewise, all of the isozymes of *E. carotovora* subsp. *atroseptica* caused cell separation and protoplast death, although some difference was noted in the degree of cell separation (Quantick *et al.*, 1983). PL isozymes of *E. carotovora* subsp. *atroseptica* also exhibited varying degrees of thermal stability, substrate specificity and sensitivity to chlorogenic acid, a normal constituent of tuber cells (Quantick *et al.*, 1983). These observations collectively suggest that all of the PL isozymes may contribute to virulence of soft-rot *Erwinia* spp. This hypothesis is not, however, supported by the properties of the enzymes from the genetically modified strains of *E. chrysanthemi*. PelE of *E. chrysanthemi* (EC16) accounted for about 50–60% of the total pectate lyase activity in potato tuber tissue and in culture under induced conditions, and compared to PelB or PelC, induced rapid electrolyte loss in potato tuber tissue (Thurn and Chatterjee, 1985b). Moreover, Roeder and Collmer (1985a) constructed a derivative of the strain 1237 by marker exchange mutagenesis that was deficient in PelC. This mutant, like the parent, macerated potato tuber tissue, suggesting that PelC was not necessary in soft-rot pathogenesis. Similar marker exchange mutagenesis of other *pel* genes, currently in progress, should indicate a differential role in pathogenicity of various pectate lyases.

For a better overall understanding of PL isozymes, knowledge is also required of the structure of each of the isozymes. This would entail purification of the isozymes to homogeneity and examination of such structural features as mass, amino acid composition and sequence, and the presence of carbohydrate residues, lipids or phosphates. Some have been purified (Basham and Bateman, 1975). The use of chromatofocusing (Davis *et al.*, 1984) and affinity systems (Lee and West, 1981; Davis *et al.*, 1984; Inoue *et al.*, 1984; Kegoya *et al.*, 1984) in purification protocols already formulated should reduce the number of purification steps and increase the yield of purified preparations for a detailed structural investigation. The very similar molecular weight and pI of certain pectate lyases (see above), however, present serious problems in fractionating them by the conventional protein purification methods. As an example, while Thurn *et al.* (1985) could purify the extracellular PelE (pI 9.9; mol. wt. 50 KD) of EC16 to homogeneity as indicated by SDS-PAGE and IEF in gels, the same degree of purity with PelB (pI 7.8; mol. wt. 43 KD) or PelC (pI 8.2; mol. wt. 45 KD) was not achieved. This vexing problem can now be resolved by purifying pectate lyases from *E. coli* strains carrying a specific pel^+ gene (see below for the cloning of *pel*).

(iv) Regulation. In *E. carotovora* subsp. *carotovora* and *E. chrysanthemi*, PL is induced by polygalacturonate and repressed by glucose (Moran and Starr, 1969; Chatterjee *et al.*, 1979; Chatterjee and Starr, 1980; Stack *et al.*, 1980). The level of PL is somewhat higher in glycerol than in glucose. Thus, in *Erwinia*, three patterns of PL synthesis are evident: (1) a high level induced with polygalacturonate; (2) an intermediate (basal) level with glycerol, gluconate or galacturonate; and (3) a low repressed (albeit detectable) level with glucose.

Indirect evidence suggests that the repressive effect of glucose on PL synthesis in *E. carotovora* subsp. *carotovora* is modulated by the intracellular concentration of *c*AMP, since (1) exogenous *c*AMP reversed the repressive effects of glucose (Hubbard *et al.*, 1978), (2) the intracellular concentration of *c*AMP decreased during glucose repression and increased in induced cultures (Hubbard *et al.*, 1978), and (3) in a *c*AMP deficient mutant, the repressive effect was overcome by the addition of exogenous *c*AMP (Mount *et al.*, 1979). Keen *et al.* (1984) and Collmer *et al.* (1985a) also observed glucose repression of PL production in *E. coli* strains carrying PL genes of *E. chrysanthemi*. Moreover, in *E. chrysanthemi*, a mutation (*mcu*) (Chatterjee and Starr, 1977) that prevented the utilization of various substrates, including polygalacturonate, in the presence or absence of *c*AMP, mapped on the chromosome roughly in the same region where the locus *crp* maps on the chromosomes of *E. coli* (Bachmann, 1983) or *S. typhimurium* (Sanderson and Roth, 1978). [The *crp* locus codes for *c*AMP receptor protein (CRP).] This mutation (*mcu*) possibly caused a defect in the *c*AMP receptor protein thus rendering the catabolic operons noninducible. In *E. chrysanthemi* (B374), Hugouvieux-Cotte-Pattat *et al.* (1985) obtained a class (Cri$^-$) of mutants in which the synthesis of pectate lyases as well as proteases and cellulases was no longer susceptible to glucose effect (i.e., the mutants were resistant to catabolite repression); pectate lyases, however, remained inducible by polygalacturonate catabolic products (see below for pectate lyase induction). The mutations mapped between *ile* and *thr* loci on the *E. chrysanthemi* chromosome.

The molecular mechanism underlying the *c*AMP effect on PL in *Erwinia* spp. remains unclear. Two possibilities must be considered: (i) the binding of the *c*AMP–CRP complex to the promoter region which may subsequently influence the binding of RNA polymerase, and (ii) the interaction of the *c*AMP–CRP complex with regulatory proteins, as has been seen in the arabinose operon. Therefore, the stimulation by *c*AMP of PL production in both *E. carotovora* subsp. *carotovora* and *E. chrysanthemi* (Tsuyumu, 1979; Stack *et al.*, 1980; Collmer and Bateman, 1981; Chatterjee *et al.*, 1981b) could result from one or more of the following effects: (1) transcriptional stimulation resulting from the binding of a *c*AMP–CRP to the promoter region; (2) transcriptional stimulation as a result of the interaction between *c*AMP–CRP and a (positive) regulator protein; and/or (3) enhanced transport of the inducer molecule and its intracellular accumulation. Genetic and physiological investigations on pectolytic enzymes should allow us to distinguish between these possibilities.

The kinetics of PL production in *E. carotovora* subsp. *carotovora* and *E. chrysanthemi* in the presence of polygalacturonate showed a maximum differential rate of enzyme production after an extended lag period ranging from 60 to 180 min, depending on the cultural conditions and bacterial strains (Chatterjee *et al.*, 1979, 1981b). These results suggested that the impermeable substrate, polygalacturonate, was broken down by the basal (extracellular) pectolytic enzymes of these bacteria, the catabolic products were then transported into the cells, and these induced PL synthesis. This view is supported by the pattern of PL induction with polygalacturonate (PGT) in the presence or absence of EDTA and with unsaturated digalacturonate (UDG) and saturated digalacturonate (SDG), the catabolic products of PGT (see Fig. 17) (Tsuyumu, 1977, 1979; Collmer and Bateman, 1981; Chatterjee *et al.*, 1981b). PL synthesis commenced sooner in cells growing in a medium containing PGT than in cells growing in the presence of PGT and EDTA (Tsuyumu, 1977). Moreover, the EDTA effect was abolished with the addition of Ca^{2+}. Since Ca^{2+} is required for PL activity, the findings suggested that a breakdown product(s) of PGT was required for the induction of PL with PGT. This hypothesis was further supported by the induction of PL by SDG and UDG. Induction of PL in *E. carotovora* subsp. *carotovora* (EC-1) occurred sooner with UDG than with PGT (Tsuyumu, 1977, 1979). In a probable *E. carotovora* subsp. *atroseptica* strain EC (= strain AC9006), the differential rate of PL synthesis was higher with SDG and PGT than with UDG or glycerol (Chatterjee *et al.*, 1981b). Maximal rate of PL synthesis in the *Erwinia* strain AC9006 usually commenced after 90–180 min of growth, and the substrates (PGT, SDG, UDG, or glycerol) did not alter the lag period significantly (Chatterjee *et al.*, 1981b). Tsuyumu (1977, 1979), however, reported that higher levels of PL production started earlier with UDG than with PGT. Whether this apparent discrepancy resulted from strain differences or differences in the composition of growth medium remains to be determined. It should be noted, however, that Chatterjee *et al.* (1981b) determined the total PL activity, whereas Tsuyumu (1979) investigated substrate effects on intracellular activity. Thus an alternative explanation for the apparent discrepancy might be a preferential stimulation of the intracellular PL by UDG.

In the *Erwinia* strain EC (= AC9006) mentioned above, the addition of *c*AMP to PGT-, SDG- or glycerol-grown cultures did not appreciably alter the differential rate of PL synthesis or the lag period, when compared to cultures grown in the absence of *c*AMP, but did result in increased total PL following a prolonged incubation (6–16 h) of cells with these substrates. In contrast, UDG used in conjunction with *c*AMP caused a two-fold higher rate of PL synthesis and abolished the lag period (Chatterjee *et al.*, 1981b). A similar response of *E. carotovora* subsp. *carotovora* (strain EC-1) with *c*AMP and UDG was noted by Tsuyumu (1979).

In *E. chrysanthemi* (EC16) PL synthesis was stimulated by SDG, UDG, and PGT; higher rates of enzyme synthesis commenced sooner with these substrates than with glycerol (Chatterjee *et al.*, 1981b). SDG caused a dramatic increase in PL synthesis as evidenced by a differential rate eight-fold higher than that of PGT grown cultures and by a shorter lag period. Although PL synthesis in the presence of UDG commenced without an apparent lag, the rate of synthesis remained low throughout the duration of the experiment (Chatterjee *et al.*, 1981b). *c*AMP stimulated the effects of PGT and UDG in EC16 by stimulating the rate of PL synthesis. The effect was most striking with UDG where *c*AMP caused a 15-fold higher rate of synthesis than noted with UDG in the absence of the nucleotide. The high differential rate of PL synthesis with SDG was not stimulated further by the addition of *c*AMP which was similar to that of UDG + *c*AMP-grown cultures.

Collmer and Bateman (1981) found that the isomeric product, 3-deoxy-D-glycero-2,5-hexodiulosonate (DGH) (also known as 2,5-diketo-3-deoxygluconate = DKII), of the unsaturated monomer, 4-deoxy-L-*threo*-5-hexoseulose uronic acid (DTH = DKI), stimulated PL production in an OGL-deficient *E. chrysanthemi* mutant. Analysis of transposon insertion mutants of *E. chrysanthemi* defective in OGL production (Thurn *et al.*, 1983; Chatterjee *et al.*, 1985b) confirmed the observations of Collmer and Bateman (1981). It is therefore plausible that the 30–60 min lag period in the induced synthesis of PL with UDG or SDG (Tsuyumu, 1977, 1979; Chatterjee *et al.*, 1981b) was probably required for the formation of DTH from the dimers by the enzyme OGL (Fig. 17) and the conversion of DTH to DGH. However, in the presence of both *c*AMP and the dimers, there was little or no lag in the synthesis of PL in *E. carotovora* subsp. *carotovora* and *E. chrysanthemi* (Tsuyumu, 1979; Chatterjee *et al.*, 1981b). At present this finding is subject to various interpretations, such as (1) the stimulation of OGL synthesis by the *c*AMP–CRP complex, thereby favoring the formation of DTH and DGH, (2) the stimulation of the entry of the diuronides into bacterial cells, or (3) the formation of a positive effector molecule, a ternary complex of regulator protein, inducer, and *c*AMP.

Recent findings with Tn*5* insertion mutants deficient in OGL or 3-deoxy-D-glycero-2,5-hexodiulosonate dehydrogenase (KDUD) showed that an intermediate(s), most likely DGH, of the deoxyketouronate pathway (Fig. 17) induced the production of PL, PG, OGL and KDUD (Tyrell *et al.*, 1983; Chatterjee *et al.*, 1985b) suggesting "product induction" of pectic depolymerizing enzymes, at least in *E. chrysanthemi* (EC16). The findings of Condemine *et al.* (1984) suggest that a similar situation may also prevail in other strains of this bacterium.

The reader should be alerted that PL production in *E. chrysanthemi* is also influenced by such physiological factors as water activity (Mildenhall *et al.*, 1981), temperature and aeration (Thurn, Tyrell and Chatterjee, unpublished data).

As is evident, many uncertainties remain in our understanding of the regulation of PL in *Erwinia* spp. The advances in our knowledge of the enzymes and the intermediates in the polygalacturonate catabolic pathways, the cloning of *pel* genes (see below), and the occurrence of mutants constitutive for PL production (Hugouvieux-Cotte-Pattat *et al.*, 1985) should facilitate formulation of appropriate physiological experiments and isolation of additional mutants which might prove useful in probing regulatory mechanisms.

(v) Genetics. Genetical research on pectolytic enzymes was hampered by the lack of a convenient method for the screening of one or the other pectolytic activities and the unavailability of gene transfer systems to characterize mutants. Both of these deficiencies have now been alleviated to a great extent. In conventional procedures for assaying pectolytic activity on agar media at a neutral pH, polymer degradation is brought about by both *trans*-eliminative and hydrolytic activities. Under these conditions, mutational alteration in one or the other activity would be phenotypically masked. Early studies on pectolytic enzymes of *E. carotovora* subsp. *carotovora* (Nasuno and Starr, 1966; Moran *et al.*, 1968) showed a maximal hydrolytic activity at pH 5.5 and optimal *trans*-eliminative cleavage of polygalacturonate in the pH range of 8–9. The vastly different pH optima offered the possibility of distinguishing between these two enzymatic activities. Hildebrand (1971) later exploited differential pH optima in screening pseudomonads and other bacteria for the production of various pectolytic enzymes. Likewise, Chatterjee and Starr (1977) used polygalacturonate agar medium at pH values of 8.0–8.3 to score PL activity. A summary of the procedures used to score for various extracellular enzymatic activities of *E. chrysanthemi* is described in Fig. 18.

It should be pointed out that the polygalacturonate agar (Chatterjee and Starr, 1977; Starr *et al.*, 1977) used in the isolation of *E. chrysanthemi* PL mutants may not work for bacteria (such as *E. carotovora* subsp. *atroseptica* and *carotovora*) that produce endo-PG (*peh*) in addition to endo-PL (*pel*). In this case, it may be essential to obtain PG-deficient mutants first. PL-deficient mutants can then be obtained using Peh$^-$ bacteria as the parent. In the isolation of Peh$^-$ mutants, several modifications of polygalacturonate agar may prove useful. These could include supplementing the medium with EGTA (to chelate Ca^{2+}), adding NaCl and lowering the pH of the medium to 6.0 or below by using a buffer such as Tris-maleate. Such a medium should allow detection of PG with little or no interference from PL.

Exo- and endo-types of cleavage by lyases or hydrolases can be differentiated by a double agar layer procedure (Tsuyumu and Goto, 1978) which measures the production of acid-soluble pectic acid as well as the formation of reducing ends from pectin or polygalacturonate by isolated colonies. In addition to the isolation of structural gene mutations, the double-layer procedure should prove valuable in obtaining mutants altered in regulatory or

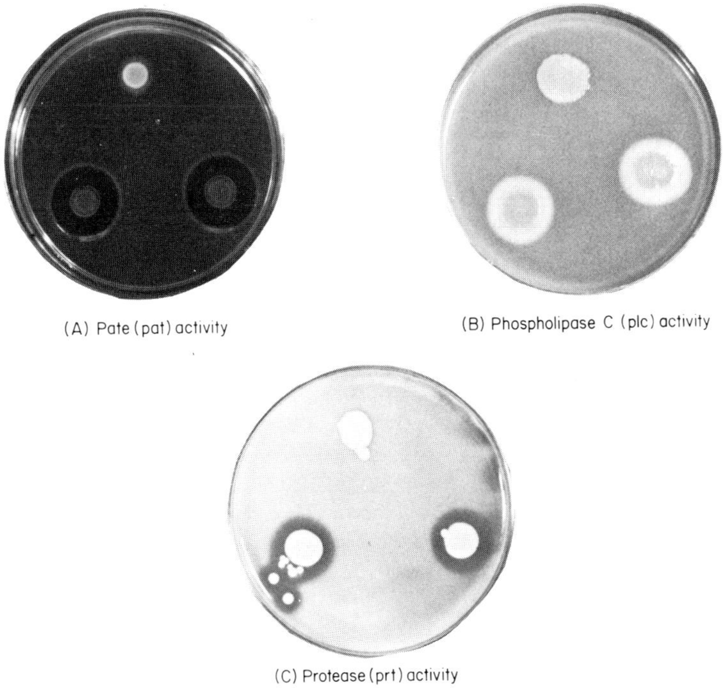

Fig. 18. Procedures for the screening of extracellular pectate lyase (A), phospholipase C (B), or protease (C) activities. For pectate lyase activity, bacteria were grown for 16–18 hr on PYA medium (Chatterjee and Starr, 1977) and the plates were flooded with 6 *N* HCl. Clearing (halo) around the colonies indicated the breakdown of polygalacturonate by pectate lyase. Phospholipase C activity was scored by inoculating bacteria to lecithin agar (Chrisope *et al.*, 1976) and by visual examination for a precipitate around colonies after 16–24 hr incubation. Protease activity was assessed by growing bacteria on nutrient gelatin agar plates for 48–72 hr and then flooding the plates with acidic mercuric chloride. Clearing around the colony results from the breakdown of gelatin by protease activity. Results shown were obtained with Tn*5*-insertion mutants of *E. chrysanthemi* defective in extracellular enzyme production and with a parent strain which produces all three extracellular enzymes. (A. K. Chatterjee, K. K. Thurn and D. J. Tyrell, unpublished results.)

catalytic properties. The usefulness of these and other modifications in the isolation of *E. carotovora* pv. *carotovora* mutants remains to be tested.

PL deficient or Pel$^-$ mutants of *E. chrysanthemi* (Chatterjee and Starr, 1977, 1978b; Thurn and Chatterjee, 1984) fell into two classes: class 1 mutants had considerably more intracellular activity but much less extracellular activity than the parent strains. These mutants are defective in the export of pectolytic enzymes (see Section IV.C.2.b.ii.). Similar export defective mutants have been obtained in *E. carotovora* subsp. *carotovora* strain Ecc193 (Chatterjee *et al.*, 1985a). Independent single site mutations were linked to biosynthetic markers (*gua*, *lys*, *pur*) and caused a defective export of PL, PG, and CL (Zink and Chatterjee, 1984a; Chatterjee *et al.*, 1985a). Class II mutants had much reduced levels of both intra- and extracellular activities (Ferguson and Chatterjee, 1980). In preliminary experiments quantitative, but not qualitative, differences were found in the PL isozymes in certain class II mutants and their parent strains (Ferguson and Chatterjee, 1980). At present it is unclear if these mutations are in the structural gene or a regulator locus.

As stated above, genetic analysis of PL isozymes has not been feasible using the conventional mutagenesis and recombinational genetic approaches. Recently, however, the genetic analysis of pectolytic enzymes has been facilitated by cloning of genes for pectate lyases of *E. chrysanthemi* (Keen *et al.*, 1984; Collmer, 1985a,b; Kotoujansky *et al.*, 1985; Reverchon *et al.*, 1985; Van Gijesegem *et al.*, 1985a) and pectate lyases and polygalacturonases of *E. carotovora* subsp. *carotovora* (Allen *et al.*, 1984; Zink and Chatterjee, 1984b, 1985; Allen *et al.*, 1985; Roberts *et al.*, 1985; Willis *et al.*, 1985) (Table 8). The physical, genetic and biochemical analyses of RP4::mini Mu (pULB113) primes carrying *E. chrysanthemi* (B374) pel^{+} genes indicated a tight linkage between *pelA*, *pelD* and *pelE* and between *pelB* and *pelC* (Van Gijsegem *et al.*, 1985a). Moreover, one pULB113-prime, pULB262, harboring about 130 kb of the B374 DNA, carried all five pel^{+} genes indicating a clustering of *pel* on the chromosome. The linkage analysis of the various primes revealed the following organization on the B374 chromosome of *pel* and *cel* (*cel* = a gene locus for cellulase production; see below): (*cel kdgk uxu*)-*metB-ile-cya-pelBC-pelADE-purE* (Van Gijsegem *et al.*, 1985a). The data from *in vitro* cloning also confirmed this genetic organization of *pel* in the Paris strain 3937 (Kotoujansky *et al.*, 1985). By screening a genomic library of 3937 in the lambda L47–1 vector, phage clones were obtained that produced various combinations of pectate lyases, including a clone λ41 which produced both PelB and PelC, and another clone, λ81 which produced PelA, PelD and PelE. In contrast to these strains, EC16 produced four pectate lyases: PelA, PelB, PelC and PelE (see above). The initial findings of Keen *et al.* (1984) suggested the occurrence in EC16 of two *pel* genes specifying the pI 9.8 enzyme and the pI 8.8 enzyme. The products of pPL3, pPL7, pPL34, pPL74 and pPL742 constructed by Keen *et al.* (1984) were analyzed on isoelectric focusing gels and additional EC16 DNA carrying cosmid hybrids constructed (F. Barras, K. K. Thurn, and A. K. Chatterjee, in prep.). The pel^{+} plasmids specified the following enzymes: pPL7, pPL74 and pAC308, PelA and PelE; pPL742, PelE; pPL74::Tn*5*, PelA; pPL3 and pAKC 307, PelB and PelC; and pPL34, PelC. These findings, along with the enzymatic profiles of further subclones of some of the plasmids, indicate an organization of EC16 *pel* genes similar to those found in the other strains, except for the absence of *pelD* and its product. Several *pel* genes also appear clustered in the Cornell strain 1237 (Collmer *et al.*, 1985a; A. Collmer, personal communication).

With the isolation of structural genes for the various *E. chrysanthemi* pectate lyases, it is now possible to examine their export, regulation and function. Indeed, progress in these areas is rapidly occurring in various laboratories. For example, the nucleotide sequence of most of *pelE* is now known (N. T. Keen, personal communication) and the sequence data for other *pel* genes should soon be forthcoming. Biochemical properties, in particular the differences in catalytic properties and tissue macerating ability of each of the pectate lyases, are being studied; this knowledge may provide some insights into the functions of the specific PLs in host–pathogen interactions. In addition, each of the *pel* genes is being inactivated either by transposon insertion or by generating deletions *in vitro* and inserting drug-resistant (selective) markers in order to construct by marker exchange *E. chrysanthemi* strains deficient in the production of a specific enzyme (Roeder and Collmer, 1985a). The pathogenic properties of such strains should elucidate the differential role of the pectate lyases.

Compared to *E. chrysanthemi*, less progress has occurred in the genetics of pectolytic enzymes of *E. carotovora* subsp. *carotovora*. However, Ecc71 and other strains are known to produce several pectate lyases in addition to an endopolygalacturonase; the pI values of the major pectate lyases are very different from those of *E. chrysanthemi* pectate lyases (Reid and Collmer, 1985; Willis *et al.*, 1985). Two pectate lyase genes and an endopolygalacturonase gene of Ecc71 have been cloned (Zink and Chatterjee, 1985). Willis *et al.* (1985, in prep.) subsequently isolated four additional *pel* genes from a cosmid gene library. Moreover, a tight linkage between the endopolygalacturonase gene and a *pel* gene was detected in Ecc71, and also apparently in strain EC14 (Roberts *et al.*, 1985). A clustering of pectate lyase genes was also found on a 7-kb DNA fragment of the strain EC (Lei *et al.*, 1985).

The studies with *E. chrysanthemi* and *E. carotovora* subsp. *carotovora* are a beginning of the genetic analysis of pectolytic enzymes, and a foundation upon which much additional research can be done. Future developments are likely to occur in such areas as (1) the organization of structural and regulator genes into operon(s) or regulon(s), (2) the structural properties of pectate lyases in relation to their secretion and interactions with plant components, and (3) the modes of expression of the polygalacturonate catabolic genes.

3. *Pectin Lyases*

As stated above, endo-pectin lyase (E.C.4.2.2.1.0) degrades pectin by *trans*-eliminative cleavage (Albersheim and Killias, 1962; Chesson, 1980). This enzyme also cleaves highly (>98%) methylesterified polygalacturonate, commonly known as Link pectin (Morell *et al.*, 1934). While pectin lyase (PNL) is commonly found in fungi (Fogarty and Ward, 1974; Chesson, 1980) it is rarely detected in bacteria. Its involvement, either by itself or in conjunction with other pectic depolymerases, in the elicitation of tissue macerating diseases is probable since it causes maceration of plant tissues (Byrde and Fielding, 1968; Ishii, 1976) and permeability changes in plant cells (Hislop *et al.*, 1979).

There are several reports (Kamimiya *et al.*, 1972, 1974; Itoh *et al.*, 1980, 1982) of pectin lyase production in Japanese strains of *E. carotovora* subsp. *carotovora* in the presence of the DNA-damaging agents, mitomycin C, nalidixic acid and bleomycin or upon UV irradiation. Pectin lyase production and cellular lysis were subsequently observed in mitomycin C-treated cultures of *E. carotovora* subsp. *atroseptica* and *carotovora*, and *E. chrysanthemi* (Tsuyumu and Chatterjee, 1984). The level of pectin lyase was generally higher in *E. carotovora* subsp. *carotovora* than in other *Erwinia* spp. A similar induction was observed with nalidixic acid or UV light in randomly selected strains of these *Erwinia* spp. From these findings we conclude that pectin lyase production in response to DNA-damaging agents is a trait common to most soft-rot causing *Erwinia* spp.

Several studies (Kamimiya *et al.*, 1974; Itoh *et al.*, 1982; Tsuyumu and Chatterjee, 1984; Tsuyumu *et al.*, 1985) have demonstrated that pectin and pectate lyase activities are associated with different polypeptides. This conclusion was supported by: (i) the lack of inducibility of pectin lyase by pectic materials that induce pectolytic enzymes (see below); (ii) the lack of inducibility of the pectolytic enzymes by DNA-damaging agents; (iii) the fractionation of enzymatic activities by standard protein purification procedures; (iv) the differences in catalytic and structural properties of the purified enzymes; and (v) the absence of immunological cross-reaction. Both of these enzymes were, however, produced in infected plant tissues.

(a) Regulation of Pectin Lyase Production. Following treatment with DNA-damaging agents, cellular lysis and pectin lyase production occur in most *Erwinia* strains. Findings (Kamimiya *et al.*, 1972; Itoh *et al.*, 1980, 1982; Tsuyumu and Chatterjee, 1984) strongly suggest that cellular lysis following mitomycin C or UV treatment results from the induction of bacteriocins or temperate phages. In *E. carotovora* subsp. *carotovora*, the coinduction of bacteriocins (carotovoricins, CTV) and pectin lyase has been shown (Itoh *et al.*, 1980; Zink *et al.*, 1985a,b). In the lysogenic strain (KS612) of *E. chrysanthemi*, a coordinate induction of the temperate phage, Erch-12 and pectin lyase was also detected. In some *E. chrysanthemi* strains, mitomycin C-induced production of structures resembling bacteriophage tails (defective

phages) or bacteriocins were noted. In this context it is noteworthy that bacteriocinogeny (Echandi and Moyer, 1979) or lysogeny (Paulin and Nassan, 1978; Chatterjee and Brown, 1980; Resibois *et al.*, 1984) are fairly common in *E. chrysanthemi*.

The genetic basis for the coordinate induction of pectin lyase and bacteriocins of temperate phages is unknown. Itoh *et al.* (1980) postulated that the pectin lyase gene(s) was located within or adjacent to the genome(s) of a defective phage (bacteriocin) or temperate phage, and hence were coordinately regulated by a common repressor. An alternate hypothesis is that pectin lyase is one of a group of inducible functions belonging to the regulatory unit known as SOS functions. In *Escherichia coli*, UV irradiation induces numerous functions (Witkin, 1976; Little and Mount, 1982; Walker, 1984, 1985) including prophage induction, DNA degradation, impairment of cell division and filament formation, and *recA*$^+$ product synthesis. Other agents that also inhibit DNA replication mimic some of the effects induced by UV irradiation. It is now known that the *recA*$^+$ product controls homologous recombination and the SOS functions in *E. coli* including the induction of lamdoid phages and colicins. The RecA is induced by mitomycin C, nalidixic acid, UV irradiation, γ-irradiation, and thymine starvation (see Tsuyumu and Chatterjee, 1984, for specific references). Under these inducive conditions, it is converted to a form that mediates the inactivation of the *lexA* repressor controlling a coordinate expression of SOS functions.

The *recA* function is required in the induction of pectin lyase and carotovoricin in *E. carotovora* subsp. *carotovora* Ecc71 (Zink *et al.*, 1985b). A RecA::Tn*5* Ecc strain, constructed by marker exchange, resembled RecA$^-$ strains of *E. coli* in that it was highly sensitive to methyl methanesulfonate, nitroquinoline oxide, and UV light. The *recA* mutation did not affect the production of pectate lyase, polygalacturonase, cellulase and protease, or the ability to cause soft-rot of potato tubers. However, in this RecA$^-$ strain, unlike the RecA$^+$ parent strain, PNL and CTV were not induced by the DNA damaging agents or detected in infected potato tuber tissue. The RecA$^+$ phenotype, including the inducibility of PNL and CTV, could be restored in the mutants in *trans* by the *recA*$^+$ gene from either Ecc71 or *E. coli*. While these observations demonstrate a role of the RecA product, the molecular basis for the effect awaits elucidation.

*(b) Isolation and Characterization of Tn*5 *Insertion Mutants of* E. carotovora *subsp.* carotovora *Defective in Pectin Lyase (PNL) and Carotovoricin (CTV).* As part of a systematic genetic analysis of pectin lyase and carotovoricin in *E. carotovora* subsp. *carotovora* strain Ecc71, Tn*5* insertion mutants were isolated using the R plasmid vehicle, pJB4JI (Zink *et al.*, 1984; McEvoy *et al.*, 1985). The mutants were analyzed for basal and induced levels of PNL and CTV, lysis in the

presence of mitomycin C and other DNA-damaging agents, presence of Mu DNA, and the physical location of Tn*5* insertions. The representative data allow the following conclusions: (i) *pnl*, *ctv* and the gene for lysis (*lss*) are most likely components of independent transcriptional units, and (ii) these genes in strain Ecc71 probably reside on the chromosome and not on an 8.2-kb cryptic plasmid. In contrast, in *E. carotovora* subsp. *carotovora* strain 379, bacteriocin (= carotovoricin ?) production was apparently specified by a large (*ca.* 100 kb) conjugative plasmid (Ward and Copeman, 1985); whether or not it also specified PNL is not yet known. The properties of Tn*5* insertion mutants (McEvoy *et al.*, 1985), however, suggest that either the *pnl* and *ctv* genes share a common regulator molecule or each of these genes is controlled by specific regulators; in either event the regulator molecule(s) are subject to RecA processing.

(c) Significance of Pectin Lyase Production in the Ecology of Soft-rot Erwinia. The ecological significance of the unusual regulation, i.e., pectin lyase production in many soft-rot bacteria in response to stress caused by DNA-damaging agents, remains obscure. Several lines of evidence, however, suggest the *recA*-mediated expression of *pnl* in plants. For example, production of both PNL and CTV was detected in potato tuber tissue with Rec^+ *E. carotovora* subsp. *carotovora* strains but not with a Rec^- strain (Zink *et al.*, 1985b). Also, production of the temperate phage Erch-12 (Chatterjee and Brown, 1980) and PNL in a Rec^+ *E. chrysanthemi* (EC183) was observed in infected plant tissue and in culture medium supplemented with plant extracts believed to contain putative DNA-damaging agents (Tsuyumu *et al.*, 1984). Thus, the available data suggest that the expression of *Erwinia* genes for the pectolytic enzymes (pectate lyase and polygalacturonase) and the pectinolytic enzyme (PNL) is modulated by host factors. When in the presence of uronate derivatives, these bacteria produce the pectolytic enzymes that act as virulence factors. However, when the host presents DNA-damaging agents, the bacteria respond by inducing *recA* and activating its product, thereby allowing the expression of *pnl*.

Noteworthy in the context of PNL induction in plants is the observation that many plants produce secondary metabolites and an array of stress compounds (viz. phytoalexins; Bailey and Mansfield, 1982; Darvill and Albersheim, 1984; Kuć and Rush, 1985), some of which are DNA damaging (Ames, 1983; Surico *et al.*, 1985). (Phytoalexins are low molecular weight antimicrobial compounds produced by plants.) Moreover, the products (pectic fragments) of pectolytic activities (endo-pectate lyase of *E. carotovora* subsp. *carotovora* and endo-polygalacturonase of *Rhizopus stolonifer*) are elicitors of phytoalexins (Lee and West, 1981; Bruce and West, 1982; Davis *et al.*, 1984; West *et al.*, 1985) as well as proteinase inhibitors (Ryan *et al.*, 1985). Albersheim and associates (Davis *et al.*, 1984) also found that the elicitor-active molecules derived from citrus pectin were less potent compared to those from polygalacturonate. The differences were ascribed to: (a) the low activity of pectate lyase on citrus pectin compared to polygalacturonate, thereby producing from pectin fewer oligogalacturonides of correct size (9–13 galacturonosyl residues), the elicitors of pterocarpan phytoalexins; or (b) lower elicitor activity of methyl-esterfied oligogalacturonide fragments produced from citrus pectin as opposed to nonesterified oligogalacturonides produced from pectate. The latter hypothesis would suggest that pectate lyase (and polygalacturonase) of *E. carotovora* subsp. *carotovora*, compared to pectin lyase, is more effective in producing elicitors of phytoalexin synthesis.

Although this hypothesis, as well as the induction in plants of putative DNA-damaging agents by pectic fragments and other elicitors, awaits confirmation, we propose the following model. Once within the host tissue, *E. carotovora* subsp. *carotovora* and other soft-rot *Erwinia* produce pectolytic enzymes that degrade the pectic substances and generate uronate derivatives. The latter compounds in conjunction with other putative elicitors activate the host defense system(s) which in turn gives rise to "stress-metabolites" that are growth inhibitory or DNA-damaging or both. The bacteria respond to these stress compounds by inducing *recA* and activating the *recA* product. This response allows the pathogen to repair DNA damage and concomitantly produce PNL, an additional virulence factor. This hypothesis takes into account the findings that: (a) damage-inducible pectin lyase production is a trait common to most soft-rot *Erwinia*; (b) while pectin lyase causes damage to plant tissues, the pectolytic enzyme, pectate lyase, is probably the primary virulence determinant; (c) some of the products of pectolytic activities elicit a host defense system; and (d) plant tissues, under normal and stress conditions, contain DNA-damaging agents. The hypothesis also postulates a novel mechanism by which a pathogen can overcome an inimical environment presented by the host. These hypotheses are being tested by examining the response of genetically-modified bacteria in normal and stressed plants or plant tissues and by characterizing plant metabolites and their effects on genetically-altered bacteria.

4. *Cellulases and Cellobiases*

Cellulose, hemicellulose and pectin are the major structural components of the plant cell wall. Thus, the breakdown of cellulose could be significant in bacterial pathogenicity, particularly by those bacteria that cause tissue macerating diseases. While pectinases are of paramount importance in the elicitation of soft-rot disease, whether or not cellulases contribute to bacterial virulence is not known. While a number of phytopathogenic bacteria produce cellulases (see below for the details of the enzymes), until recently little systematic work has been done with enzymes of these bacteria. It is no surprise, therefore, that most of our fundamental knowledge of the cellulases and the complexities of the cellulolytic processes comes from the work done in fungi, notably in *Trichoderma* (Ghose, 1977; Halliwell, 1979; Ryu and Mandels, 1980; Fogarty, 1983; Shoemaker *et al.*, 1983) and such nonphytopathogenic bacteria as *Bacterioides succinogenes* (Schellhorn and Forsberg, 1984), *Cellulomonas fimi* (Beguin and Eisen, 1978; Gilkes *et al.*, 1984a,b; Langsford *et al.*, 1984; Wakarchuk *et al.*, 1984), and *Clostridium thermocellum* (Ait *et al.*, 1979; Ng and Zeikus, 1981; Johnson *et al.*, 1982; Cornet *et al.*, 1983; Lamed *et al.*, 1983; Beguin *et al.*, 1985). These and other systems are reviewed by Ljungdahl and Eriksson (1985).

Aside from their possible significance in host-pathogen associations, cellulolytic enzymes of phytopathogenic and plant-associated bacteria merit attention from the aspect of biomass conversion. Cellulose is the only renewable carbon source that is available in bulk, comprising one-half of 100 billion tons of photosynthetically fixed carbon (Ryu and Mandels, 1980). An efficient biological conversion of this enormous biomass, harnessed through solar energy and available worldwide, potentially could provide soluble products through energy conserving industrial processes. Because of this enormous potential, there has been a surge of interest in the cellulolytic enzymes of the soft-rot *Erwinia* spp., and we summarize below some of the recent findings.

(a) Cellulases. The enzymatic breakdown of cellulose involves a multienzyme (cellulase) complex consisting primarily of endo-β-glucanases, exo-β-glucanases and cellobiases. The endo-β-glucanases cleave the polymer in a random manner, thereby rapidly reducing the chain length (i.e., the degree of polymerization) but increasing the reducing groups rather slowly. In contrast, the exo-β-glucanases remove glucose or cellobiose residues from the nonreducing end, producing a rapid increase in reducing groups. Cellobiases are β-glucosidases that hydrolyze the β (1→4) glucosidic bond in cellobiose; they may also hydrolyze β (1→2)-, β (1→3)-, and β (1→6)-linkages (see Ryu and Mandels, 1980 for details). [A number of phytopathogenic bacteria possess β-glucosidase activity (Hildebrand and Schroth, 1964; Hayward, 1977), although apparently not cellulolytic.]

Cellulose is an unbranched polymer of β (1→4) linked glucose units. Despite this simple structure, cellulose can assume a crystalline structure which contributes to its insolubility as well as resistance to enzymatic or chemical damage. In contrast, amorphous or derived cellulose is more readily hydrated and also more susceptible to enzymatic cleavage. Reese *et al.* (1950) postulated the concept of the C_1 enzyme and C_x enzyme to differentiate between the enzymatic complexes in the breakdown of cellulose. For example, the solubilization of crystalline cellulose requires complete or true cellulases (i.e., the C_1 enzyme) consisting of all necessary components. On the other hand, amorphous or derived cellulose can be degraded by true cellulases as well as by individual endo- or exo-glucanases, i.e. by incomplete cellulases (C_x enzyme).

E. chrysanthemi and other soft-rot *Erwinia* produce cellulases (Beraha and Garber, 1971; Andro *et al.*, 1984; Boyer *et al.*, 1984a,b; Chatterjee *et al.*, 1985a; Thurn and Chatterjee, 1985a); some *E. chrysanthemi* strains apparently can also degrade crystalline cellulose by an unknown mechanism (see Kotoujansky *et al.*, 1985). The endoglucanase produced by the Marseille strain 3665 of *E. chrysanthemi* has been purified and characterized (Boyer *et al.*, 1984b). The production of this enzyme was stimulated under conditions that reduced the growth rate, and was subject to catabolite repression (Boyer *et al.*, 1984a). The maximum specific activity of *E. chrysanthemi* endoglucanase, produced in synthetic or complex medium under limited aeration, was similar to that of *Clostridium thermocellum* (Ait *et al.*, 1979; Boyer *et al.*, 1984a). A hybrid cosmid from a 3665 genomic library which conferred to *E. coli* a Cel^+ phenotype has been isolated by Barras *et al.* (1984a). The cellulase activity in Cel^+ *E. coli* clones was considerably lower than in the *E. chrysanthemi* strain 3665 (Barras *et al.*, 1984a). Barras *et al.* (1984a) speculated that the poor expression in *E. coli* of the *Erwinia* cel^+ may have been due to the absence of a positive regulatory gene that may not have been simultaneously cloned along with the structural gene. However, recent results showed that the cloned gene product was not the same as had previously been characterized by Boyer *et al.* (1984b). That would imply the existence of at least two cellulases in strain 3665 (F. Barras, pers. comm.). The *cel* genes from other strains have also been cloned by Kotoujansky *et al.* (1985) and Van Gijsegem *et al.* (1985a).

These developments are highly significant in terms of the role of endoglucanases (C_x enzymes) in pathogenicity of *E. chrysanthemi* as well as in elucidating the regulatory mechanisms in cellulase production. However, the biotechnological potential of *Erwinia* cellulases is not yet apparent. Several issues remain unresolved. For example, is there a C_1 (or true cellulase) complex in these bacteria for an efficient degradation of crystalline cellulose? Can the *Erwinia* enzymes (genes) stimulate the efficiency of other cellulolytic systems? If the *Erwinia* cellulases fail to significantly degrade native and crystalline forms of cellulose, is it due to the instability or inefficiency of one or more C_1 components?

(b) Cellobiases. As stated above, one of the major products of endo- and exoglucanase activities on cellulose is cellobiose, which is converted to glucose by the cellobiase or β-glucosidase activity (see Woodward and Wiseman, 1982, for the properties and applications of fungal and bacterial β-D-glucosidases). The genetics and regulation of cellobiose catabolism have been studied in *E. chrysanthemi* (Barras *et al.*, 1984b; Barras *et al.* (1985) FEMS Microbiol. Letters 30, 209–12). The findings indicate the presence of a specific cellobiose utilization system. The genetic data revealed the presence of at least two *clb* genes encoding cellobiose permease and cellobiase hydrolase. They are probably organized as an operon in the *arg-ile* region of the *E. chrysanthemi*

chromosome. Studies with the cloned *clb* genes in *E. coli* and in *E. chrysanthemi* indicate (i) that *clb* expression is controlled by the *c*AMP–CRP system; (ii) that cellobiose is transported via the phosphotransferase (PTS) system (see Postma and Roseman, 1976; and Saier, 1977 for the details of the PTS); and (iii) that a *gyrA* mutation affects the expression of *clb* (Barras *et al.*, 1984b, 1985; Barras *et al.* (1985). J. Bacteriol. In Press.). While these findings are not directly related to pathogenicity of *E. chrysanthemi*, the knowledge should nevertheless facilitate genetic as well as physiological manipulation for an enhanced efficiency of the total cellulolytic complex. In addition, the findings are significant in that they indicate an overall similarity in catabolic systems and their regulation in phytopathogenic bacteria and their nonphytopathogenic counterparts.

In concluding this section, we would like to stress that the role(s) of the total extracellular enzyme complex of soft-rot bacteria in the development of disease awaits elucidation. As indicated previously, cellulase, protease and phospholipase activities may well function primarily as catabolic enzymes, degrade polymeric substances of host origin and generate residues that are utilized by the pathogen as sources of carbon, nitrogen or energy. However, in macerating strains of *Pseudomonas fluorescens*, protease(s) appears to be a major enzyme in dissolution of potato slices (Wang and Kelman, 1982). The cloning of a macerating factor (Schlemmer *et al.*, 1985) should enable elucidation of its structure, function and regulation. Likewise, endo-pectin lyase and endo-polygalacturonase are believed to play important roles in the pathogenicity of various microorganisms (Byrde and Fielding, 1968; Ulrich, 1975; Ishii, 1976; Hislop *et al.*, 1979). Since most previous studies have been done with partially purified enzymes, and not by using genetically modified bacteria, we cannot rule out the possibility that these enzymes, acting in conjunction with such primary virulence determinants as toxins or pectate lyases, augment bacterial pathogenicity. Noteworthy in this context is the occurrence of generally (five- to 20-fold) higher levels of xylanase activity in the corn stalk rot (CSR) strains of *E. chrysanthemi* compared to the levels in non-CSR strains of the bacterium pathogenic on other plants (Braun and Kelman, 1985). This and other observations suggested that the xylanase activity of the CSR strains contributed, at least in part, to the maceration of corn tissue. This hypothesis awaits genetic verification. Despite that, it raises several significant issues, for example, the primary or secondary (ancillary) role of xylanase in pathogenicity of CSR strains; and the role of both the quality and quantity of extracellular enzyme complexes in host range (host-specificity) of soft-rot bacteria. The findings with xylanases and pectin lyases clearly demonstrate the importance of critical studies of the extracellular enzyme complex of various soft-rot bacteria.

D. Toxins

Many bacteria produce metabolites deleterious to specific hosts when administered in small doses. Such metabolites are called toxins, which are classified into two broad groups, exotoxins and endotoxins. A more precise definition of toxins is not possible because of their various physicochemical properties and biological activities. Many phytopathogenic bacteria produce extracellular glycopeptides, peptides, polysaccharides, or small metabolites such as amino acid analogs (Patil, 1974; Yoder, 1980; Mitchell, 1981; Strobel, 1977, 1982; Mitchell and Hart, 1983) that are phytotoxic; there is considerable doubt that the polysaccharides and low molecular weight carboxylic acids are toxic per se (Noda *et al.*, 1980; Yoder, 1980; Daly, 1981; Perreau *et al.*, 1982; Vidaver, 1982; Robeson and Cook, 1985). There is also a report (Hignett and Quirk, 1979) of a lipopolysaccharide–protein complex or endotoxin of *P. syringae* pv. *morsprunorum* toxic to cherry leaves. Thus, the most studied toxins produced by phytopathogenic bacteria are extracellular and elicit a phytotoxic response.

The concept that toxins are involved in bacterial pathogenicity has existed since the germ theory of disease. With many animal and human pathogens such as *E. coli*, *Shigella dysenteriae*, *Vibrio cholerae*, and *C. diphtheriae*, the validity of this hypothesis has been rigorously established. Likewise, at least for several pathovars of *P. syringae*, there is strong evidence that toxins are important in the development of plant disease. In other bacteria, such as *Corynebacterium* spp., the evidence is less compelling. Likewise, although a potent antibiotic has been detected in cultures of *X. albilineans* (Birch and Patel, 1985), its role in the leaf scald disease of sugarcane is not yet known. For fastidious bacteria, little is known about the mechanisms of pathogenicity. Davis and Lee (1982) and Daniels (1983) reviewed this area and concluded that only for *Spiroplasma citri* is there any evidence for toxin production. While two chemically different, nonspecific toxins of >400 Da have been produced *in vitro* by *S. citri*, it is not clear whether or not either of them (or other toxic substances) is actually involved in pathogenesis. A broad-bean cut-leaf bioassay was developed by Daniels (1979) to assess toxic metabolite production by *S. citri* and other plant pathogenic bacteria. The assay is based on the appearance of a black polyphenol oxidation product.

Yoder (1980) discussed various criteria that are commonly used to evaluate the role of toxins in pathogenesis. The criteria included host specificity, presence in infected plants, production at a key step in disease development, induction of typical disease symptoms, correlation of virulence with the quantity of toxin produced *in vitro*, and effects of genetic differences of the pathogen and the host. He pointed out, and we agree, that genetic approaches are likely to provide definitive evidence for the role of toxins. In addition to establishing an association between production and pathogenicity in the bacteria, genetic approaches should also enable determination of the specific function of toxins in pathogenesis, namely their role in the establishment of the pathogen within a host and in symptom production. Genetic approaches are also required to clarify the qualitative and quantitative aspects of toxin production both *in vitro* and *in vivo*.

Much has been made of the general lack of host specificity of bacterial phytotoxins (see, for example, Patil, 1974; Rudolph, 1976; Daly and Uritani, 1979). However, the specificity of bacterial phytopathogens need not reside in a toxic metabolite(s) they produce, as best shown by nontoxin-producing pathogenic *P. syringae* pv. *phaseolicola* (Mitchell, 1978). Toxins of phytopathogenic bacteria are but one of a series of pathogenic determinants, and other characteristics of the pathogen and host, such as cell surface and host metabolic properties, could well determine host specificity. For example, extracellular polysaccharides of *Pseudomonas* and *Xanthomonas* spp. produce water-soaking symptoms only on hosts susceptible to the pathogen (El-Banoby and Rudolph, 1979a,b; see Section IV.A.1. for additional details). And ethylene production is stimulated by the toxin coronatine (Ferguson and Mitchell, 1985), at least in leaf discs.

It is also important to distinguish between diseases caused by *in vivo* microbial activity in addition to toxinogenesis and diseases caused by the uptake of microbial toxins. Contrary to examples of animal and human diseases, such as staphylococcal food poisoning and botulism in humans, we know of no instance where the mere uptake of preformed bacterial and fungal toxins by plants results in disease under natural conditions. Thus symptoms of plant diseases caused by bacteria seem to require joint or sequential interactions of various pathogenic factors of which toxins, enzymes, or cytolytic factor(s) constitute only a part. Host specificity of a pathogen may be due to one or several of the many factors involved in pathogenesis.

In the literature (e.g., see Yoder, 1980) one encounters such terms as primary and secondary determinants of pathogenicity (or virulence). The origin of these terms can probably be traced to animal and human pathology. Secondary or primary determinants of virulence have some utility in those diseases where either invasiveness or toxinogenesis is the predominant factor. However, many bacterial pathogens, including a number that cause diseases in plants, are both invasive and toxinogenic. In these instances, where various pathogenic factors act in a cooperative or sequential manner, it is difficult, and probably deceptive, to classify pathogenic factors as either primary or secondary.

It is, however, useful to differentiate between pathogenic factors and virulence factors. These factors can be differentiated both qualitatively and quantitatively, as suggested by Yoder (1980). Pathogenicity describes the qualitative ability of an organism to cause disease and virulence describes the quantitative aspects or severity of the pathological changes incited by a pathogen. While bacterial determinants that cause pathological changes in the host are also required in virulence, qualitative or quantitative changes in pathogenic determinants could alter the severity of the pathology. Toxins produced by phytopathogenic bacteria probably function both as pathogenic factors, producing toxin-specific damage, and virulence factors, exacerbating other pathological processes.

Many phytopathogenic bacteria produce phytotoxic metabolites (see Mitchell, 1981, 1984, and also Bashan *et al.*, 1980; Buchanan and Starr, 1980; Perreaux *et al.*, 1982; Bauer and Beer, 1983; Mitchell *et al.*, 1983). However, until recently, research was limited to documentation of the occurrence of such substances, elucidation of their modes of action, and host specificity or the lack thereof. Little is known about the genetics, biosynthesis, regulation, and the role of toxins in bacterial plant pathogenicity. Genetic and physiological investigations have been hampered due to the lack of methods to readily quantify toxins produced *in vivo* or *in vitro*, and the lack of appropriate collections of mutants. Some of these problems have recently been alleviated for phytotoxic peptides, thereby making it feasible to systematically investigate the genetics and physiology of toxinogenesis, particularly in phytopathogenic pseudomonads. Current knowledge of the phytotoxins of fluorescent pseudomonads is summarized in Table XIV. The considerable research on peptide antibiotics produced by *Bacillus* spp., and the knowledge acquired in these systems, could well serve as a guide to research on phytotoxic peptides. Based on such studies, we will consider, using phaseolotoxin, syringomycin, and syringotoxin as examples of peptide toxins, the concepts of primary and secondary metabolites as they apply to phytotoxins, possible modes of synthesis of peptide phytotoxins, and the genetics and physiology of toxinogenesis. While our treatment of the subject is highly speculative, we hope that by providing such hypotheses work may be stimulated in areas of phytotoxin research that have been grossly neglected in the past. We remind the reader that there are available several review articles (Patil, 1974; Strobel, 1974, 1977; Yoder, 1980; Strobel, 1982) and book chapters (Durbin, 1972; Daly, 1976a,b; Rudolph, 1976; Durbin and Steele, 1979; Daly, 1981; Mitchell, 1981; Panopoulos and Staskawitz, 1981; Shaw, 1981, Mitchell, 1984) on the chemistry, genetics, and modes of action of bacterial phytotoxins, so we will deal with these topics only insofar as they are pertinent to the subject discussed.

1. Composition and Structure of Peptide Toxins

(a) Syringomycin and Syringotoxin. Following its discovery in 1966 by DeVay and associates (Sinden, 1966; DeVay *et al.*, 1968), several attempts have been made to purify syringomycin (SR) and to determine its composition. Based on a

Table XIV. Properties of some phytotoxins produced by fluorescent pseudomonads[a]

Toxin	Producing bacterium	Structure	(Bio)assay	Mode of action	Genetic determinants
Coronatine	*P. syringae* pv. *atropurpurea; P. syringae* pv. *glycinea*	1-carboxycyclopropyl-1-amide; coronafacic acid and coronamic acid linked by amide linkage	1. Hypertrophic growth of potato tuber tissue 2. Chlorosis in bean leaves 3. Immunological[b] 4. Growth inhibition of *Micrococcus* spp.[c]	?	An 8.3-kb plasmid in *P. syringae* pv. *glycinea*[bc]
Tabtoxin	*P. syringae* pv. *tabaci; P. s.* pv. *coronafaciens; P. s.* pv. *garcae*	Peptide: tabtoxinine-β-lactam-threonine tabtoxinine-β-lactam-serine	1. Chlorosis or necrosis in plants 2. Glutamine-dependent growth inhibition of *E. coli*	Inhibition of glutamine synthetase	? chromosomal
Phaseolotoxin	*P. syringae* pv. *phaseolicola*	[N^{δ} (N^{γ}-sulfodiaminophosphinyl)]-ornithyl-alanyl-homoarginine	1. Bean leaves 2. Arginine (citrulline)-dependent growth inhibition of *E. coli*	Inhibition of ornithine carbamoyl transferase	chromosomal
Tagetitoxin	*P. syringae* pv. *tagetis*	Cyclic hemithioketal structure ($C_{11}H_{18}NO_{13}PS$)[g]	Apical chlorosis[g]	Interference with chloroplast development[g]	?
Syringomycin	*P. syringae* pv. *syringae*	?Hexapeptide: ser: phe: x[d]: arg: (2:1:2:1)	1. Necrosis of maize leaf tissue 2. Growth inhibition of *Geotrichum candidum*	Inhibition of mitochondrial functions[e]	chromosomal
Syringotoxin	*P. syringae* pv. *syringae* (citrus ecotype)	? Pentapeptide: thr: ser: gly: orn: x[f]: (1:1:1:1:1)	1. Necrosis in citrus petiole 2. Growth inhibition of *G. candidum*	Inhibition of mitochondrial functions[e]	chromosomal

[a] For details see text and also Durbin (1981).
[b] Specific immunological assays have been developed by J. V. Leary and associates.
[c] J. V. Leary (personal communication).
[d] The findings of Hemming and Strobel (1983) suggest that the unknown material could be δ-*N*-hydroxyornithine.
[e] See Surico and DeVay (1982) for additional details.
[f] Possibly OH-orn; see text.
[g] See Mitchell and Hart (1983) for details.

partially purified toxin, Sinden *et al.* (1971) suggested that it was a small peptide. Subsequently, using a homogeneous preparation of SR, Gross *et al.* (1977) found that the toxin was probably a hexapeptide composed of serine, phenylalanine, an unknown basic ninhydrin-positive material, and arginine in a 2:1:2:1 ratio. The findings of Hemming and Strobel (1983) suggest that the unknown material could be δ-*N*-hydroxyornithine. A minimum molecular weight of SR was estimated at 922 by using the molecular weight of lysine for that of the unknown amino acid (Gross *et al.*, 1977). Although the toxin did not seem to contain amino sugars (*N*-acetylglucosamine, galactosamine, mannosamine), reducing sugars, or amino sulfur-containing compounds, the presence of additional component(s) was suggested by the low nitrogen content (an actual recovery of 12.9% N_2 as opposed to an expected theoretical value of 16.7%) upon combustion analysis. The toxin was found to be stable in acidic solutions, but rapidly lost biological activity in alkaline solutions above pH 8.0 (Sinden *et al.*, 1971). The base inactivated toxin was also less mobile upon polyacrylamide gel electrophoresis (Gross *et al.*, 1977). Despite such findings acid hydrolysates of ^{14}C-labeled SR showed the presence only of ninhydrin-reactive substances (Gross *et al.*, 1977). The loss of biological activity of SR upon exposure to alkaline conditions deserves further scrutiny. Since both active and base-inactivated SR have an identical amino acid composition (Gross *et al.*, 1977), the occurrence of additional base-labile groups or intramolecular bonds, rather than cleavage of peptide bonds, is suggested. It should be noted that the base-labile structural feature is necessary for biological activity. Additional studies are required to resolve these and other questions (Mitchell, 1981) as well as to elucidate the structure of the toxin.

Syringotoxin (ST) is produced by strains of *P. syringae* pv. *syringae* causing citrus canker (Gross and DeVay, 1977; Gonzalez *et al.*, 1981a). The toxin produced by strain Ps268 contained threonine, serine, glycine, ornithine, and an unidentified basic ninhydrin-positive substance in a 1:1:1:1:1 ratio (Gross *et al.*, 1977). Based upon its elution pattern in an amino acid analyzer, the unknown residue appeared to be the same as that in SR. The differences in composition of ST and SR are also reflected in the different mobilities of the toxins in polyacrylamide gels and different affinities for the adsorbent CM 52 (Gross *et al.*, 1977). The molecular weight of the pentapeptide was estimated at 685. However, the presence in ST of several additional minor ninhydrin-positive substances was also suggested. One such component, in preparations of ST, and in a sample of SR from the strain B-427 of *P. syringae* pv. *syringae*, was subsequently identified as α-alanine (Gonzalez *et al.*, 1981a).

Early reports on the possible mode of action of syringomycin described the binding of syringomycin to DNA and its subsequent inhibition of RNA polymerase (Penner *et al.*, 1969) and a rapid detergent-like lysis of cellular membranes (Backman and DeVay, 1971). More recently, syringomycin and syringotoxin were found to produce numerous alterations in the structure and function of isolated maize mitochondria. These alterations included uncoupling of oxidative phosphorylation, stimulation of succinate, malate/glutamate, and NADH respiration, stimulation of ATPase activity, inhibition of ATP formation, swelling, and a decrease in mitochondrial membrane potential (Surico and DeVay, 1982).

(b) Phaseolotoxin. Two structures have been proposed for the biologically unique toxin, associated with halo blight of field bean. The structure of (N^{δ}-phosphosulfamyl)-ornithylalanylhomoarginine (Fig. 19) has been proposed by Mitchell (1976). Recently, Moore *et al.* (1984) proposed a revised structure, in which the tripeptide is attached to a N^{δ} (N^{∇}-sulfodiamino-phosphinyl) moiety

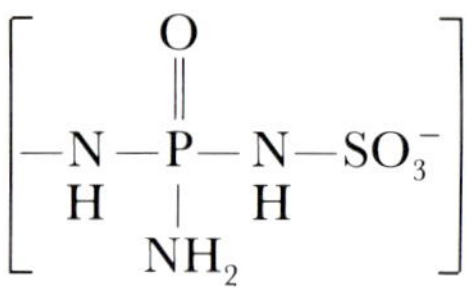

. For historical reasons, the following discussion of the structure–function relationships and biosynthesis of phaseolotoxin makes use of the N^{δ}-phosphosulfamyl ornithine structure; it is equally applicable to the revised structure.

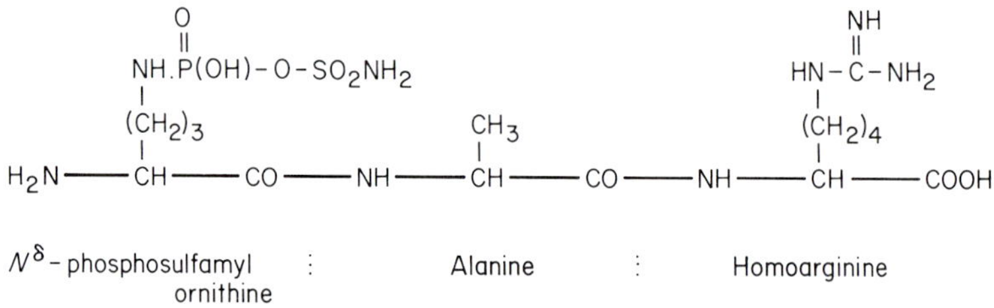

Fig. 19. Structure of phaseolotoxin. (After Mitchell, 1976.)

Mitchell (1985) found that culture supernatants of *P. syringae* pv. *phaseolicola* had biological activity, with the induction of chlorosis and ornithine accumulation. At least 80% of this was due to the N^{δ}-(phosphosulfamyl)-tripeptide. Rudolph and Rasche (1981) essentially confirmed some of these results. Moreover, the chlorosis-inducing property and ornithine accumulation was associated with (N^{δ}-phosphosulfamyl)-ornithylalanylhomoarginine and its degradation products, (N^{δ}-phosphosulfamyl)-ornithylalanine and (N^{δ}-phosphosulfamyl)-ornithine. Other degradation products such as ornithylalanylhomoarginine, ammonium phosphosulfamate, and ammonium sulfamate did not induce chlorosis even when administered at a 10–fold higher concentration (Mitchell, 1976). It was, therefore, concluded that biological activity of phaseolotoxin resided in the N^{δ}-phosphosulfamyl ornithine residue. According to Moore *et al.* (1984), octicidin ($=\text{ornithine}{-}\text{HN}{-}\overset{\overset{\large O}{\|}}{\underset{\underset{\large NH_2}{|}}{P}}{-}\underset{H}{N}{-}SO_3^-$) was the major toxin in bean leaves infected with *P. s.* pv. *phasedicola*.

It is useful to examine briefly the structural features of phaseolotoxin in relation to its biological activity. Early studies (Patil *et al.*, 1970) showed that *P. syringae* pv. *phaseolicola* toxin inhibited the enzyme ornithine carbamoyl transferase (OCT) of bean plants. This conclusion has been confirmed (Mitchell and Bieleski, 1977; Ferguson and Johnston, 1980; Staskawicz *et al.*, 1980).

The structural similarity between carbamyl phosphate and the N^{δ}-phosphosulfamyl group led Mitchell (1976) to suggest that if the latter compound has SO_2 replaced by C=O it should act as a competitive inhibitor of OCT. This hypothesis was supported by the findings (Hoogenraad, 1978; Mori *et al.*, 1977) that a chemically synthesized analog, N^{δ}-(phosphoacetyl)-L-ornithine (δ-PALO) inhibited OCT competitively with respect to carbamyl phosphate and noncompetitively with respect to L-ornithine. Hoogenraad (1978) concluded that δ-PALO and L-ornithine combined with different forms of the enzyme, whereas δ-PALO and carbamyl phosphate competed for the same form of the enzyme. It is noteworthy that the affinity of δ-PALO for OCT was about 250-fold higher than that of carbamyl phosphate. The potent biological activity of (N^{δ}-phosphosulfamyl)-ornithine and its structural similarity with the analog inhibitor, δ-PALO, suggested that both compounds inhibit the same way. It would be of interest to compare inhibition constants for octicidine and δ-PALO.

Mitchell (1978) found that all strains of *P. syringae* pv. *phaseolicola* that caused chlorosis of dwarf bean produced phaseolotoxin while strains that caused necrotic spots, but not chlorosis, did not produce phaseolotoxin. Mitchell and Bieleski (1977) also produced data strongly suggesting the production of phaseolotoxin in plants. All these findings lead to the conclusion that chlorosis in tissues infected by *P. syringae* pv. *phaseolicola* is caused by phaseolotoxin.

2. *Quantitation*

Phytotoxic peptides produced by *P. syringae* pv. *phaseolicola* and *P. syringae* pv. *syringae* possess broad-spectrum antibiotic activities toward both eukaryotic and prokaryotic cells (Mitchell, 1981; Yoder, 1981). This nonspecific toxicity is the basis of several bioassays for the toxins produced by these phytopathogenic bacteria. For both syringomycin and syringotoxin, the extremely sensitive fungus *Geotrichum candidum* has been widely used as the test organism. The inoculum size and physiological state of the test organism affect quantitative reproducibility, so that it might be useful to substitute sensitive bacteria (Sinden *et al.*, 1971) for the fungus.

Prototrophic *E. coli* strain K12 has been used as the test organism for phaseolotoxin assays (Staskawicz and Panopoulos, 1979). Staskawicz and Panopoulos (1979) established a dose–response curve (Fig. 20a) with purified phaseolotoxin preparations, and could detect 10–12 pg of the toxin. Since the basis for this assay is the requirement for arginine resulting from inhibition of OCT activity, the assay has to be performed in the absence of arginine and citrulline. As *P. syringae* pv. *phaseolicola* wild-type strains are prototrophic, this assay can be used to detect toxin producing *P. syringae* pv. *phaseolicola* strains. A similar agar assay procedure has also been developed for the quantitation of the β-lactam toxin, tabtoxin, produced by *P. syringae* pv. *tabaci* (Gasson, 1980), apparently based upon the inhibition of glutamine synthesis by the toxin in the test bacterium, *E. coli*.

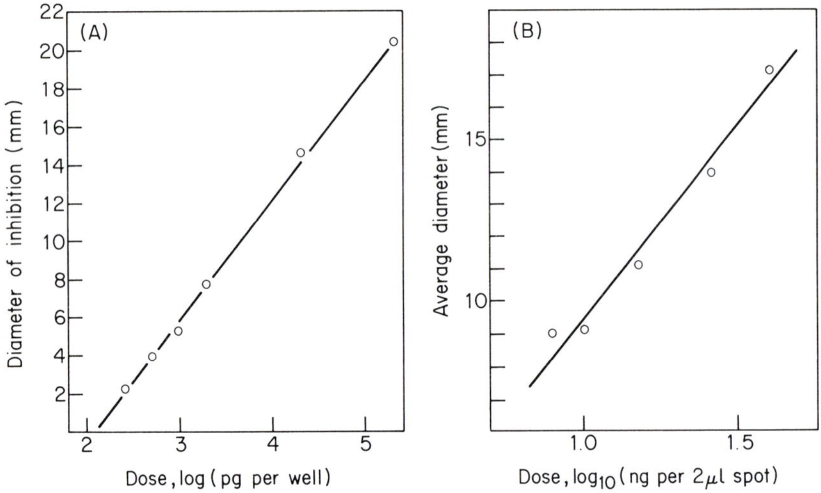

Fig. 20. (A) Bioassays of phaseolotoxin. Dose–response curve for inhibition of *Escherichia coli* strain N100 by phaseolotoxin. Twenty-microliter volumes of serial dilutions of phaseolotoxin made in phosphate buffer were delivered into 5-mm-diameter wells cut into the minimal agar medium after it was overlaid with the indicator strain. The petri dishes contained 10 ml of agar medium plus 4 ml of overlay agar. Each point represents the average of three replicates; line of best fit is $Y = 6.35X - 13.2$ in which X is log [picograms] of phaseolotoxin and Y is millimeters inhibition. (Reproduced from Staskawicz and Panopoulos, 1979.) (B) Dose–response relationship between diameter of chlorotic spot and dose of phaseolotoxin (plotted on a $\log_{10}$ scale). Standard phaseolotoxin solutions were prepared in $\times 3$ concentrate of control growth medium; 2-μl droplets were applied to the primary leaves of bean plants and the leaf was pricked through the droplet. After 56 hr the long and short axes of each chlorotic spot were measured, then averaged to give the diameter. The line of best fit is $y = 12.0x - 2.5$ [y = diameter, in mm; $x = \log_{10}$ (dose in ng)]. The equation can be rearranged to give the amount of phaseolotoxin in terms of the overall diameter of the chlorotic spot. (Reproduced from Mitchell, 1978.)

In these foregoing procedures growth inhibition of a test organism is assayed on agar media that can support growth of both the toxin-producing cells and the toxin-sensitive test organism. Thus, in addition to quantitation of toxins present in culture supernatants or fractionated samples, these methods allow direct testing for biologically active material synthesized and excreted by bacterial colonies growing on agar media. These procedures are useful in testing the inheritance of toxin genes in genetic recombinants and in detecting mutants defective in extracellular toxin production or transport such as toxin-negative mutants as well as super-producing mutants. Using such procedures, nontoxinogenic strains have been obtained for various pathovars of *P. syringae* (Panopoulos and Staskawicz, 1981; Romes and Patil, 1984; Morgan and Chatterjee, 1985).

Such colony assays do not measure intracellular toxin or the production of analogs that lack biological activity. These limitations do not pose a serious problem since such strains would be identified as Tox^-, and the determination of the genetic and biochemical basis of the Tox^- phenotype would require further work.

In addition to agar plate assays, the toxins can be quantitated using plant tissue. For syringomycin, leaf and shoot assays (Gross and DeVay, 1977) provide convenient means of quantitating biological activity. Various plant species (peach, maize, cowpea) have been used. Phaseolotoxin can be assayed on primary leaves of dwarf bean plants (Mitchell and Bieleski, 1977; Mitchell, 1978) and gives a linear dose–response reationship (Fig. 20b). The plant bioassay appears less sensitive and less precise than that using bacteria or fungi (see Fig. 20a and b).

The phosphorus and sulfur content of phaseolotoxin can be exploited to radiolabel the toxin with ^{32}P or ^{35}S and quantitate it by counting. Using $^{35}SO_4^{2-}$, Mitchell (1978) prepared radioactive phaseolotoxin, purified it in three to four steps, and determined the quantity of toxin present in culture supernatants of various *P. syringae* pv. *phaseolicola* strains. It may be possible to simplify the fractionation protocol further and thereby rapidly quantitate the radiolabeled toxin synthesized by whole cells or by cell-free extracts. The radiolabel-based procedures are potentially much more accurate for quantifying toxins, though less specific, than bioassay procedures.

No methods have been described for the quantitation of SR or ST by using radioactive precursors or by chemical assays. The ability to radiolabel SR or ST should be facilitated by using a synthetic medium for the efficient production of these toxins (Gross, 1985) in conjunction with the published fractionation procedures (Sinden *et al.*, 1971; Gross *et al.*, 1977; Gonzalez *et al.*, 1981a).

3. *Regulation*

Few systematic studies have been undertaken to elucidate the effect of cultural conditions on the production of peptide phytotoxins. Moreover, published studies are mostly semiquantitative in nature. Thus, the factors that regulate toxin biosynthesis and the underlying molecular mechanisms remain largely unexplored. It is apparent that production of syringomycin, syringotoxin, and phaseolotoxin is favored during postexponential growth of bacteria on complex media. This pattern of synthesis is typical of so-called "secondary metabolites" (Hopwood and Merrick, 1977; Demain, 1981) or "idiolites" (Walker, 1978).

Syringomycin, syringotoxin, and phaseolotoxin production normally occurs during postexponential growth. It remains to be determined if enzymes (synthetases) that catalyze toxin biosynthesis are absent, inactive, or limited in activity during exponential growth due to the deficiency of precursors. (We assume that the synthesis of phytotoxic peptides in pseudomonads is nonribosomal.) In the case of several peptide antibiotics produced by *Bacillus* spp., the production of peptide synthetases has been detected in cells during the late logarithimic phase of growth (Drew and Demain, 1977; Katz and Demain, 1977). A similar situation might occur in pathovars of *P. syringae*.

Variations in the balance of nutrients can influence toxin synthesis (Table XV) (e.g., see Shaw, 1981). For example, the production of syringomycin is stimulated by casamino acids (Gross and DeVay, 1977). The recent findings

Table XV. Effects of substrates and cultural conditions on the production of syringomycin and phaseolotoxin[a]

Substrate or growth condition	Effect on the production of	
	Syringomycin	Phaseolotoxin
Glucose	Stimulation (2)[b]	Stimulation at low concentration (?)[c] (3,4)
Glycerol	?[c]	Stimulation (4)
Peptone	Depression (2)	?
Casamino acids[d]	Stimulation (2,5)	?
Glutamate	?	Stimulation; production initiated early in growth phase (4)
Cysteine	?	Stimulation (4)
Yeast extract	?	Stimulation (4)
Phosphate	Depression (1,2,5)	?; no apparent effect (3,4)
Temperature: optimal	23–25°C (1,2)	18°C (4)
Aeration	Depression (2)	?; stimulation (3,4)

[a] Syringomycin is produced by certain strains of *Pseudomonas syringae* pv. *syringae*; phaseolotoxin by *P. syringae* pv. *phaseolicola*.

[b] Reference: (1) DeVay *et al.* (1968); (2) Gross and DeVay (1977); (3) Mitchell (1978); (4) Rudolph *et al.* (1976); (5) Gross (1985).

[c] ?, Unknown or uncertain; data are suggestive.

[d] Histidine was the only amino acid that stimulated syringomycin production in a synthetic medium (5).

of Gross (1985) indicate that in a synthetic medium, histidine was the only amino acid which allowed efficient syringomycin production when used as a sole nitrogen source. Phaseolotoxin production commenced sooner with the addition of glutamic acid to a synthetic medium (Rudolph *et al.*, 1978). The stimulatory effect of glutamic acid on phaseolotoxin production (Rudolph *et al.*, 1978) might be attributed to the formation of such precursors as ornithine, homoarginine, and the N^{δ} (N^{V}-sulfodiamino-phosphinyl) moiety. Or the added glutamate may be a nitrogen source in excess of that required for growth. We remind the reader that glutamic acid and glutamine are key nitrogen donors in the biosynthesis of various amino acids and pyrimidines. Thus, later in the growth cycle when primary metabolic systems have slowed down, glutamic acid and the amino acids derived therefrom may be unnecessary for growth and available for the synthesis of phaseolotoxin. In the presence of cysteine higher levels of phaseolotoxin were synthesized (Rudolph *et al.*, 1978), although cysteine is not a component of the phaseolotoxin tripeptide (Fig. 19).

Inorganic ions can also affect toxin production. Phosphate caused suppression of syringomycin production but apparently had little effect on phaseolotoxin production (see, for example, Gross and DeVay, 1977; Rudolph *et al.*, 1978). There are no data on kinetics of phosphate inhibition of syringomycin production or on the reversal of the inhibition upon phosphate depletion. The data presented by DeVay *et al.* (1968) and Gross and DeVay (1977) and Gross (1985) indicate that the phosphate effect is concentration dependent. The requirement of iron for efficient syringomycin production has also been shown (Gross, 1985). Zinc was involved in the processing of tabtoxin to its active form, tabtoxinine-β-lactam (Durbin and Uchytil, 1985).

The production of many antibiotics belonging to different biosynthetic groups (e.g., peptide antibiotics, polyene macrolides, tetracyclines) is regulated by phosphate (Martin and Demain, 1976, 1980). Regulation might occur by many different pathways or, alternatively, a common regulatory effector might control different biosynthetic processes. Demain and associates (Drew and Demain, 1977; Martin and Demain, 1980) have considered three possibilities in explaining phosphate inhibition of antibiotic synthesis: (1) role of ATP as an effector molecule in antibiotic synthesis; (2) adenylate energy charge (Atkinson and Walton, 1967) as the regulatory parameter controlling energy-yielding and energy-consuming metabolic processes; and (3) inhibition of enzymes such as antibiotic synthetases or phosphatases. Any one or a combination of these conditions could account for the phosphate effect on syringomycin production in *P. syringae* pv. *syringae*. However, far more sophisticated mechanisms may be operating. Bacteria have elaborate mechanisms for "sensing" metabolite concentrations (and concentration gradients) and generating internal signals in response, which control their behavior.

The production of syringomycin and phaseolotoxin also appear to be affected by aeration (Table XV). Higher levels of syringomycin production in still cultures, compared with aerated cultures, could be attributed either to shifts in metabolic systems in the presence of low levels of oxygen or to inactivation of synthetases by oxygen. In aerated cultures, a carbon source such as glucose is probably rapidly metabolized with the generation of ATP and other high-energy compounds. Under these conditions the cell metabolism may be directed toward processes leading to cell growth. The amino acids (phenylalanine, serine, arginine) produced may be directed to protein synthesis and not to the peptide toxin.

The available data do not indicate glucose repression of toxin synthesis under aerated conditions. On the contrary, glucose stimulated the synthesis of syringomycin, and high levels of the toxin were obtained with 1.5% glucose (Gross and DeVay, 1977). In experiments involving phaseolotoxin production (Mitchell, 1976), glucose concentration of the medium was dropped to 0.2% from 1%. However, Rudolph *et al.* (1978) found no significant difference in toxin levels when cells were grown either on glucose or glycerol.

The production of syringomycin, syringotoxin, and phaseolotoxin is stimulated under low incubation temperatures coincident with field observations of pronounced disease severity in early spring. Phaseolotoxin production by *P. syringae* pv. *phaseolicola* does not occur at 25°C or higher; 18°C appears to be optimal. The biochemistry and genetics of the temperature effect await elucidation.

The results of *in vitro* studies allow certain predictions to be made concerning toxin production in plants. Bacteria grow at a slower rate in plants than *in vitro* on rich artificial media (e.g., Smidt and Vidaver, 1982). The slower growth, possibly resulting from nutrient imbalance or limitation, or partially effective inhibitors, results in a slower metabolic rate, which could be physiologically conducive to toxin production. It is thus possible that the two-phase kinetics of toxin production seen under *in vitro* conditions on rich media is actually an artifact, and that in plants toxin production occurs simultaneously. Such coupling between growth and toxin production may have teleological significance. The altered physiology of host cells (e.g., release of nutrients) resulting from the activities of toxins of such pathogens as *P. syringae* pv. *phaseolicola* and *P. syringae* pv. *syringae* could provide substrates catabolized by the pathogens and suppress host defense systems facilitating pathogen multiplication. Findings on the multiplication and/or translocation of toxinogenic and nontoxinogenic strains of *P. syringae* pv. *syringae* (Gonzalez and Vidaver, 1979) and *P. syringae* pv. *phaseolicola* (Patil *et al.*, 1974) are generally consistent with this hypothesis. In contrast, in a nutritionally rich environment on artificial media, toxin production is not essential for the growth of the organism. So, phytopathogenic bacteria such as *P. syringae* pv. *phaseolicola* or *P. syringae* pv. *syringae* may have evolved to produce toxins only when the specific growth rate falls below a certain level due to physiological constraints, in contrast, for example, to *P. solanacearum* or *X. campestris* pv. *phaseoli*, which are associated with high temperature and presumed greater growth rate. Chemostat data are needed to test this hypothesis. The use of plant cells as substrates in such studies might prove fruitful in defining parameters regulating toxinogenesis *in vivo*.

4. Biosynthesis

The enzymes and the steps in the synthesis of peptide toxins and other classes of prospective phytotoxins (e.g., amino acid analogs, glycopeptides, polysaccharides, sugar derivatives) await elucidation. Therefore, using the knowledge of peptide synthesis acquired in other bacterial systems, such as *Bacillus* spp., we outline possible modes of synthesis of syringomycin, syringotoxin and phaseolotoxin. This purely speculative venture may serve as a guide to the research involving synthesis of peptide toxins of phytopathogenic bacteria.

Available knowledge suggests that phytotoxic peptides and antibiotic peptides have a number of properties in common. For example, both are much smaller molecules than proteins, and contain unusual amino acids (such as ornithine, homoarginine, or D-amino acids) not found in proteins. In the case of peptide antibiotics, a family of related peptides, rather than a single substance, is produced by an organism. Although (N^{δ}-phosphosulfamyl)-ornithylalanylhomoarginine was the major phytotoxic compound, Mitchell and Parsons (1977) detected (N^{δ}-phosphosulfamyl)-ornithylserylhomoarginine in the culture filtrate of *P. syringae* pv. *phaseolicola*, suggesting the production of related substances. *P. syringae* pv. *syringae* strains, however, produce either syringomycin or syringotoxin, or neither, but not both. Moreover, there is no evidence to suggest bacterial production of different forms of syringomycin that are all biologically active. It is possible that the synthesis of phytotoxic peptides by phytopathogenic bacteria occurs at a higher fidelity than with peptide antibiotics synthesized by nonphytopathogenic bacteria. The evolution of a high fidelity peptide-synthesizing system might be due to selection pressure exerted by the host on the bacterium to consistently produce a biologically active material.

The significance of the tripeptide structure of phaselotoxin is not known. According to Mitchell and Bieleski (1977), (N^{δ}-phosphosulfamyl)-ornithylalanylhomoarginine, was broken down sequentially to (N^{δ}-phosphosulfamyl)-ornithylalanine, (N^{δ}-phosphosulfamyl)-ornithine, phosphosulfamate, and sulfamate in plants. Of these breakdown products, (N^{δ}-phosphosulfamyl)-ornithine was the main component present and was also as toxic as the N^{δ}-phosphosulfamyl tripeptide. The findings of Moore *et al.* (1983) also indicate that octicidin (see page 158) is the chemical entity that accumulates in plants. These observations suggest the occurrence of the following sequence of events in plants. The bacterium synthesizes the tripeptide toxin which is transported across the bacterial cell envelope and excreted out into the host tissue. Plant enzymes (peptidases) degrade the molecule to octicidine which not only produces a toxic response but is also transported within the plant. The tripeptide structure might impart a degree of stability to the molecule, preventing a rapid breakdown of the crucial linkage between the N^{δ}-(N^{∇}-sulfodiamino-phosphinyl) moiety and ornithine (Fig. 19); this linkage is required for biological activity (Mitchell and Bieleski, 1977). Alternatively, the tripeptide structure may be required in the excretion of the toxin.

Studies on the synthesis of peptide antibiotics have revealed that many of these biosynthetic systems are quite distinct from those synthesizing proteins. The distinguishing features of peptide-synthesizing systems are: (1) insensitivity to chloramphenicol or puromycin; (2) high sensitivity *in vitro* to SH poisons and, possibly, a requirement for thiols to keep the enzyme thiols reduced; (3) presence of 4′-phosphopantethein in the active enzyme complex; (4) *in vitro* synthetic activity in enzymes (proteins) in the molecular weight region of 10^5 to 5×10^5; (5) the requirement for multiple enzyme components; and (6) insensitivity to deoxyribonucleases and ribonucleases. This is because *t*RNA, *m*RNA, and ribosomes are not directly involved in the synthesis of peptide antibiotics. [Such compounds are required in the synthesis of antibiotic (peptide) synthetases that activate and polymerize amino acids into peptides.] Peptide synthetases may be assayed and recognized by their catalysis of SH reagent-inhibitable ATP/PP_i exchange reactions specifically dependent upon one or more of the phytotoxic peptides, analogous to the requirements for the cell-free synthesis of peptide antibiotics by *Bacillus* spp. (Katz and Demain, 1977; Lipmann, 1980).

There is still uncertainty regarding the composition of syringotoxin, as well as the amino acid sequence of both syringomycin and syringotoxin. Understanding the biosynthesis of these toxins requires identification of the nature of the unknown ninhydrin-positive material present in the toxin hydrolysate of syringotoxin and syringomycin, determining the absence of other residues besides amino acids, and the amino acid sequence. With such knowledge a system might be developed for cell-free synthesis of the toxins that would permit examination of the properties of toxin synthetase(s).

The development of a cell-free syringomycin- or syringotoxin-synthesizing system would constitute a major advance in probing the biosynthesis of the toxin(s) and in elucidating the effects of regulatory parameters. Specifically, such systems would facilitate investigations on: (1) activators and inhibitors of toxin synthetases; (2) specificity and fidelity of the toxin-synthesizing system; (3) *in vitro* complementation between genetically modified synthetases; and (4) comparisons between the toxin molecules synthesized in a cell-free system and the extracellular toxins produced by whole cells of *P. syringae* pv. *syringae*.

In contrast to syringomycin, the composition and structure of phaseolotoxin are now known (Mitchell, 1976, 1978; Mitchell and Parsons, 1977; Moore *et al.*, 1984). Thus it is now possible to proceed directly with investigations on the biosynthesis of this structurally unusual molecule in *P. syringae* pv. *phaseolicola*. Biosynthesis probably occurs in several steps: one group of steps would assemble the N^{δ}-(N^{∇}-sulfodiamino-phosphinyl) moiety and place it on the ornithine, while another group of steps would activate and assemble the amino acids; the latter steps might involve formation of amino acyl-enzyme thioesters and their sequential polymerization by a mechanism analogous to that for the peptide antibiotics discussed above. A homoarginine-dependent exchange reaction would likely be distinctive.

5. *Phytotoxin Immunity*

Along with biosynthesis, the question of immunity must be addressed. Microbial cells producing non-specific toxin are likely to possess mechanisms which allow them to avoid the effects of their own toxins. In the case of phaseolotoxin, producing cells were found to possess an ornithine carbamoyltransferase which was resistant to the toxin (Staskawicz *et al.*, 1980). This type of immunity does not appear to be present in *Pseudomonas syringae* pv. *tabaci*. Strains of this organism produce tabtoxin whose hydrolytic product, tabtoxinine-β-lactam, inhibits glutamine synthetase from pea (Thomas *et al.*, 1983). Glutamine synthetase from *P. syringae* pv. *tabaci* also does not appear to be resistant to tabtoxinine-β-lactam (Thomas and Durbin, 1985).

Non-toxigenic strains of *Pseudomonas syringae* pv. *phaseolicola* are more sensitive to phaseolotoxin than producing strains (Staskawicz *et al.*, 1980). In contrast, non-toxigenic mutants of a syringotoxin-producing strain of *P. syringae* pv. *syringae* were not altered in their sensitivity to syringotoxin (Morgan and Chatterjee, unpublished).

6. *Genetics*

Despite their obvious value, until recently the genetics of toxinogenesis in phytopathogenic bacteria had not been examined systematically. In the past, the lack of personnel trained in genetic techniques, the lack of genetic systems and, in most cases, the inability to score the toxinogenic phenotype have largely contributed to this deficiency. With the development of microbiological assay procedures for syringomycin (Sinden *et al.*, 1971) and phaseolotoxin (Staskawicz and Panopoulos, 1979), syringotoxin (Gross and DeVay, 1977) and tabtoxin (Gasson, 1980), it has been possible to look for mutant bacteria altered in toxinogenic phenotype and to study toxin genes by physical means and by gene cloning (see below). It should be noted that colony assays on agar plates score for the presence or absence of extracellular toxins, and this limitation should be considered in characterizing (and describing) putative toxin mutants.

We must also recognize that mutations in genes for toxin biosynthesis, precursor formation, and toxin transport components might lead to the nontoxinogenic phenotype. Thus, in investigating the genetics of toxinogenesis, distinctions should be made between those mutations that alter toxin synthesis in the presence of precursors and those mutations that prevent toxin synthesis by altering the biosynthesis of precursors. Ideally, mutants could be classified according to the function of the steps that are altered. In addition, mutations in biosynthetic pathways of certain primary metabolites, such as those involved in the synthesis of glutamate and assimilation of sulfate, could affect toxin biosynthesis and may give rise to nontoxinogenic phenotypes. These might be identified, however, by using suitable supplemented media.

Nonantibiotic or low antibiotic producers have been detected in auxotrophic mutants (see Hopwood and Merrick, 1977). Likewise, T. C. Currier and M. K. Morgan (personal communication) have obtained Ade^-, Arg^-, Met^-, and Trp^- mutants of *P. syringae* pv. *syringae*, some of which produce low or no detectable levels of syringomycin. Cysteine, and to a lesser extent serine auxotrophy, reduce levels of toxin production in a syringotoxin producing strain of *P. syringae* pv. *syringae* (Morgan and Chatterjee, 1985). The nontoxinogenic phenotype of such auxotrophic mutants may be attributed to the deficiency of the intracellular pools of precursors required in toxin synthesis. Again, media supplementation might identify them.

The possibility should be considered of the occurrence of regulatory mutations leading to no toxin, some toxin production or greater than wild-type production. These are Tox^-, Tox^{leaky}, and Tox^{hyper} phenotypes, respectively. Regulatory mutations resulting in Tox^- or Tox^{leaky} phenotypes could be due to regulation either of precursor formation or of peptide synthetases. The Tox^{hyper} phenotype could only occur from mutational alteration both in the level of all precursors and the quantity of toxin synthetases or a membrane defect(s) if there is a high level of intracellular toxin in wild-type strains.

In *P. syringae* pv. *phaseolicola* (Romeo and Patil, 1984; Peet *et al.*, 1985) and *P. syringae* pv. *syringae* (Morgan and Chatterjee, 1985), two classes of Tn*5* insertion mutants were obtained: Class I, those that produced no toxin as indicated by bioassays; and Class II, those that produced considerably lower levels of toxin. The physical analysis of a collection of Tn*5* insertion mutations in *P. syringae* pv. *syringae* (Morgan and Chatterjee, 1985) suggested that the Class I mutants probably had Tn*5* insertions in structural genes for toxin-synthetases or other enzymes involved in toxinogenesis, and that the Class II mutants resulted from Tn*5* insertions either in a gene that regulated the synthetase levels or in other genes affecting the efficiency of toxinogenesis. A similar situation may also prevail in *P. syringae* pv. *phaseolicola*.

The observations that plasmids specify some of the virulence traits of animal and human pathogenic bacteria (for example, Elwell and Shipley,

1980) led to the concept that toxinogenesis might also be determined by plasmids of phytopathogenic bacteria. This hypothesis was bolstered by the early success in *A. tumefaciens* analyses and the occurrence of one or more plasmids in most phytopathogenic bacteria (see Section III.D.). Plasmids proved to be essential not only for tumorigenic *A. tumefaciens* and the root proliferation pathogen *A. rhizogenes* (Moore *et al.*, 1979; White and Nester, 1980) but also for some strains of the gall-forming *P. syringae* pv. *savastanoi* (Smidt and Kosuge, 1978; Comai and Kosuge, 1980; Comai *et al.*, 1982; also see Section IV.E.). The production of the phytotoxin, coronatine, was correlated with the presence of a 58-MDa plasmid in *Pseudomonas syringae* pv. *atropurpurea* (Sato *et al.*, 1983) and with a 8.8-kb plasmid in *P. syringae* pv. *glycinea* strains (Willis *et al.*, 1986). The 8.8-kb plasmid was cloned in *E. coli* and *E. coli* transformants containing this recombinant plasmid produced a compound which appeared to be identical to coronatine (Willis *et al.*, 1986). In other toxin-producing pathovars of *P. syringae*, initial correlations (Gantotti *et al.*, 1979; Gonzalez and Vidaver, 1979) with plasmid presence, curing, and loss of toxin production, have not been definitive. In fact, plasmidless and toxinogenic strains of all of these pathovars can be as pathogenic as plasmid-carrying strains (Gonzalez and Vidaver, 1978). In *P. syringae* pv. *syringae*, plasmid transfer, after transposon marking of pCG131, to a Tox^- recipient, did not result in toxin production (Gonzalez *et al.*, 1984). Currier and Morgan (1983) also cured several *P. syringae* pv. *syringae* strains of all resident plasmids; the cured strains remained fully toxinogenic. However, there remained the possibility that plasmids specifying toxin production could have transposed into the chromosome in cured strains. In both *P. syringae* pv. *syringae* (Currier and Morgan, 1983) and *P. syringae* pv. *tabaci* (Piwowarski and Shaw, 1982) this has been shown not to be the case by DNA:DNA hybridization studies. Likewise, Jamieson *et al.* (1981) were not able to associate the loss of a plasmid with spontaneous loss of toxinogenicity in *P. syringae* pv. *phaseolicola*. Since plasmids do not appear to code for phytotoxins such as tabtoxin, phaseolotoxin and syringomycin, investigations have now been directed towards chromosomal determinants.

In Tn*5* insertion mutants of *P. syringae* pv. *syringae* altered in the production of ST, the loss of toxin production was associated with Tn*5* insertions into chromosomal *Eco*RI fragments of about 10.5, 17.8, and 19.3 kb. Of 15 toxin-negative mutants, 10 contained Tn*5* insertions into a 17.8-kb fragment, three contained insertions into an apparently adjacent 10.5 kb, and two contained insertions into an unlinked 19.3-kb fragment (Morgan and Chatterjee, 1985). Preliminary mapping of the insertions revealed a fairly even dispersion of insertion along the expanse of 10.5- and 17.8-kb fragments (Morgan and Chatterjee, in progress). Tn*5* insertion fragments from each of the three size classes have been cloned for use as probes for wild-type toxin sequences. When used as probes of genomic DNAs from other *P. syringae* strains and pathovars in Southern blot analyses, strong homology was detected between the three ST probes and genomic DNAs from *P. syringae* pv. *syringae* which produced SR. Very little or no homology was detected with *P. syringae* pvs. *phaseolicola*, *pisi* or *tabaci* strains (Morgan and Chatterjee, in progress). These results suggest that these probes can be used to isolate SR as well as ST genes and that the synthesis machinery of ST and SR are probably quite similar. The lack of homology between the ST probes and *P. syringae* pvs. *phaseolicola* and *tabaci* suggest that the enzymes which synthesize the peptide moiety of the toxins in these two pathovars are not closely related to those involved in the synthesis of ST or SR. The frequency at which the syringotoxin-negative Tn*5* mutant arose (0.28%) suggests the involvement of approximately 770 kDa of protein, assuming random insertion of Tn*5*. This amount of protein is comparable with peptide antibiotic synthetase systems of *Bacillus* spp. which comprise multienzyme complexes totaling 250–770 kDa of protein. In addition to the toxin-negative mutants, two classes of mutants with reduced levels of toxin production were obtained which produced 3–4% or 50% of wild-type toxin levels (Morgan and Chatterjee, 1985).

Mutants altered in the production of phaseolotoxin have also been isolated. Both toxin-negative (Romeo and Patil, 1984) and mutants with extremely low levels of toxin activity (Peet *et al.*, 1984; Romeo and Patil, 1984) have been described. Wild-type sequences have been cloned which restore toxin activity in some of these mutants (Clements *et al.*, 1985; Peet *et al.*, 1985).

E. Indoleacetic Acid

1. Biosynthesis and Regulation

The auxin indoleacetic acid (IAA) is normally produced in plants and stimulates plant growth (Bonner and Varner, 1976). Certain bacteria, such as *P. syringae* pv. *savastanoi*, *C. fascians*, and *A. tumefaciens*, produce IAA (see Nester and Kosuge, 1981; Vidaver and Starr, 1981; Liu *et al.*, 1982), which apparently elicits a hormonal imbalance in plants. The abnormalities are seen as knots, fasciations, or galls. Hypertrophic and hyperplastic responses are seen as abnormal cell enlargements, initiation of root initials, and enhanced meristematic and cambial activity (Nester and Kosuge, 1981).

Our current knowledge of the biosynthetic mechanisms and the genetics of IAA production in bacteria mostly comes from the work of Kosuge and associates. While some of these findings are discussed below, further detail can be obtained elsewhere (Nester and Kosuge, 1981; Kosuge *et al.*, 1983).

The biosynthesis of IAA in *P. syringae* pv. *savastanoi* proceeds from tryptophan by the sequential activities of tryptophan monooxygenase and indoleacetamide hydrolase as shown in Fig. 21 (Smidt and Kosuge, 1978; Comai

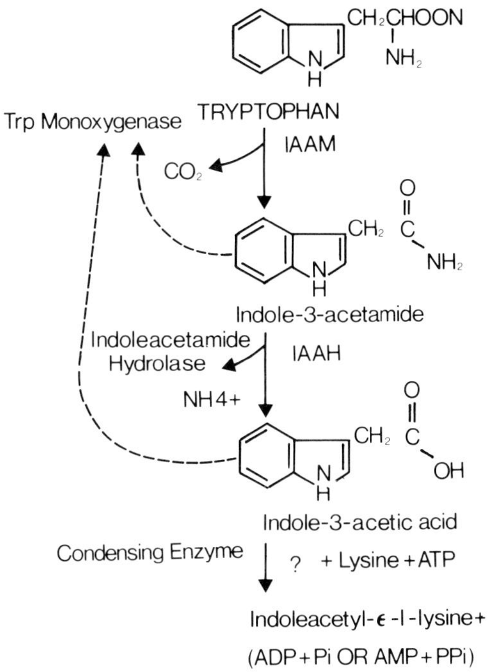

Fig. 21. Enzymes and intermediates in the conversion of L-tryptophan to IAA and indoleacetyl-ε-L-lysine in *Pseudomonas syringae* pv. *savastanoi*. Broken arrows indicate feedback inhibition of tryptophan monooxygenase by intermediates.

and Kosuge, 1980, 1982). Only tryptophan monooxygenase, the first enzyme of the pathway, has been well characterized (Kosuge *et al.*, 1966; Hutzinger and Kosuge, 1968; Kuo and Kosuge, 1969). The condensation of IAA to indoleacetyl-ε-L-lysine occurs at the expense of ATP (Fig. 21). The physiological significance of this reaction remains unclear.

Biosynthesis of IAA in various microorganisms may proceed by alternate pathways (see, for example, Kosuge *et al.*, 1966). One such pathway, reported for *A. tumefaciens* (Liu *et al.*, 1982), is

$$\text{Tryptophan} \xrightarrow[NH_2]{\text{transaminase}} \text{indole-3-pyruvic acid} \xrightarrow[CO_2]{\text{decarboxylase}} \text{indole-3-acetic acid}$$

The production of IAA in *P. syringae* pv. *savastanoi* appears to be regulated by a combination of factors: (1) feedback inhibition, (2) the tryptophan quantity or pool, and (3) transcription of the *iaa* genes (operon). In cell-free systems, indole-3-acetamide inhibited tryptophan monooxygenase; likewise, indole-3-acetic acid inhibited the monooxygenase but not indoleacetamide hydrolase. Thus, the feedback regulation of IAA biosynthesis appears to operate at the level of the first biosynthetic enzyme, tryptophan monooxygenase. However, such feedback inhibition, observed in cell-free systems, might not be operative in whole cells under physiological conditions. IAA may either be rapidly excreted, preventing the accumulation of a high intracellular pool of the product(s), or can be converted to indoleacetyl-ε-L-lysine; the latter compound may not feedback inhibit tryptophan monooxygenase, although this possibility remains to be determined.

The regulatory effects of precursor pools on the production of bacterial metabolites were discussed in Section IV.D.3. Both physiological and genetic evidence (Kuo and Kosuge, 1969; Smidt and Kosuge, 1978) indicate such a role for the tryptophan pool in IAA production in *P. syringae* pv. *savastanoi*. For example, the addition of L-tryptophan to culture medium greatly stimulated the production of IAA by the bacterium (Kuo and Kosuge, 1969). Moreover, the data of Kuo and Kosuge (1969) indicated that the stimulation of IAA synthesis by tryptophan resulted from an intracellular pool rather than an enhanced synthesis of IAA biosynthetic enzymes. This observation was subsequently supported by genetic evidence of Smidt and Kosuge (1978). They obtained α-methyltryptophan (MT; an analog of tryptophan)-resistant mutants, some of which hyperproduced IAA while others did not synthesize IAA. We might note that in *E. coli* 5-methyltryptophan resistance results in changes in anthranilate synthetase, the first enzyme in the biosynthesis of tryptophan from chorismate; this enzyme is subject to feedback inhibition by tryptophan. Consequently, such *E. coli* mutants are resistant to tryptophan inhibition, and accumulate tryptophan in culture media. Smidt and Kosuge (1978) hypothesized that a similar situation prevailed in *P. syringae* pv. *savastanoi*, so that the overproduction of IAA in α-methyltryptophan-resistant mutants was ascribed to an increased tryptophan pool which therefore favored IAA synthesis. The lack of IAA production in certain analog-resistant mutants appeared to result from a deficiency of tryptophan monooxygenase activity (Comai and Kosuge, 1982; Smidt and Kosuge, 1978). This enzyme has now been characterized (Hutcheson and Kosuge, 1985).

2. *Genetics*

IAA production in *A. tumefaciens* (Liu and Kado, 1979; Liu *et al.*, 1982) and in certain strains of *P. syringae* pv. *savastanoi* (Comai and Kosuge, 1980; Comai *et al.*, 1982) is a plasmid-determined trait (see Section III.D. for the details on the IAA plasmid of *P. syringae* pv. *savastanoi*). By using gene cloning and transpositional mutagenesis techniques, Comai and Kosuge (1982) obtained preliminary evidence suggesting an operon type organization of *iaa* genes and the existence of a promoter sequence. They postulated the following direction of transcription: Promoter⟶ *iaaM*⟶ *iaaH*⟶

Whether or not there also is a regulator gene(s) remains to be determined. The organization of the chromosomal IAA genes in other strains of *P. syringae* pv. *savastanoi* awaits elucidation.

The evidence for Ti plasmid-coded production of IAA in *A. tumefaciens* (Liu *et al.*, 1982) includes: (1) lack of IAA production in Ti plasmid-free strains; (2) restoration of IAA production upon the introduction of nopaline or octopine type Ti plasmids; and (3) inactivation by Tn*5* mutagenesis of a gene (*iaaP*) required in IAA synthesis. Sequence studies with the *iaaM* gene of *P. syringae* pv. *savastanoi* suggest homology with that of the T-DNA (TMS-1) of *A. tumefaciens* (Yamada *et al.*, 1985). This suggests evolutionary conservation of pathogenic determinants for this property.

The elegant work of Kosuge and associates now paves the way for detailed genetic and molecular investigations of the *iaa* genes in *P. syringae* pv. *savastanoi*. An obvious research goal is to confirm the operon type organization of the *iaa* genes, including the gene that specifies the production of the condensing enzyme. Another gene, for IAA-lysine synthetase (*iaaL*), was cloned and hyper-expressed in *E. coli* (Glass and Kosuge, 1985). The hypothesis presented was that conversion of IAA to IAA-lysine may regulate the IAA pool size in the bacterium and thus modulate virulence by affecting gall size. Other questions of interest are the coordinate regulation of the biosynthetic enzymes. In clarifying the regulation of IAA biosynthesis, knowledge is also required of tryptophan biosynthesis and its control in *P. syringae* pv. *savastanoi*. However, the regulation by tryptophan may be studied in an *E. coli* background where much is now known of the expression of the *trp* operon (Yanofsky, 1981) and for which strains defective in *trp* biosynthetic and regulatory genes are available. Such studies are clearly feasible since the *iaaM* gene from *P. syringae* pv. *savastanoi* cloned into an R plasmid, was expressed in an *E. coli* host (Comai and Kosuge, 1982).

V SUMMARY AND PERSPECTIVES

We have presented a review and interpretation of current knowledge of most known and prospective factors that affect pathogenicity of phytopathogenic bacteria, with the notable exception of the crown gall system, amply reviewed elsewhere (Nester and Kosuge, 1981; Van Montagu and Schell, 1982; Roberts, 1982; Hooykaas and Schilperoort, 1984; Nester *et al.*, 1984). Some potential bacterial virulence determinants were not discussed, primarily because of the lack of firm genetic data. Some correlations of observed compounds with pathogenicity include plant growth regulators such as cytokinins (Mitchell, 1981; Nester and Kosuge, 1981; Vidaver, 1982) produced by *Agrobacterium tumefaciens* and *Corynebacterium fascians*, the gene for which has been cloned and found homologous to that in *A. tumefaciens* (Mellano and Cooksey, 1985) and ethylene by at least *Pseudomonas solanacearum* (see Primrose, 1979), the Kudzu strains of *P. syringae* pv. *phaseolicola* (Goto *et al.*, 1985). Other biochemical factors of potential importance include a wide variety of enzymes, such as proteases, cellulases, phospholipases, and unidentified plant-macerating enzymes (Bateman and Basham, 1976), several characterized toxins (see Rudolph, 1976; Mitchell, 1981), and high-molecular-weight extracellular peptidyl glycans (see Daly, 1981; Strobel, 1982; Vidaver, 1982). Structural features likely to be important in pathogenicity or virulence need to be examined in further detail, especially pili and fimbriae (Ottow, 1975; Sokatch, 1979; Sequeira, 1984) and the cytoplasmic membrane of *Spiroplasma* spp. (Wroblewski *et al.*, 1977; Mudd *et al.*, 1979). Any or all of these factors may be shown to be significant in disease, provided that appropriate genetic studies are done. It is clear that the techniques and tools of modern genetics are rapidly becoming used and offer the prospects of providing definitive answers to questions on the identification and role of pathogenic determinants. Recombinant DNA technology has been of particular success in analyzing plasmid gene functions in *A. tumefaciens* and several human and animal pathogens. Similar studies with *Corynebacterium*, *Erwinia*, *Pseudomonas*, *Streptomyces*, and *Xanthomonas* spp. are underway. A near-term success in elucidating chromosomal gene functions is anticipated in some of these bacteria because of the available genetic tools such as transposons, conjugative plasmids, and cloning vectors. The fastidious nature of other phytopathogenic bacteria, such as *Spiroplasma* and xylem-limited bacteria, is likely to make genetic studies more difficult (but see Daniels, 1983, and Mouches *et al.*, 1983) as is the absence of reports for any gene transfer systems. While most genetic studies with mycoplasma-like organisms and other phloem inhabitors await their isolation and characterization, some progress on elucidating structure and function is likely with the successful cloning of DNA from MLO-enriched extracts (Kirkpatrick *et al.*, 1985).

Future work in the genetics of bacterial plant pathogens is expected to assist plant and microbial ecologists in understanding the relationship of bacteria to particular environments, plant breeders and plant pathologists in defining and manipulating resistance determinants in plants, and chemists in formulating compounds for improved plant disease control. Finally, we hope that this article will encourage more work in this area by both prospective and seasoned plant pathologists.

REFERENCES

Acker, G., Wartenberg, K. and Knapp, W. (1980). *Zentralblatt für Bakteriologie, Parasiten-Kunde, Infektionskrankheiten und Hygiene* (Abteilung I, Originale A) **247,** 229–240.

Adams, G. A. and Young, R. (1965). *Can. J. Biochem.* **43,** 1499–1512.

Adams, M. H. (1959). "Bacteriophages". Interscience Publishers, Inc., New York.

Ait, N., Creuzet, N. and Forget, P. (1979). *J. Gen. Microbiol.* **113,** 399–402.

Akiyama, Y., Eda, S., Kato, K. and Tanaka, H. (1984). *Carbohydr. Res.* **133,** 289–296.

Akiyoshi, D. E., Morris, R. O., Hinz, R., Mischke, B. S., Kosuge, T., Garfinkel, D. J., Gordon, M. P. and Nester, E. W. (1983). *Proc. Natl. Acad. Sci., USA* **80,** 407–411.

Alberghina, A., Mazzucchi, U. and Pupillo, P. (1973). *Phytopathol. Z.* **78,** 204–213.

Albersheim, P. (1966). *Meth. Enzymol.* **8,** 628–631.

Albersheim, P., Neukom, H. and Deuel, H. (1960). *Arch. Biochem. Biophys.* **90,** 46–51.

Al-Issa, A. N. and Sigee, D. C. (1982). *Phytopathol. Z.* **104,** 104–114.

Allen, A. K. and Neuberger, A. (1973). *Biochem. J.* **135,** 307–314.

Allen, C. and Lacy, G. H. (1985). *Phytopathology* **75,** 1326, #384 (Abstr.).

Allen, C., Roberts, D. P., Ford, M., Stromberg, V. K., Berman, P. M., Lacy, G. H. and Mount, M. S. (1984). *Phytopathology* **74,** 816, #A214 (Abstr.).

Allen, C., Roberts, D. P. and Lacy, G. H. (1985). *Proc. 6th Intern. Conf. Plant Pathogenic Bacteria*, Maryland (in press).

Alton, N. K. and Vapnek, D. (1979). *Nature* **282,** 864–869.

Ames, B. N. (1983). *Science* **221,** 1256–1264.

Ames, G. F.-L. and Nikaido, K. (1976). *Biochemistry* **15,** 616–623.

Ames, G. F.-L., Spudich, E. N. and Nikaido, H. (1974). *J. Bacteriol.* **117,** 406–416.

Anderson, A. J. (1984). *Appl. Environ. Microbiol.* **48,** 31–35.

Anderson, D. and Mills, D. (1983). *Phytopathology* **73,** 824, #A459 (Abstr.).

Anderson, D. M. and Mills, D. (1985). *Phytopathology* **75,** 104–108.

Andreoli, P., Overbeeke, N., Veltkamp, E., van Embden, J. and Nijkamp, H. (1978). *Mol. Gen. Genet.* **160,** 1–11.

Andrews, J. H. (1976). *Biol. Revs.* **51,** 211–253.

Andro, T., Chambost, J.-P., Kotoujansky, A., Cattaneo, J., Bertheau, Y., Barras, F., Van Gijsegem, F. and Coleno, A. (1984). *J. Bacteriol.* **160,** 1199–1203.

Anwar, H., Brown, M. R. W., Cozens, R. M. and Lambert, P. A. (1983). *J. Gen. Microbiol.* **129,** 499–507.

Arber, W., Iida, S., Jutte, H., Caspers, P., Meyer, J. and Hanni, C. (1978). *Cold Spring Harbor Symp. Quant. Biol.* **43,** 1197–1208.

Arbuthnott, J. P. and Smyth, C. J. (1979). In "Adhesion of Microorganisms to Surfaces" (D. C. Ellwood, J. Melling and P. Rutter, eds), pp. 165–198. Academic Press, Orlando and London.

Archer, D. B., Best, J. and Barber, C. (1981). *J. Gen. Microbiol.* **126,** 511–514.

Ashwell, G. (1962). *Meth. Enzymol.* **5,** 190–208.

Aspinall, G. O. (1970). "Polysaccharides". Pergamon Press, Oxford.

Aspinall, G. O., Craig, J. W. T. and Whyte, J. L. (1968a). *Carbohydr. Res.* **7,** 442–452.

Aspinall, G. O., Gestetner, B., Molloy, J. A. and Uddin, M. (1968b). *J. Chem. Soc.* (c), 2554–2559.

Atkinson, D. E. and Walton, G. M. (1967). *J. Biol. Chem.* **242,** 3239–3241.

Austrian, R., Bernheimer, H. P., Smith, E. E. B. and Mills, G. T. (1959). *J. Exp. Med.* **110,** 585–602.

Avery, O. T., MacLeod, C. M. and McCarty, M. (1944). *J. Exp. Med.* **79,** 137–158.

Ayers, A. R., Goodman, R. N., Ayers, S. B. and White, J. A. (1977). *Proc. Am. Phytopathol. Soc.* **4,** 107, #116 (Abstr.).

Ayers, A. R., Ayers, S. B. and Goodman, R. N. (1979). *Appl. Environ. Microbiol.* **38,** 659–666.

Bachhuber, M., Brill, W. J. and Howe, M. M. (1976). *J. Bacteriol.* **128,** 749–753.

Bachmann, B. J. (1983). *Microbiol. Rev.* **47,** 180–230.

Bachmann, B. J. and Low, K. B. (1980). *Microbiol. Rev.* **44,** 1–56.

Bachmann, B. J., Low, K. B. and Taylor, A. L. (1976). *Bacteriol. Rev.* **40,** 116–167.

Backmann, P. A. and DeVay, J. E. (1971). *Physiol. Plant Pathol.* **1,** 215–233.

Bagdasarian, M., Lurz, R., Ruckert, B., Franklin, F. C. H., Bagdasarian, M. M., Frey, J. and Timmis, K. N. (1981). *Gene* **16,** 237–247.

Bagdasarian, M. M., Amann, E., Lurz, R., Ruckert, B. and Bagdasarian, M. (1983). *Gene* **26,** 273–282.

Bagley, S. T. and Starr, M. P. (1979). *Curr. Microbiol.* **2,** 381–386.

Bahl, C. P., Wu, R., Brousseau, R., Sood, A. K., Hsiung, H. M. and Narang, A. (1978). *Biochem. Biophys. Res. Commun.* **81,** 695–703.

Baigent, N. L., DeVay, J. E. and Starr, M. P. (1963). *N.Z.J. Sci.* **6,** 75–100.

Bailey, J. A. and Manfield, J. E. (eds) (1982). "Phytoalexins" Blackie, London.

Baker, C. J., Neilson, M. J., Sequeira, L. and Keegstra, K. G. (1984). *Appl. Environ. Microbiol.* **47,** 1096–1100.

Banerjee, P. C., Vanage, R. I., Chakrabarty, A. M. and Maitra, P. K. (1985). *J. Bacteriol.* **161,** 458–460.

Barksdale, L. (1955). *C. R. Acad. Sci. Paris* **240,** 1831–1833.

Barksdale, L. and Arden, S. B. (1974). *Ann. Rev. Microbiol.* **28,** 265–299.

Barras, F., Boyer, M.-H., Chambost, J.-P. and Chippaux, M. (1984a). *Mol. Gen. Genet.* **197,** 513–514.

Barras, F., Chambost, J.-P. and Chippaux, M. (1984b). *Mol. Gen. Genet.* **197,** 486–490.

Barth, P. T. (1979). *Plasmid* **2,** 130–136.

Barth, P. T., Datta, N., Hedges, R. and Grinter, N. (1976). *J. Bacteriol.* **125,** 800–810.

Barton-Willis, P. A., Wang, M. C., Holliday, M. J., Long, M. R. and Keen, N. T. (1984). *Physiol. Plant Pathol.* **25,** 387–398.

Basham, H. G. and Bateman, D. F. (1975). *Phytopathology* **65,** 141–153.

Bashan, Y., Okon, Y. and Henis, Y. (1980). *Physiol. Plant Pathol.* **17,** 111–119.

Bassford, P., Jr., Diedrich, D. L., Schnaitman, C. L. and Reeves, P. (1977). *J. Bacteriol.* **131,** 608–622.

Bateman, D. F. (1976). *In* "Biochemical Aspects of Plant–Parasite Relationships" (J. Friend and D. R. Threlfall, eds), pp. 79–103. Academic Press, London and Orlando.

Bateman, D. F. and Basham, H. G. (1976). *In* "Physiological Plant Pathology" (R. Heitefuss and P. H. Williams, eds), pp. 316–355. Springer-Verlag, New York.

Bateman, D. F. and Millar, R. L. (1966). *Ann. Rev. Phytopathol.* **4,** 119–146.

Bauer, D. W. and Beer, S. V. (1983). *Phytopathology* **73,** 362 (Abstr.).

Bauer, D. W. and Beer, S. V. (1985). *Proc. 6th Intern. Conf. Plant Path. Bacteria,* Maryland (in press).

Bayer, M. E. (1979). *In* "Bacterial Outer Membranes: Biogenesis and Functions" (M. Inouye, ed.), pp. 167–202. John Wiley, New York.

Bayer, M. E. and Thurow, H. (1977). *J. Bacteriol.* **130,** 911–936.

Bayer, M. E. and Starkey, T. W. (1972). *Virology* **49,** 236–256.

Bayer, M. E., Koplow, J. and Goldfine, H. (1975). *Proc. Natl. Acad. Sci. USA* **72,** 5145–5149.

Bayer, M. E., Thurow, H. and Bayer, M. H. (1979). *Virology* **94,** 95–118.

Becker, A. and Gold, M. (1975). *Proc. Nat. Acad. Sci. USA* **72,** 581–585.

Beckwith, J. R., Signer, E. R. and Epstein, W. (1966). *Cold Spring Harbor Symp. Quant. Biol.* **31,** 393–401.

Beers, R. F., Jr. and Bassett, E. G. (1977). "Recombinant Molecules: Impact on Science and Society". Raven Press, New York.

Beguin, P. and Eisen, H. (1978). *Eur. J. Biochem.* **87,** 525–531.

Beguin, P., Cornet, P. and Aubert, J.-P. (1985). *J. Bacteriol.* **162,** 102–105.

Beher, M. G., Schnaitman, C. A. and Pugsley, A. P. (1980) *J. Bacteriol.* **143,** 906–913.

Bellofatto, V., Shapiro, L. and Hodgson, D. A. (1984). *Proc. Natl. Acad. Sci. USA* **81,** 1035–1039.

Bender, C. L. and Cooksey, D. A. (1985). *Phytopathology* **75,** 1325, #379 (Abstr.).

Ben-Gurion, R. and Shafferman, A. (1981). *Plasmid* **5,** 183–187.

Bennett, P. M. and Richmond, M. H. (1976). *J. Bacteriol.* **126,** 1–6.

Bennett, P. M. and Richmond, M. H. (1978). *In* "The Bacteria: A Treatise on Structure and Function" (L. N. Ornston and J. R. Sokatch, eds), Vol. VI, pp. 1–69. Academic Press, London and Orlando.

Bennett, P. M., Grinsted, J. and Richmond, M. (1977). *Mol. Gen. Genet.* **154,** 205–211.

Bennett, P. M., Grinsted, J., Choi, C. L. and Richmond, M. H. (1978). *Mol. Gen. Genet.* **159,** 101–106.

Bennett, R. A. (1980). *J. Gen. Microbiol.* **116,** 351–356.

Bennett, R. A. and Billing, E. (1975). *J. Appl. Bacteriol.* **39,** 307–315.

Bennett, R. A. and Billing, E. (1978). *Ann. Appl. Biol.* **89,** 41–45.

Bennett, R. A. and Billing, E. (1980). *J. Gen. Microbiol.* **116,** 341–349.

Benson, S. A., Hall, M. N. and Silhavy, T. J. (1985). *Ann. Rev. Biochem.* **54,** 101–134.

Benyon, J. L., Beringer, J. E. and Johnston, A. W. B. (1980). *J. Gen. Microbiol.* **120,** 421–429.

Benzinger, R. (1978). *Microbiol. Rev.* **42,** 194–236.

Benzinger, R. and Hartman, P. E. (1962). *Virology* **18,** 614–626.

Beraha, L. and Garber, E. D. (1971). *Phytopathol. Z.* **70,** 335–344.

Beraha, L., Garber, E. D. and Billeter, B. A. (1974). *Phytopathol. Z.* **81,** 15–22.

Berg, C. M., Whalen, W. A., and Archambault, L. B. (1983). *J. Bact.* **155,** 1009–1014.

Berg, D. E. and Berg, C. M. (1983). *Biol. Technol.* **1,** 417–435.

Berg, D. E., Davies, J., Allet, B. and Rochaix, J.-D. (1975). *Proc. Nat. Acad. Sci. USA* **72,** 3628–3632.

Bergmans, H. E. N., Kooijman, D. H. and Hoekstra, W. P. M. (1980). *FEMS Microbiol. Lett.* **9,** 211–214.

Beringer, J. E., Beynon, J. L., Buchanan-Wollaston, A. V. and Johnston, A. W. B. (1978). *Nature* **276,** 633–634.

Berman, P. M., Mount, M. S. and Lacy, G. H. (1983). *Phytopathology* **73,** 1342. (Abstr.).

Bernheimer, H. P., Wermundsen, I. E. and Austrian, R. (1967). *J. Bacteriol.* **93,** 320–333.

Bertheau, Y., Madgidi-Hervan, E., Kotoujansky, A., Nguyen-The, C., Andro, T. and Coleno, A. (1984). *Anal. Biochem.* **139,** 383–389.

Bertoni, G., Niepold, F. and Mills, D. (1985). *Phytopathology* **75,** 1381, #815 (Abstr.).

Bevan, M. W. and Chilton, M.-D. (1982). *Ann. Rev. Genet.* **16,** 357–384.

Biek, D. and Roth, J. R. (1980). *Proc. Natl. Acad. Sci. USA* **77,** 6047–6051.

Billing, E. (1960). *Nature* **186,** 819–820.

Billing, E. (1974). *J. Appl. Bacteriol.* **37,** 643–648.

Billing, E. (1984). *Acta Horticult.* **151,** 249–253.
Birch, R. G. and Patil, S. S. (1985). *J. Gen. Microbiol.* **131,** 1069–1075.
Birge, E. A. (1981). "Bacterial and Bacteriophage Genetics". Springer-Verlag, New York.
Biswas, G. D., Sox, T. and Sparling, P. F. (1977). *J. Bacteriol.* **129,** 983–992.
Blobel, G. and Dobberstein, B. (1975a). *J. Cell Biol.* **67,** 835–851.
Blobel, G. and Dobberstein, B. (1975b). *J. Cell Biol.* **67,** 852–862.
Bolivar, F. and Backman, K. (1979). *Meth. Enzymol.* **68,** 245–267.
Bolivar, F., Rodriguez, R. L., Greene, P. J., Betlach, M. C., Heyneker, H. L., Boyer, H. B., Crosa, J. H. and Falkow, S. (1977). *Gene* **2,** 95–113.
Bonner, J. and Varner, J. E. (1976). "Plant Biochemistry", 3rd. Edn. Academic Press, London and Orlando.
Borek, E. and Ryan, A. (1973). *Prog. Nucl. Acids. Res. Mol. Biol.* **13,** 249–300.
Botstein, D. and Klecker, N. (1977). *In* "DNA Insertion Elements, Plasmids, and Episomes" (A. I. Bukhari, J. A. Shapiro and S. L. Adhya, eds), pp. 185–204. Cold Sping Harbor Laboratory, Cold Spring Harbor, New York.
Boucher, C. A. and Sequeira, L. (1978). *Can. J. Microbiol.* **24,** 69–72.
Boucher, C., Bergeron, B., Barate de Bertalmio, M. and Dénarié, J. (1977). *J. Gen. Microbiol.* **98,** 253–263.
Boucher, C., Message, B., Debieu, D. and Zischek, C. (1981). *Phytopathology* **71,** 639–642.
Boucher, C., Barberis, P., Demery, D. and Trigalet, A. (1985a) *In* "Advances in Molecular Genetics of the Bacteria–Plant Interaction" (A. A. Szalay and R. P. Legocki, eds), pp. 165–168. Media Services, Cornell University, Ithaca, NY.
Boucher, C., Martinel, A., Barberis, P., Alloing, G. and Zischek, C. (1985b). *Proc. 6th Intern. Conf. Plant Path. Bacteria*, Maryland (in press).
Boyer, H. W. and Nicosia, S. (1978). "Genetic Engineering". Elsevier/North-Holland Biomedical Press, Amsterdam.
Boyer, M. H., Chambost, J. P., Magnan, M. and Cattaneo, J. (1984a). *J. Biotech.* **1,** 229–239.
Boyer, M. H., Chambost, J. P., Magnan, M. and Cattaneo, J. (1984b). *J. Biotech.* **1,** 241–252.
Bradbury, J. F. (1984). *In* "Bergey's Manual of Systematic Bacteriology" (N. R. Krieg and J. G. Holt, eds), Vol. 1, pp. 199–210. Williams & Wilkins Co., Baltimore.
Bradbury, W. C., Mills, S. D., Preston, M. A., Barton, L. J. and Penner, J. L. (1984). *Anal. Biochem.* **137,** 129–133.
Bradley, D. E., Douglas, C. J. and Peschon, J. (1984). *Can. J. Microbiol.* **30,** 676–681.
Braun, E. J. (1982). *Phytopathology* **72,** 159–166.
Braun, E. J. and Kelman, A. (1985). *Proc. 6th Intern. Conf. Plant Path. Bacteria*, Maryland (in press).
Braun, V. (1975). *Biochim. Biophys. Acta* **415,** 335–377.
Braun, V. (1978). *In* "Relations Between Structure and Function in the Prokaryotic Cell". Symp. 28 Soc. Gen. Microbiol. (R. Y. Stanier, H. J. Rogers and J. B. Ward, eds), pp. 111–138. Cambridge University Press, Cambridge.
Braun, V. and Hantke, K. (1974). *Ann. Rev. Biochem.* **43,** 89–121.
Brenner, D. J., Fanning, G. R. and Steigerwalt, A. G. (1974). *Int. J. Syst. Bacteriol.* **24,** 197–204.
Brenner, D. J., Fanning, G. R. and Steigerwalt, A. G. (1977). *Int. J. Syst. Bacteriol.* **27,** 211–221.

Brenner, D. J., Fanning, G. R., Leete Knutson, J. K., Steigerwalt, A. G. and Krichevsky, M. I. (1984). *Int. J. Syst. Bacteriol.* **34,** 45–55.
Breton, A. M. (1984). *FEMS Microbiol. Lett.* **22,** 85–88.
Brewin, N. J., Beringer, J. E., Buchanan-Wollaston, A. V., Johnston, A. W. B. and Hirsch, P. R. (1980). *J. Gen. Microbiol.* **116,** 261–270.
Brinton, C. C., Jr. (1965). *Trans. N. Y. Acad. Sci.* **27,** 1003–1054.
Broda, P. (1979). "Plasmids". W. H. Freeman & Co., San Francisco.
Brosius, J. (1984a). *Gene* **27,** 151–160.
Brosius, J. (1984b). *Gene* **27,** 161–172.
Brown, M. A., Ziegle, J. S., and Chatterjee, A. K. (1981). *Phytopathology* **71,** 863 (Abstr).
Brown, M. R. W. and Wood, S. M. (1972). *J. Pharm. Pharmcol.* **24,** 215–218.
Bruce, R. J. and West, C. A. (1982). *Plant Physiol.* **69,** 1181–1188.
Bryan, J., Saeed, N., Fox, D. and Sastry, G. R. K. (1982). *Arch. Microbiol.* **131,** 271–277.
Buchanan, G. E. and Starr, M. P. (1980). *Curr. Microbiol.* **4,** 63–68.
Buchanan-Wollaston, V. (1979). *J. Gen. Microbiol.* **112,** 135–142.
Buddenhagen, I. and Kelman, A. (1964). *Ann. Rev. Phytopathol.* **2,** 203–230.
Bukhari, A. I., Shapiro, J. A. and Adhya, S. L. (1977). "DNA Insertion Elements, Plasmids, and Episomes". Cold Spring Harbor Laboratory, Cold Spring Harbor, New York.
Buxton, R. S. and Drury, L. S. (1983). *FEMS Microbiol. Lett.* **7,** 287–290.
Byrde, R. J. W. and Fielding, A. H. (1968). *J. Gen. Microbiol.* **52,** 287–297.
Cadieux, J. E., Kuzio, J., Milazzo, F. H. and Kropinski, A. M. (1983). *J. Bacteriol.* **155,** 817–825.
Cadmus, M. C., Rogovin, S. P., Burton, K. A., Pittsley, J. E., Knutson, C. A. and Jeanes, A. (1976). *Can. J. Microbiol.* **22,** 942–948.
Callow, J. A. (ed.) (1983). "Biochemical Plant Pathology". John Wiley, New York.
Calos, M. P. and Miller, J. H. (1980). *Cell* **20,** 579–595.
Campbell, A. (1981). *Ann. Rev. Microbiol.* **35,** 55–83.
Campbell, A., Berg, D. E., Lederberg, E. M., Starlinger, P., Botstein, D., Novick, R. P. and Szybalski, W. (1979). *Plasmid* **2,** 466–473.
Campos, J. M., Geisselsoder, J. and Zusman, D. R. (1978). *J. Mol. Biol.* **119,** 167–178.
Carlson, R. R. and Vidaver, A. K. (1982). *Int. J. Syst. Bacteriol.* **32,** 315–326.
Carlton, B. C., and Brown, B. J. (1981). *In* "Manual of Methods for General Bacteriology" (P. Gerhardt, ed.), pp. 222–242. American Society for Microbiology, Washington, D.C.
Carroll, R. B. and Lukezic, F. L. (1972). *Phytopathology* **62,** 555–564.
Casadaban, M. J. and Chou, J. (1984). *Proc. Natl. Acad. Sci. U.S.A.* **81,** 535–539.
Casadaban, M. J. and Cohen, S. N. (1979) *Proc. Natl. Acad. Sci. U.S.A.* **76,** 4530–4533.
Casadaban, M. J., Chou, J. and Cohen, S. N. (1980). *J. Bacteriol.* **143,** 971–980.
Castilho, B. A., Olfson, P. and Casadaban, M. J. (1984). *J. Bacteriol.* **158,** 488–495.
Cerf, A. and Desmettre, P. (1982). *C. R. Acad. Sci. Paris* **295,** 587–592.
Chakrabarty, A. M. (1976). *Ann. Rev. Genet.* **10,** 7–30.
Chakrabarty, A. M. (1980). *In* "Plasmids and Transposons: Environmental Effects and Maintenance Mechanisms" (C. Stuttard and K. R. Rozee, eds), pp. 21–30. Academic Press, Orlando and London.
Chakrabarty, A. M., Mylroie, J. R., Friello, D. A. and Vacca, J. G. (1975). *Proc. Natl. Acad. Sci. USA* **72,** 3647–3651.

Chambon, P. (1981). *Sci. Am.* **244,** 60–71.
Chandler, M., Silver, L. and Caro, L. (1977). *J. Bacteriol.* **131,** 421–430.
Chang, M. T. and Doyle, M. P. (1984). *Infect. Immun.* **43,** 472–476.
Chater, K. F., Hopwood, D. A., Kieser, T. and Thompson, C. J. (1982). *Curr. Top. Microbiol. Immunol.* **96,** 69–95.
Chatterjee, A. K. (1978). *In* "Proc. 4th Int. Conf. Plant Path. Bact." (Station de Pathologie Végétale et Phytobactériologie, ed.), pp. 3–16. Institut National de la Recherche Agronomique, Angers.
Chatterjee, A. K. (1980). *J. Bacteriol.* **142,** 111–119.
Chatterjee, A. K. (1985) *In* "Advances in the Molecular Genetics of the Bacteria–Plant Interaction" (A. A. Szalay and R. P. Legocki, eds), pp. 153–158. Media Services, Cornell University, Ithaca, NY.
Chatterjee, A. K. and Brown, M. A. (1980). *J. Bacteriol.* **143,** 1444–1449.
Chatterjee, A. K. and Brown, M. A. (1981). *Curr. Microbiol.* **6,** 269–273.
Chatterjee, A. K. and Gibbins, L. N. (1971). *J. Bacteriol.* **105,** 107–112.
Chatterjee, A. K. and Starr, M. P. (1973a). *J. Bacteriol.* **116,** 1100–1106.
Chatterjee, A. K. and Starr, M. P. (1973b). *Infect. Immun.* **8,** 563–572.
Chatterjee, A. K. and Starr, M. P. (1977). *J. Bacteriol.* **132,** 862–869.
Chatterjee, A. K. and Starr, M. P. (1978a). *In* "Proc. 4th Int. Conf. Plant Path. Bact." (Station de Pathologie Végétale et Phytobactériologie, ed.), pp. 81–88. Institut National de la Recherche Agronomique, Angers.
Chatterjee, A. K. and Starr, M. P. (1978b). *In* "Proc. 4th Int. Conf. Plant Path. Bact." (Station de Pathologie Végétale et Phytobactériologie, ed.), pp. 89–94. Institut National de la Recherche Agronomique, Angers.
Chatterjee, A. K. and Starr, M. P. (1980). *Ann. Rev. Microbiol.* **34,** 645–676.
Chatterjee, A. K., Ross, H. and Sanderson, K. E. (1976). *Can. J. Microbiol.* **22,** 1549–1560.
Chatterjee, A. K., Buss, R. F. and Starr, M. P. (1977). *Antimicrob. Agents Chemother.* **11,** 897–905.
Chatterjee, A. K., Buchanan, G. E., Behrens, M. K. and Starr, M. P. (1979). *Can. J. Microbiol.* **25,** 94–102.
Chatterjee, A. K., Brown, M. A., Ziegle, J. S. and Thurn, K. K. (1981a). *In* "Proc. 5th Int. Conf. Plant Path. Bact." (J. C. Lozano, ed.), pp. 389–402. Centro Internacional de Agricultura Tropical, Cali.
Chatterjee, A. K., Thurn, K. K., and Tyrell, D. J. (1981b). *In* "Proc. 5th Int. Conf. Plant Path. Bact." (J. C. Lozano, ed.), pp. 252–262. Centro Internacional de Agricultura Tropical, Cali.
Chatterjee, A. K., Ziegle, J. S. and Thurn, K. K. (1982). *Phytopathology* **72,** 934, #40 (Abstr.).
Chatterjee, A. K., Thurn, K. K. and Feese, D. A. (1983). *Appl. Environ. Microbiol.* **45,** 644–650.
Chatterjee, A. K., Ross, L. M., McEvoy, J. L. and Thurn K. K. (1985a). *Appl. Environ. Microbiol.* **50,** 1–9.
Chatterjee, A. K., Thurn, K. K. and Tyrell, D. J. (1985b). *J. Bacteriol.* **162,** 708–714.
Cheng, S. C. and Modrich, P. (1983). *J. Bacteriol.* **154,** 1005–1008.
Chesson, A. (1980). *J. Appl. Bacteriol.* **48,** 1–45.
Chilton, M.-D., Farrand, S. K., Levin, R. and Nester, E. W. (1976). *Genetics* **83,** 609–618.
Chilton, M.-D., Tepfer, D., Petit, A., David, C., Casse-Delbart, F. and Tempe, J. (1982). *Nature* **295,** 432–434.

Chow, L. T. and Broker, T. R. (1978). *J. Bacteriol.* **133,** 1427–1436.

Chow, L. T. and Bukhari, A. I. (1977). *In* "DNA Insertion Elements, Plasmids, and Episomes" (A. I. Bukhari, J. A. Shapiro and S. L. Adhya, eds), pp. 295–306. Cold Spring Harbor Laboratory, Cold Spring Harbor, New York.

Chrisope, G. L., Fox, C. W. and Marshall, R. T. (1976). *Appl. Environ. Microbiol.* **31,** 784–786.

Christofi, N., Wilson, M. I. and Old, D. C. (1979). *J. Appl. Bacteriol.* **46,** 179–183.

Chumley, F. G., Menzel, R. and Roth, J. R. (1979). *Genetics* **91,** 639–655.

Civerolo, E. L. (1985). *Phytopathology* **75,** 524–528.

Clarke, L. and Carbon, J. (1975). *Proc. Natl. Acad. Sci. USA* **72,** 4361–4365.

Clements, D. E., Romeo, C. J., Kamdar, H. V. and Patil, S. S. (1985). *Phytopathology* **75,** 1355, #611 (Abstr.).

Close, T. J., Zaitlin, D. and Kado, C. I. (1984). *Plasmid* **12,** 111–118.

Clowes, R. C. and Hayes, W. (1968). "Experiments in Microbial Genetics". John Wiley, New York.

Codner, R. C. (1971). *J. Appl. Bacteriol.* **34,** 147–160.

Cohen, S. N. (1975). *Sci. Am.* **233,** 24–33.

Cohen, S. N., Chang, A. C. Y. and Hsu, L. (1972). *Proc. Natl. Acad. Sci. USA* **69,** 2110–2114.

Cohen, S. N., Casadaban, M. J., Chou, J. and Tu, C.-P. D. (1979). *Cold Spring Harbor Symp. Quant. Biol.* **43,** 1247–1255.

Collins, J. (1979). *Curr. Top. Microbiol. Immunol.* **78,** 121–170.

Collins, M. D. and Jones, D. (1980). *J. Appl. Bacteriol.* **48,** 459–470.

Collins, M. L. P., Mallon, D. E. and Niederman, R. A. (1979). *J. Bacteriol.* **139,** 1089–1092.

Collmer, A. and Bateman, D. F. (1981). *Proc. Natl. Acad. Sci. USA* **78,** 3920–3924.

Collmer, A., Berman, P. and Mount, M. S. (1982a). *In* "Phytopathogenic Prokaryotes" (M. Mount and G. Lacy, eds), Vol. 1, pp. 395–422. Academic Press, London and Orlando.

Collmer, A., Whalen, C. H., Beer, S. V. and Bateman, D. F. (1982b). *J. Bacteriol.* **149,** 626–634.

Collmer, A., Schoedel, C., Roeder, D. L., Ried, J. L. and Rissler, J. F. (1985a). *J. Bacteriol.* **161,** 913–920.

Collmer, A., Schoedel, C., Roder, D. L. and Ried, J. L. (1985b). *In* "Advances in Molecular Genetics of the Bacteria–Plant Interaction" (A. A. Szalay and R. P. Legocki, eds), pp. 159–163. Media Services, Cornell University, Ithaca, NY.

Comai, L. and Kosuge, T. (1980). *J. Bacteriol.* **143,** 950–957.

Comai, L. and Kosuge, T. (1982). *J. Bacteriol.* **149,** 40–46.

Comai, L. and Kosuge, T. (1983). *J. Bacteriol.* **154,** 1162–1167.

Comai, L., Surico, G. and Kosuge, T. (1982). *J. Gen. Microbiol.* **128,** 2157–2163.

Comai, L., Schilling-Cordaro, C., Mergia, A. and Houck, C. M. (1983). *Plasmid* **10,** 21–30.

Condemine, G., Hugouvieux-Cotte-Pattat, N. and Robert-Baudouy, J. (1984). *J. Gen. Microbiol.* **130,** 2839–2844.

Cook, R. J. and Baker, K. F. (1983). "Nature and Practice of Biological Control of Plant Pathogens". American Phytopathological Society, St. Paul.

Coplin, D. L. (1979). *J. Gen. Microbiol.* **113,** 181–184.

Coplin, D. L. and Rowan, R. G. (1978). *In* "Proc. 4th Int. Conf. Plant Path. Bact." (Station de Pathologie Végétale et Phytobactériologie, ed.), pp. 67–73. Institut National de la Recherche Agronomique, Angers.

Coplin, D. L., Sequeira, L. and Hanson, R. S. (1974). *Can. J. Microbiol.* **20,** 519–529.

Coplin, D. L., Rowan, R. G., Chisholm, D. A. and Whitmoyer, R. E. (1981a). *Appl. Environ. Microbiol.* **42,** 599–604.

Coplin, D. L., Frederick, R. D., Tindal, M. H. and McCammon, S. L. (1981b). *In* "Proc. 5th Int. Conf. Plant Path. Bact." (J. C. Lozano, ed.), pp. 379–388. Centro Internacional de Agricultura Tropical, Cali.

Coplin, D. L., Meaney, C., Bradshaw-Rouse, J. J. and McCammon, S. L. (1982). *Phytopathology* **72,** 1002, #577 (Abstr.).

Coplin, D. L., Frederick, R. D. and Majerczak, D. (1984). *Phytopathology* **74,** 876, #A694 (Abstr.).

Coplin, D. L., Frederick, R. D. and Majerczak, D. (1985). *Proc. 6th Intern. Conf. Plant Path. Bacteria*, Maryland (in press).

Corey, R. R. and Starr, M. P. (1957a). *J. Bacteriol.* **74,** 141–145.

Corey, R. R. and Starr, M. P. (1957b). *J. Bacteriol.* **74,** 146–150.

Cornelis, G. (1980). *J. Gen. Microbiol.* **117,** 243–247.

Cornelis, G., Ghosal, D. and Saedler, H. (1978). *Mol. Gen. Genet.* **160,** 215–224.

Cornet, P., Tronik, D., Miller, J. and Aubert, J.-P. (1983). *FEMS Microbiol. Lett.* **16,** 137–141.

Cosloy, S. D. and Oishi, M. (1973). *Proc. Natl. Acad. Sci. USA* **70,** 84–87.

Costerton, J. W. (1979). *Ann. Rev. Microbiol.* **33,** 459–479.

Costerton, J. W., Ingram, J. M. and Cheng, K.-J. (1974). *Bacteriol. Rev.* **38,** 87–110.

Costerton, J. W., Irvin, R. T. and Cheng, K.-J. (1981a). *CRC Crit. Rev. Microbiol.* **8,** 303–338.

Costerton, J. W., Irvin, R. T. and Cheng, K.-J. (1981b). *Ann. Rev. Microbiol.* **35,** 299–324.

Covarrubias, L. and Bolivar, F. (1982). *Gene* **17,** 79–89.

Cronan, J. E., Jr. (1979). *In* "Bacterial Outer Membranes: Biogenesis and Functions" (M. Inouye, ed.), pp. 35–65. John Wiley, New York.

Csonka, L. N., Howe, M. M., Ingraham, J. L., Pierson, L. S., III and Thurnbough, C. L., Jr. (1981). *J. Bacteriol.* **145,** 299–305.

Cuppels, D. A. (1985). *Proc. 6th Intern. Conf. Plant Path. Bacteria*, Maryland (in press).

Curiale, M. S. and Mills, D. (1977). *J. Bacteriol.* **131,** 224–228.

Curiale, M. S. and Mills, D. (1982). *J. Bacteriol.* **152,** 797–802.

Currier, T. C. and Morgan, M. K. (1981). *In* "Proc. 5th Int. Conf. Plant Path. Bact." (J. C. Lozano, ed.), pp 420–426. Centro Internacional de Agricultura Tropical, Cali.

Currier, T. C. and Morgan, M. K. (1982). *J. Bacteriol.* **150,** 251–259.

Currier, T. C. and Morgan, M. K. (1983). *Can. J. Microbiol.* **29,** 84–89.

Currier, T. C. and Nester, E. W. (1976). *Anal. Biochem.* **76,** 431–441.

Curtiss, R., III. (1969). *Ann. Rev. Microbiol.* **23,** 69–136.

Curtiss, R., III. (1981). *In* "Manual of Methods for General Bacteriology" (P. Gerhardt, ed.), pp. 243–265. American Society for Microbiology, Washington, D.C.

Dahlback, B., Hermansson, M., Kjelleberg, S. and Norkrans, B. (1981). *Arch. Microbiol.* **128,** 267–270.

Daly, J. M. (1976a). *In* "Encyclopedia of Plant Physiology" (A. Pirson and M. H. Zimmerman, eds), Vol. 4, pp. 27–50. Springer-Verlag, Berlin.

Daly, J. M. (1976b). *In* "Encyclopedia of Plant Physiology" (A. Pirson and M. H. Zimmerman, eds), Vol. 4, pp. 450–479. Springer-Verlag, Berlin.

Daly, J. M. (1981). *In* "Toxins in Plant Disease" (R. D. Durbin, ed.), pp. 331–394. Academic Press, London and Orlando.

Daly, J. M. (1984). *Ann. Rev. Phytopathol.* **22,** 273–307.

Daly, J. M. and Deverall, B. J. (eds) (1983). "Toxins and Plant Pathogenesis". Academic Press, London and Orlando.

Daly, J. M. and Uritani, I. (1979). "Recognition and Specificity in Plant Host–Parasite Interactions". Japan Scientific Societies Press, Tokyo.

Daniels, M. J. (1979). *J. Gen. Microbiol.* **114,** 323–328.

Daniels, M. J. (1983). *Ann. Rev. Phytopathol.* **21,** 29–43.

Daniels, M. J., Barber, C. E., Turner, P. C., Cleary, W. G. and Sawczyc, M. K. (1984a). *J. Gen. Microbiol.* **130,** 2447–2455.

Daniels, M. J., Barber, C. E., Turner, P. C., Sawczyc, M. K., Byrde, R. J. W. and Fielding, A. H. (1984b). *EMBO J.* **3,** 3323–3328.

Daniels, M. J., Barber, C. E., Turner, P. C. and Sawczyc, M. K. (1985). *In* "Advances in Molecular Genetics of the Bacteria–Plant Interaction" (A. A. Szalay and R. P. Legocki, eds), pp. 169–172. Media Services, Cornell University, Ithaca, NY.

Danilevich, V. N., Stepanshin, Y. G., Volozhantsev, N. V. and Golub, E. I. (1978). *Mol. Gen. Genet.* **161,** 337–339.

Darvill, A. G. and Albersheim, P. (1984). *Ann. Rev. Plant Physiol.* **35,** 243–275.

Darvill, A., McNeil, M., Albersheim, P. and Delmer, D. P. (1980). *In* "The Biochemistry of Plants" (P. K. Stumpf and E. E. Conn, eds), Vol. 1, pp. 91–162. Academic Press, London and Orlando.

Darzins, A., Nixon, L. L., Vanags, R. I. and Chakrabarty, A. M. (1985). *J. Bacteriol.* **161,** 249–257.

Dassa, E. and Boquet, P.-L. (1981). *Mol. Gen. Genet.* **181,** 192–200.

Date, T., Inuzuka, M. and Tomoeda, M. (1977). *Biochemistry* **16,** 5579–5584.

Davis, B. D. and Tai, P.-C. (1980). *Nature* **283,** 433–437.

Davis, B. D., Dulbecco, R., Eisen, H. N. and Ginsberg, H. S. (1980). "Microbiology", 3rd Edn. Harper & Row, Hagerstown.

Davis, K. R., Lyon, G. D., Darvill, A. G. and Albersheim, P. (1984). *Plant Physiol.* **74,** 52–60.

Davis, M. J., Whitcomb, R. F. and Gillespie, A. G., Jr. (1981). *In* "The Procaryotes" (M. P. Starr, H. Stolp, H. G. Trüper, A. Balows and H. G. Schlegel, eds), pp. 2172–2188. Springer-Verlag, New York.

Davis, M. J., Gillaspie, A. K., Jr., Vidaver, A. K. and Harris, R. W. (1984). *Int. J. Syst. Bacteriol.* **34,** 107–117.

Davis, R. E. and Lee, I.-M. (1982). *In* "Phytopathogenic Procaryotes" (M. Mount and G. Lacy, eds), Vol. 1, pp. 491–513. Academic Press, London and Orlando.

Davis, R. W., Botstein, D. and Roth, J. R. (1980). "Advanced Bacterial Genetics, A Manual for Genetic Engineering". Cold Spring Harbor Laboratory, Cold Spring Harbor, New York.

Dazzo, F. B. and Brill, W. J. (1977). *Appl. Environ. Microbiol.* **33,** 132–136.

Dazzo, F. B. and Hubbell, D. H. (1975). *Appl. Microbiol.* **30,** 1017–1033.

Deasey, M. C., Stapleton, M. J. and Matthysse, A. G. (1985). *Proc. 6th Intern. Conf. Plant Path. Bacteria*, Maryland (in press).

Débarbouillé, M. and Schartz, M. (1979). *J. Mol. Biol.* **132,** 521–534.

De Boer, S. H. (1980). *Can. J. Microbiol.* **26,** 567–571.

De Boer, S. H. (1985). *Proc. 6th Intern. Conf. Plant Path. Bacteria*, Maryland (in press).

De Boer, S. H., Copeman, R. J. and Vruggink, H. (1979). *Phytopathology* **69,** 316–319.

de Bruijn, F. J. and Lupski, J. R. (1984). *Gene* **27,** 131–149.
de Graaff, J., Kreuning, P. C. and van de Putte, P. (1973). *Mol. Gen. Genet.* **123,** 283–288.
de Graaff, J., Kreuning, P. C. and Stouthamer, A. H. (1974). *Antonie van Leeuwenhoek J. Microbiol. Serol.* **40,** 161–170.
de Graff, F. K., Krenn, B. E. and Klaasen, P. (1984). *Infect. Immun.* **43,** 508–514.
Dehority, B. A. (1969). *J. Bacteriol.* **99,** 189–196.
Demain, A. L. (1981). *Science* **214,** 987–995.
Derylo, M., Glowacka, M., Lorkiewicz, Z. and Russa, R. (1975). *Mol. Gen. Genet.* **140,** 175–181.
DeVay, J. E., Lukezic, F. L., Sinden, S. L., English, H. and Coplin, D. L. (1968). *Phytopathology* **58,** 95–101.
Devillers-Thiery, A., Kindt, T., Scheele, G. and Blobel, G. (1975). *Proc. Natl. Acad. Sci. USA* **72,** 5016–5020.
De Vos, P. and De Ley, J. (1983). *Int. J. Syst. Bacteriol.* **33,** 487–509.
Dianese, J. C. and Schaad, N. W. (1982). *Phytopathology* **72,** 1284–1289.
Diaz-Maurino, T. and Perkins, H. R. (1974). *J. Gen. Microbiol.* **80,** 533–539.
Dickenson, C. H. and Preece, T. F. (1976). "Microbiology of Aerial Plant Surfaces". Academic Press, London and Orlando.
Dickey, R. S. (1979). *Phytopathology* **69,** 324–329.
Dickey, R. S. (1981). *Phytopathology* **71,** 23–29.
Dickinson, C. H. and Lucas J. A. (1982). "Plant Pathology and Plant Pathogens", 2nd Edn. Blackwell Scientific Publications, Oxford.
DiRienzo, J. M., Nakamura, K. and Inouye, M. (1978). *Ann. Rev. Biochem.* **47,** 481–532.
Ditta, G., Stanfield, S., Corbin, D. and Helinski, D. R. (1980). *Proc. Natl. Acad. Sci. USA* **77,** 7347–7351.
Djordjevic, M. A., Zurkowski, W. and Rolfe, B. G. (1982). *J. Bacteriol.* **151,** 560–568.
Doesburg, J. J. (1965). "Pectic Substances in Fresh and Preserved Fruits and Vegetables". I.B.V.T. Communication No. 25, Wageningen.
Dolph, P. J. and Coplin, D. L. (1985). *Proc. 6th Intern. Conf. Plant Path. Bacteria,* Maryland (in press).
Dopfer, H., Stackebrandt, E. and Fiedler, F. (1982). *J. Gen. Microbiol.* **128,** 1697–1708.
dos Santos, R. M. D. B. and Dianese, J. C. (1985). *Phytopathology* **75,** 581–587.
Doudoroff, M. and Palleroni, N. J. (1974). *In* "Bergey's Manual of Determinative Bacteriology," 8th Edn (R. E. Buchanan and N. E. Gibbons, eds), pp. 217–243. Williams & Wilkins Co., Baltimore.
Douglas, C. J., Halperin, W. and Nester, E. W. (1982). *J. Bacteriol.* **152,** 1265–1275.
Dowson, W. J. (1957). "Plant Diseases Due to Bacteria". Cambridge University Press, Cambridge.
Drake, J. W. (1970). "The Molecular Basis of Mutation". Holden-Day, San Francisco.
Drake, J. W. and Baltz, R. H. (1976). *Ann. Rev. Biochem.* **45,** 11–37.
Drew, S. W. and Demain, A. L. (1977). *Ann. Rev. Microbiol.* **31,** 343–356.
Drigues, P., Lafforgue, D. and Asselineau, J. (1982). *C. R. Acad. Sci. Paris* **294,** 443–445.
Drummond, M. (1979). *Nature* **281,** 343–347.
Dudman, W. F. (1959). *Nature* **184,** 1969–1970.
Duncan, M. J. (1981). *J. Gen. Microbiol.* **122,** 61–67.
Durbin, R. D. (1972). *In* "Phytotoxins in Plant Diseases" (R. K. S. Wood, A. Ballio and A. Graniti, eds), pp. 19–34. Academic Press, London and Orlando.
Durbin, R. D. (1981). "Toxins in Plant Diseases". Academic Press, London and Orlando.

Durbin, R. D. and Steele, J. A. (1979). *In* "Recognition and Specificity in Plant Host Parasite Interactions" (J. M. Daly and I. Uritani, eds), pp. 115–131. Japan Scientific Societies Press, Tokyo.
Durbin, R. D. and Uchytil, T. F. (1985). *Proc. 6th Intern. Conf. Plant Path. Bacteria*, Maryland (in press).
Duvick, J. P. (1982). "Interaction of *Pseudomonas solanacearum* with tobacco cell walls and with a wall-associated agglutinin". Ph.D. Thesis, University of Wisconsin, Madison. 174 pp.
Duvick, J. P. and Sequeira, L. (1984a). *Appl. Environ. Microbiol.* **48,** 192–198.
Duvick, J. P. and Sequeira, L. (1984b). *Appl. Environ. Microbiol.* **48,** 199–205.
Dye, D. W. (1981). *N.Z.J. Agr. Res.* **24,** 223–229.
Dye, D. W. and Kemp, W. J. (1977). *N.Z.J. Agri. Res.* **20,** 563–582.
Dye, D. W. and Lelliott, R. A. (1974). *In* "Bergey's Manual of Determinative Bacteriology," 8th Edn (R. E. Buchanan and N. E. Gibbons, eds), pp. 243–249. Williams & Wilkins Co., Baltimore.
Dye, D. W., Bradbury, J. F., Goto, M., Hayward, A. C., Lelliott, R. A. and Schroth, M. N. (1980). *Rev. Plant Path.* **59,** 153–168.
Ebel-Tsipis, J., Botstein, D. and Fox, M. S. (1972). *J. Mol. Biol.* **71,** 433–448.
Eden-Green, S. J. and Knee, M. (1974). *J. Gen. Microbiol.* **81,** 509–512.
Ehrenshaft, M. and Mills, D. (1985). *Appl. Environ. Microbiol.* **50,** 169–171.
Ehrlich, S. D. (1978). *Proc. Natl. Acad. Sci. USA* **75,** 1433–1436.
El-Banoby, F. E. and Rudolph, K. (1978). *In* Proc. 4th Int. Conf. Plant Path. Bact." (Station de Pathologie Végétale et Phytobactériologie, ed.), pp. 667–673. Institut National de la Recherche Agronomique, Angers.
El-Banoby, F. E. and Rudolph, K. (1979a). *Phytopathol. Z.* **95,** 38–50.
El-Banoby, F. E. and Rudolph, K. (1979b). *Physiol. Plant Pathol.* **15,** 341–349.
El-Banoby, F. E. and Rudolph, K. (1980). *Physiol. Plant Pathol.* **16,** 425–437.
Elia, S., Gossele, F., Vantomme, R., Swings, J. and De Ley, J. (1984). *Phytopath. Z.* **110,** 89–105.
Ellingboe, A. H. (1976). *In* "Encyclopedia of Plant Pathology, Vol. 4, Physiological Plant Pathology" (R. Heitefuss and P. H. Williams, eds), pp. 761–778. Springer-Verlag, New York.
Ellingboe, A. H. (1982). *In* "Active Defense Mechanisms in Plants" (R. K. S. Wood, ed.), pp. 179–192, Plenum Press, New York.
Elliott, C. (1951). "Manual of Bacterial Plant Pathogens", 2nd Edn. Williams & Wilkins Co., Baltimore.
Ellis, J. G., Kerr, A., Petit, A. and Tempé, J. (1982). *Mol. Gen. Genet.* **186,** 269–274.
Elwell, L. P. and Shipley, P. L. (1980). *Ann. Rev. Microbiol.* **34,** 465–496.
Engebrecht, J., Simon, M. and Silverman, M. (1985). *Science* **227,** 1345–1347.
Enomoto, M. and Stocker, B. A. D. (1974). *Virology* **60,** 503–514.
Epton, N. A. S., Sigei, D. C. and Passmoor, M. (1977). *In* "Current Topics in Plant Pathology" (Z. Király, ed.), pp. 301–303. Akademiai Kiado, Budapest.
Ercolani, G. L. (1970). *Phytopathol. Mediterr.* **9,** 145–150.
Errington, J. and Vivian, A. (1981). *J. Gen. Microbiol.* **124,** 439–442.
Eshdat, Y., Silverblatt, F. J. and Sharon, N. (1981). *J. Bacteriol.* **148,** 308–314.
Esquerré-Tugayé, M.-T. and Lamport, D. T. A. (1979). *Plant Physiol.* **64,** 314–319.
Esquerré-Tugayé, M.-T., Lafitte, C., Mazau, D., Toppan, A. and Touzé, A. (1979). *Plant Physiol.* **64,** 320–326.
Ewing, W. H. and Fife, M. A. (1972). *Int. J. Syst. Bacteriol.* **22,** 4–11.
Expert, D. and Tourneur, J. (1982). *J. Virol.* **42,** 283–291.
Expert, D. and Toussaint, A. (1985). *J. Bacteriol.* **163,** 221–227.

Fader, R. C., Duffy, L. K., Davis, C. P. and Kurosky, A. (1982). *J. Biol. Chem.* **257,** 3301–3305.

Faelen, M., Toussaint, A., Van Montagu, M., Van den Elsacker, S., Engler, G. and Schell, J. (1977). *In* "DNA Insertion Elements, Plasmids, and Episomes" (A. I. Bukhari, J. A. Shapiro and S. L. Adhya, eds), pp. 521–530. Cold Spring Harbor Laboratory, Cold Spring Harbor, New York.

Faelen, M., Resibois, A. and Toussaint, A. (1979). *Cold Spring Harbor Symp. Quant. Biol.* **43,** 1169–1177.

Faelen, M., Toussaint, A., Lefèbvre, N., Mergeay, M., Braipson-Thiry, J. and Thiry, G. (1981). *Arch. Intern. Physiol. Biochem.* **89,** 855–856.

Fahy, P. C. and Persley, G. J. (1983). "Plant Bacterial Diseases. A Diagnostic Guide". Academic Press, London and Orlando.

Falkow, S. (1975). "Infectious Multiple Drug Resistance". Pion, London.

Faris, A., Wadstrom, T. and Freer, J. H. (1981). *Curr. Microbiol.* **5,** 67–72.

Ferenczy, L. (1981). *In* "Genetics as a Tool in Microbiology" (S. W. Glover and D. A. Hopwood, eds), pp. 1–34. Cambridge University Press, Cambridge.

Ferguson, A. R. and Johnston, J. S. (1980). *Physiol. Plant Pathol.* **16,** 269–275.

Ferguson, B. I. and Mitchell, R. E. (1985). *Plant Physiol.* **77,** 969–973.

Ferguson, M. W. and Chatterjee, A. K. (1980). *Proc. Am. Phytopathol. Soc.*, 142, #200 (Abstr.).

Fett, W. F. and Jones, S. B. (1983). *Phytopathology* **73,** 365 (Abstr.).

Fiedler, F., Schaffler, M. J. and Stackebrandt, E. (1981). *Arch. Microbiol.* **129,** 85–93.

Filip, C., Fletcher, G., Wulff, J. L. and Earhart, C. F. (1973). *J. Bacteriol.* **115,** 717–722.

Finn, C. W., Jr., Silver, R. P., Habig, W. H., Hardegree, M. C., Zon, G. and Garon, C. F. (1984). *Science* **224,** 881–884.

Flor, H. H. (1955). *Phytopathology* **45,** 680–685.

Fogarty, W. M. (ed.) (1983). "Microbial Enzymes and Biotechnology". Applied Science Publishers, London.

Fogarty, W. M. and Ward, O. P. (1974). *In* "Progress in Industrial Microbiology" (D. J. D. Hockenhull, ed.), Vol. 13, pp. 59–119. Churchill Livingstone, Edinburgh.

Franke, A. E. and Clewell, D. B. (1981). *J. Bacteriol.* **145,** 494–502.

Freeman, V. J. (1951). *J. Bacteriol.* **61,** 675–688.

Friedman, A. M., Long, S. R., Brown, S. E., Buikema, W. J. and Ausubel, F. M. (1982). *Gene* **18,** 289–296.

Friedman, B. A. (1962). *Phytopathology* **52,** 328–332.

Fuerst, J. A. and Hayward, A. C. (1969). *J. Gen. Microbiol.* **58,** 227–238.

Fukushima, K., Takada, K., Motoda, R. and Ikeda, T. (1982). *FEBS Lett.* **149,** 299–303.

Fulbright, D. W. and Leary, J. V. (1978). *J. Bacteriol.* **136,** 497–500.

Gaastra, W. and de Graff, F. (1982). *Microbiol. Rev.* **46,** 129–161.

Gabriel, D. W. (1985). *In* "Advances in Molecular Genetics of the Bacteria–Plant Interaction" (A. A. Szalay and R. P. Legocki, eds), pp. 202–204. Media Services, Cornell University, Ithaca, NY.

Gahan, L. C., Sanford, P. A. and Conrad, H. E. (1967). *Biochemistry* **6,** 2755–2767.

Galanos, C., Lüderitz, O. and Westphal, O. (1969). *Eur. J. Biochem.* **9,** 245–249.

Galanos, C., Lüderitz, O., Rietschel, E. T. and Westphal, O. (1977). *In* "Biochemistry of Lipids II," International Review of Biochemistry (T. W. Goodwin, ed.), Vol. 14, pp. 239–335. University Park Press, Baltimore.

Gantotti, B. V. and Beer, S. V. (1982). *J. Bacteriol.* **151,** 1627–1629.

Gantotti, B. V., Patil, S. S. and Mandel, M. (1979). *Mol. Gen. Genet.* **174,** 101–103.
Gantotti, B. V., Kindle, K. L. and Beer, S. V. (1981). *Curr. Microbiol.* **6,** 377–381.
Garber, E. D. (1959). *Bot. Gaz.* **120,** 157–161.
Garber, E. D. (1960). *Ann. N.Y. Acad. Sci.* **88,** 1187–1194.
Garber, E. D., Shaeffer, S. G. and Goldman, M. (1956). *J. Gen. Microbiol.* **14,** 261–267.
Gardner, J. M. and Kado, C. I. (1972). *Int. J. Syst. Bacteriol.* **22,** 201–209.
Gardner, J. M. and Kado, C. I. (1976). *J. Bacteriol.* **127,** 451–460.
Garfinkel, D. J. and Nester, E. W. (1980). *J. Bacteriol.* **144,** 732–743.
Garfinkel, D. J., Simpson, R. B., Ream, L. W., White, F. F., Gordon, M. P. and Nester, E. W. (1981). *Cell* **27,** 143–153.
Garibaldi, A. and Bateman, D. F. (1971). *Physiol. Plant Pathol.* **1,** 25–40.
Garibaldi, A. and Gibbins, L. N. (1975). *Can. J. Microbiol.* **21,** 1282–1287.
Garland, P. B. and Williamson, R. (1979). "Biochemistry of Genetic Engineering". The Biochemical Society, London.
Garrett, C. M. E., Crosse, J. E. and Sletten, A. (1974). *J. Gen. Microbiol.* **80,** 475–483.
Gasson, M. (1980). *Appl. Environ. Microbiol.* **39,** 25–29.
Gawron-Burke, C. and Clewell, D. B. (1984). *J. Bacteriol.* **159,** 214–221.
Gemski, P., Lazere, J. R., Casey, T. and Wohlhieter, J. A. (1980). *Infect. Immun.* **28,** 1044–1047.
Genetello, C., Van Larebeke, N., Holsters, M., De Picker, A., Van Montagu, M. and Schell, J. (1977). *Nature* **265,** 561–563.
Ghalambor, M. A. and Heath, E. C. (1966). *J. Biol. Chem.* **241,** 3222–3227.
Ghalambor, M. A., Levine, E. M. and Heath, E. C. (1966). *J. Biol. Chem.* **241,** 3207–3215.
Ghosal, D., Sommer, H. and Saedler, H. (1979). *Nucl. Acids Res.* **6,** 1111–1122.
Ghose, T. K. (ed.) (1977). "Bioconversion of Cellulosic Substances into Energy, Chemicals and Microbial Proteins". Indian Institute of Technology, Delhi, India.
Gibbins, L. N. (1978). *In* "Proc. 4th Int. Conf. Plant Path. Bact." (Station de Pathologie Végétale et Phytobactériologie, ed.), pp. 403–431. Institut National de la Recherche Agronomique, Angers.
Gibbins, L. N., Bennett, P. M., Saunders, J. R., Grinstead, J. and Connolly, J. C. (1976). *J. Bacteriol.* **128,** 309–316.
Gibbons, R. J. (1980). *In* "Microbial Adhesion to Surfaces" (R. C. W. Berkeley, J. M. Lynch, J. Melling, P. R. Rutter and B. Vincent, eds), pp. 351–388. Ellis Horwood Chichester.
Gies, D. R., Willis, D. K., Panopoulos, N. J. and Lindow, S. E. (1985). *Phytopathology* **75,** 1320, #335 (Abstr.).
Gilkes, N. R., Langsford, M. L., Kilburn, D. G., Miller, R. C., Jr., Wakarchuk, W. W., Warren, R. A. J., Whittle, D. J. and Wong, W. K. R. (1984a). *J. Gen. Microbiol.* **130,** 1377–1384.
Gilkes, N. R., Langsford, M. L., Kilburn, D. G., Miller, R. C., Jr. and Warren, R. A. J. (1984b). *J. Biol. Chem.* **259,** 10455–10459.
Gilleland, H. E., Jr. and Murray, R. G. E. (1976). *J. Bacteriol.* **125,** 267–281.
Glaser, L. (1973). *Ann. Rev. Biochem.* **42,** 91–112.
Glass, N. L. and Kosuge, T. (1985). *Phytopathology* **75,** 1325, #378 (Abstr.).
Gmeiner, J. (1979). *Arch. Microbiol.* **121,** 177–180.
Goldberg, R. B., Bender, R. A. and Streicher, S. L. (1974). *J. Bacteriol.* **118,** 810–814.
Gonzalez, C. F. and Vidaver, A. K. (1979). *Curr. Microbiol.* **2,** 75–80.
Gonzalez, C. F. and Vidaver, A. K. (1980). *Phytopathology* **70,** 223–225.
Gonzalez, C. F. and Vidaver, A. K. (1978). *In* "Proc. 4th Int. Conf. Plant Path. Bact."

(Station de Pathologie Végétale et Phytobactériologie, ed.), pp. 31–38. Institut National de la Recherche Agronomique, Angers.
Gonzalez, C. F., DeVay, J. E. and Wakeman, R. J. (1981a). *Physiol. Plant Pathol.* **18,** 41–50.
Gonzalez, C. F., Olsen, R. H. and Vidaver, A. K. (1981b). *Phytopathology* **71,** 220.
Gonzalez, C. F., Layher, S. K., Vidaver, A. K. and Olsen, R. H. (1984). *Phytopatholoy* **74,** (in press).
Goodman, R. N. (1978). *Mycopathologia* **65,** 107–113.
Goodman, R. N., Huang, J. S. and Huang, P. Y. (1974). *Science* **183,** 1081–1082.
Goodman, R. N., Ayers, A. R. and Politis, D. J. (1978). *In* "Proc. 4th Int. Conf. Plant Path. Bact." (Station de Pathologie Végétale et Phytobactériologie, ed.), pp. 483–485. Institut National de la Recherche Agronomique, Angers.
Gorin, P. A. J. and Spencer, J. F. T. (1961a). *Can. J. Chem.* **39,** 2274–2281.
Gorin, P. A. J. and Spencer, J. F. T. (1961b). *Can. J. Chem.* **39,** 2282–2289.
Gorin, P. A. J. and Spencer, J. F. T. (1964). *Can. J. Chem.* **42,** 1230–1232.
Gorin, P. A. J., Spencer, J. F. T. and Westlake, D. W. S. (1961). *Can. J. Chem.* **39,** 1067–1073.
Gorin, P. A. J., Spencer, J. F. T., Lindberg, B. and Lindh, F. (1980). *Carbohydr. Res.* **79,** 313–315.
Gorlenko, M. V. (1965). "Bacterial Diseases of Plants", 2nd Edn, revised and enlarged (translated from the 1961 Russian original). Israel Program for Scientific Translations, Jerusalem.
Gormus, B. J. and Wheat, R. W. (1971). *J. Bacteriol.* **108,** 1304–1309.
Goto, M. (1972). *J. Appl. Bacteriol.* **35,** 505–515.
Goto, M. (1981). "Bacterial Plant Pathology". Soft Science, Inc., Tokyo.
Goto, M. and Starr, M. P. (1972). *Ann. Phytopathol. Soc. Japan* **38,** 267–274.
Goto, M., Ishida, Y., Takikawa, Y. and Hyodo, H. (1985). *Plant Cell Physiol.* **26,** 141–150.
Govindarajan, A. G. and Lindow, S. E. (1985). *Phytopathology* **75,** 1380, #805 (Abstr.).
Gowrishankar, J. and Pittard, J. (1982). *J. Bacteriol.* **150,** 1122–1129.
Gradel, C. M. and Dehority, B. A. (1972). *Appl. Microbiol.* **23,** 332–340.
Graham, T. L., Sequeira, L. and Huang, T.-S. R. (1977). *Appl. Environ. Microbiol.* **34,** 424–432.
Griffith, F. (1928). *J. Hygiene* **27,** 113–159.
Gross, D. C. (1985). *J. Appl. Bacteriol.* **58,** 167–174.
Gross, D. C. and Cody, Y. S. (1985). *Can. J. Microbiol.* **31,** 403–410.
Gross, D. C. and DeVay, J. E. (1977). *Physiol. Plant Pathol.* **11,** 13–28.
Gross, D. C. and Vidaver, A. K. (1981). *Can. J. Microbiol.* **27,** 759–765.
Gross, D. C., DeVay, J. E. and Stadtman, F. H. (1977). *J. Appl. Bacteriol.* **43,** 453–463.
Gross, D. C., Vidaver, A. K. and Keralis, M. B. (1979). *J. Gen. Microbiol.* **115,** 479–489.
Gross, M. (1985). *Proc. 6th Intern. Conf. Plant Path. Bacteria*, Maryland (in press).
Grunstein, M. and Wallis, J. (1979). *Meth. Enzymol.* **68,** 379–389.
Gryczan, T. J. (1982). *In* "The Molecular Biology of the Bacilli" (D. A. Dubnau, ed.), Vol. I, pp. 307–329. Academic Press, London and Orlando.
Guimares, W. V., Panopoulos, N. J. and Schroth, M. N. (1978). *In* "Proc. 4th Int. Conf. Plant Path. Bact." (Station de Pathologie Végétale et Phytobactériologie, ed.), pp. 53–65. Institut National de la Recherche Agronomique, Angers.
Gurlitz, R. H. G. and Matthysse, A. G. (1982). *In* "Proc. 13th Int. Cong. Microbiol.", pp. 103, #P30:2.

Guyman, L. F., Esser, M. and Shafer, W. M. (1982). *Infect. Immun.* **36,** 541–547.
Haas, J. J., Fett, W. F. and Fleming, D. J. (1985). *Proc. 6th Intern. Conf. Plant Path. Bacteria*, Maryland (in press).
Habermann, P., Klaer, R., Kuhn, S. and Starlinger, P. (1979). *Mol. Gen. Genet.* **175,** 369–373.
Hadley, R. G. and Szalay, A. A. (1982). *Mol. Gen. Genet.* **188,** 361–369.
Hahn, F. E. (1978). "Progress in Molecular and Subcellular Biology", Vol. 6. Springer-Verlag, Berlin.
Halegoua, S. and Inouye, M. (1979). *In* "Bacterial Outer Membranes: Biogenesis and Functions" (M. Inouye, ed.), pp. 67–113. John Wiley, New York.
Hall, A. N. and Jafri, S. S. A. (1980). *J. Gen. Microbiol.* **117,** 263–265.
Hall, A. N., Hogg, S. D. and Phillips, G. O. (1978). *J. Appl. Bacteriol.* **44,** 215–223.
Hall, M. N. and Silhavy, T. J. (1981). *Ann. Rev. Genet.* **15,** 91–142.
Halliwell, G. (1979). *Prog. Ind. Microbiol.* **15,** 1–60.
Hamada, S. E., Luckey, J. P. and Farrand, S. K. (1979). *J. Bacteriol.* **139,** 280–286.
Hammerschmidt, R., Lamport, D. T. A. and Muldoon, E. P. (1984). *Physiol. Plant Pathol.* **24,** 43–47.
Hanahan, D. (1983). *J. Mol. Biol.* **166,** 557–580.
Hancock, R. E. W. and Carey, A. M. (1979). *J. Bacteriol.* **140,** 902–910.
Hancock, R. E. W. and Nikaido, H. (1978). *J. Bacteriol.* **136,** 381–390.
Handa, A. K. (1984). *In* "Proc. of 2nd Int. Symp., The Molecular Genetics of the Bacteria–Plant Interaction", #126 (Abstr.).
Handa, A. K. (1985). *Proc. 6th Intern. Conf. Plant Path. Bacteria*, Maryland (in press).
Harkki, A. and Palva, E. T. (1984). *Mol. Gen. Genet.* **195,** 256–259.
Hasegawa, S. and Nagel, C. W. (1968). *Arch. Biochem. Biophys.* **124,** 513–520.
Hatanaka, C. and Imamura, T. (1974). *Agric. Biol. Chem.* **38,** 2267–2268.
Hatanaka, C. and Ozawa, J. (1969). *Agric. Biol. Chem.* **33,** 116–118.
Hatanaka, C. and Ozawa, J. (1970). *Agric. Biol. Chem.* **34,** 1618–1624.
Hatanaka, C. and Ozawa, J. (1971). *Agric. Biol. Chem.* **35,** 1617–1624.
Hatanaka, C. and Ozawa, J. (1973). *Agric. Biol. Chem.* **37,** 593–597.
Hayes, W. (1968). "The Genetics of Bacteria and Tneir Viruses". John Wiley, New York.
Hayward, A. C. (1964). *J. Appl. Bacteriol.* **27,** 265–277.
Hayward, A. C. (1976). *In* "Proc. of the 1st Int. Planning Conf. and Workshop on the Ecology and Control of Bacterial Wilt Caused by *Pseudomonas solanacearum*" (L. Sequeira and A. Kelman, eds), pp. 6–21. North Carolina State University, Raleigh.
Hayward, A. C. (1977). *J. Appl. Bacteriol.* **43,** 407–411.
Heath, E. C. (1971). *Ann. Rev. Biochem.* **40,** 29–56.
Heckels, J. E. (1977). *J. Gen. Microbiol.* **99,** 333–341.
Hedges, R. W. and Jacob, A. E. (1974). *Mol. Gen. Genet.* **132,** 31–40.
Heffron, F., Rubens, C. and Falkow, S. (1975a). *Proc. Natl. Acad. Sci. USA* **72,** 3623–3627.
Heffron, F., Sublett, R., Hedges, R., Jacob, A. and Falkow, S. (1975b). *J. Bacteriol.* **122,** 250–256.
Heitefuss, R. and Williams, P. H. (1976). "Physiological Plant Pathology", Vol. 4. Springer-Verlag, Berlin.
Helander, I., Salkinoja-Salonen, M. and Rylander, R. (1980). *Infect. Immun.* **29,** 859–862.
Helmuth, R. and Achtman, M. (1978). *Proc. Natl. Acad. Sci. USA* **75,** 1237–1241.

Hemming, B. C. and Strobel, G. A. (1983). *In* "Genetic Aspects of Plant Nutrition" (M. R. Saric and B. C. Loughman, eds), pp. 347–357. Developments in Plant and Soil Sciences, Vol. 8. Martinus Nijhoff/W. Junk, The Hague.

Hendrick, C. A. (1982). "The use of bacteriophages in the study of surface carbohydrates of *Pseudomonas solanacearum*". Ph.D. Thesis, University of Wisconsin, Madison.

Hendrick, C. A. and Sequeira, L. (1984). *Appl. Environ. Microbiol.* **48,** 94–101.

Hendrick, C. A., Haskins, W. P. and Vidaver, A. K. (1984). *Appl. Environ. Microbiol.* **48,** 56–60.

Henner, D. J. and Hoch, J. A. (1980). *Microbiol. Rev.* **44,** 57–82.

Heppel, L. A. (1971). *In* "Structure and Function of Biological Membranes" (L. I. Rothfield, ed.), pp. 223–247. Academic Press, London and Orlando.

Hernalsteens, J.-P., Van Vliet, F., De Beuckeleer, M., Depicker, A., Engler, G., Lemmers, M., Holsters, M., Van Montagu, M. and Schell, J. (1980). *Nature* **287,** 654–656.

Hershey, A. D. (1971). "The Bacteriophage Lambda". Cold Spring Harbor Laboratory, Cold Spring Harbor, New York.

Heymer, B. and Rietschel, E. T. (1977). *In* "Microbiology-1977" (D. Schlessinger, ed.), pp. 344–349. American Society for Microbiology, Washington, D.C.

Hickman, J. and Ashwell, G. (1966). *J. Biol. Chem.* **241,** 1424–1428.

Hignett, R. C. and Quirk, A. V. (1979). *J. Gen. Microbiol.* **110,** 77–81.

Hildebrand, D. C. (1971). *Phytopathology* **61,** 1430–1436.

Hildebrand, D. C. and Schroth, M. N. (1964). *Appl. Microbiol.* **12,** 487–491.

Hildebrand, D. C., Alosi, M. C. and Schroth, M. N. (1980). *Phytopathology* **70,** 98–109.

Hindennach, I. and Henning, V. (1975). *Eur. J. Biochem.* **59,** 207–213.

Hinton, J. C. D., Perombelon, M. C. M. and Salmond, G. P. C. (1985), *J. Bacteriol.* **161,** 786–788.

Hirano, S. S., Maher, E. A., Kelman, A. and Upper, C. D. (1978). *In* "Proc. 4th Int. Conf. Plant Path. Bact." (Station de Pathologie Végétale et Phytobactériologie, ed.), pp. 717–724. Institut National de la Recherche Agronomique, Angers.

Hirsch, P. R. and Beringer, J. E. (1984). *Plasmid* **12,** 139–141.

Hisamatsu, M., Abe, J., Amemura, A. and Harada, T. (1980). *Agric. Biol. Chem.* **44,** 1049–1055.

Hislop, E. C., Keon, J. P. R. and Fielding, A. H. (1979). *Physiol. Plant Pathol.* **14,** 371–381.

Hoekstra, W. P. M., Bergmans, J. E. N. and Zuidweg, E. M. (1980). *J. Bacteriol.* **143,** 1031–1032.

Hoffman, J., Lindberg, B., Glowacka, M., Derylo, M. and Lorkiewicz, Z. (1980). *Eur. J. Biochem.* **105,** 103–107.

Hofstra, H. and Witholt, B. (1984). *J. Biol. Chem.* **259,** 15182–15187.

Hofstra, H., van Tol, M. J. D. and Dankert, J. (1980). *J. Bacteriol.* **143,** 328–337.

Hohn, B. and Collins, J. (1980). *Gene* **11,** 291–298.

Hohn, B. and Murray, K. (1977). *Proc. Natl. Acad. Sci., USA* **74,** 3259–3263.

Holley, E. A. and Foster, J. W. (1982). *J. Bacteriol.* **152,** 959–962.

Holloway, B. W. (1979). *Plasmid* **2,** 1–19.

Holmes, B. and Roberts, P. (1981). *J. Appl. Bacteriol.* **50,** 443–467.

Holowachuk, E. W. and Friesen, J. D. (1982). *Mol. Gen. Genet.* **187,** 248–253.

Holsters, M., de Waele, D., Depicker, A., Messens, E., Van Montagu, M. and Schell, J. (1978). *Mol. Gen. Genet.* **163,** 181–187.

Holsters, M., Hernalsteens, J. P., Van Montagu, M. and Schell, J. (1982). *In*

"Molecular Biology of Plant Tumors" (G. Kahl and J. S. Schell, eds), pp. 269–298. Academic Press, London and Orlando.
Hoogenraad, N. J. (1978). *Arch. Biochem. Biophys.* **188,** 137–144.
Hooykaas, P. J. J. and Schilperoort, R. A. (1984). *Adv. Genet.* **22,** 209–283.
Hooykaas, P. J. J., Klapwijk, P. M., Nuti, M. P., Schilperoort, R. A. and Rorsch, A. (1977). *J. Gen. Microbiol.* **98,** 477–484.
Hooykaas, P. J. J., Van Brussel, A. A. N., Den Dulk-Ras, H., Von Slogteren and Schilperoort, R. A. (1981). *Nature* **291,** 351–353.
Hooykaas, P. J. J., Peerbolte, R., Regensburg-Tuink, A. J. G., de Vries, P. and Schilperoort, R. A. (1982a). *Mol. Gen. Genet.* **188,** 12–17.
Hooykaas, P. J. J., Snijdewint, F. G. M. and Schilperoort, R. A. (1982b). *Plasmid* **8,** 73–82.
Hopwood, D. A. (1981). *Ann. Rev. Microbiol.* **35,** 237–272.
Hopwood, D. A. and Johnston, A. W. B. (1984). *In* "The Microbe 1984. Part II. Prokaryotes and Eukaryotes". 26th Symp. Soc. Gen. Microbiol. (D. P. Kelly and N. G. Carr, eds), pp. 257–281. Cambridge University Press, Cambridge.
Hopwood, D. A. and Merrick, M. J. (1977). *Bacteriol. Rev.* **41,** 595–635.
Hotchkiss, R. D. and Gabor, M. (1970). *Ann. Rev. Genet.* **4,** 193–224.
Howard, S. P. and Buckley, J. T. (1983). *J. Bacteriol.* **154,** 413–418.
Howard, S. P. and Buckley, J. T. (1985). *J. Bacteriol.* **161,** 1118–1124.
Hsu, S.-T. and Goodman, R. N. (1978). *Phytopathology* **68,** 351–354.
Huang, J.-S. (1979). *Physiol. Plant Pathol.* **15,** 193–200.
Huang, J.-S. (1980). *Physiol. Plant Pathol.* **17,** 73–80.
Huang, P.-Y., Huang, J.-S. and Goodman, R. N. (1975). *Physiol Plant Pathol.* **6,** 283–287.
Hubbard, J. P., Williams, J. D., Niles, R. M. and Mount, M. S. (1978). *Phytopathology* **68,** 95–99.
Hugouvieux-Cotte-Pattat, N., Condemine, G., Reverchon, S. and Robert-Baudouy, J. (1985). *Proc. 6th Intern. Conf. Plant Path. Bacteria,* Maryland (in press).
Hulme, A. C. (1958). *Adv. Food Res.* **8,** 297–413.
Hungerer, D., Jann, K., Jann, B., Ørskov, F. and Ørskov, I. (1967). *Eur. J. Biochem.* **2,** 115–126.
Hurlbert, R. E. and Gross, D. C. (1983). *J. Gen. Microbiol.* **129,** 2241–2250.
Husain, A. and Kehman, A. (1958). *Phytopathology* **48,** 155–165.
Hutcheson, S. W. and Kosuge, T. (1985). *J. Biol. Chem.* **260,** 6281–6287.
Hutzinger, O. and Kosuge, T. (1968). *Biochemistry* **7,** 601–605.
Ike, Y. and Clewell, D. B. (1984). *J. Bacteriol.* **158,** 777–783.
Ikeda, H. and Tomizawa, J. (1968). *Cold Spring Harbor Symp. Quant. Biol.* **28,** 329–348.
Inoue, S., Nagamatsu, Y. and Hatanaka, C. (1984). *Agric. Biol. Chem.,* **48,** 633–640.
Inouye, M. (1979a). "Bacterial Outer Membranes: Biogenesis and Functions". John Wiley, New York.
Inouye, M. (1979b). *In* "Bacterial Outer Membranes: Biogenesis and Functions" (M. Inouye, ed.), pp. 1–12. John Wiley, New York.
Inouye, M. and Halegoua, S. (1979). *CRC Crit. Rev. Biochem.* **7,** 339–371.
Inouye, S., Takeishi, K., Lee, N., DeMartini, M., Hirashima, A. and Inouye, M. (1976). *J. Bacteriol.* **127,** 555 563.
Irvin, R. T., Chatterjee, A. K., Sanderson, K. E. and Costerton, J. W. (1975). *J. Bacteriol.* **124,** 930–941.
Ish-Horowicz, D. and Burke, J. F. (1981). *Nucl. Acids Res.* **9,** 2989–2998.

Ishii, S. (1976). *Phytopathology* **66,** 281–289.

Israilski, W. P. (1955). "Bakterielle Pflanzenkrankheiten". Deutscher Bauernverlag, Berlin.

Itoh, Y., Izaki, K. and Takahashi, H. (1980). *Agric. Biol. Chem.* **44,** 1135–1140.

Itoh, Y., Sugiura, J., Izaki, K. and Takahashi, H. (1982). *Agric. Biol. Chem.* **46,** 199–205.

Iyer, R., Darby, V. and Holland, I. B. (1978). *FEBS Lett.* **85,** 127–132.

Jacobs, S. E., Habish, H. A. and Dadd, A. H. (1965). *Ann. Appl. Biol.* **56,** 161–170.

Jamieson, A. F., Bieleski, R. L. and Mitchell, R. E. (1981). *J. Gen. Microbiol.* **122,** 161–165.

Jann, B., Reske, K. and Jann, K. (1975). *Eur. J. Biochem.* **60,** 239–246.

Jansson, P.-E., Kenne, L. and Lindberg, B. (1975). *Carbohydr. Res.* **45,** 275–282.

Jansson, P.-E., Kenne, L., Lindberg, B., Ljunggren, H., Lonngrend, I., Ruden, U. and Svensson, S. (1977), *J. Am. Chem. Soc.* **99,** 3812–3815.

Jayaswal, R. K., Bressan, R. A. and Handa, A. K. (1984). *J. Bacteriol.* **158,** 764–766.

Johnson, E. A., Sakajoh, M., Halliwell, G., Madia, A. and Demath, A. L. (1982). *Appl. Environ. Microbiol.* **43,** 1125–1132.

Johnson, S. R. and Romig, W. R. (1979). *Mol. Gen. Genet.* **170,** 93–101.

Johnsrud, L. (1979). *Mol. Gen. Genet.* **169,** 213–218.

Johnsrud, L., Calos, M. P. and Miller, J. H. (1978). *Cell* **15,** 1209–1219.

Johnston, A. W. B., Beynon, J. L., Buchanan-Wollaston, A. V., Setchell, S. M., Hirsch, P. R. and Beringer, J. E. (1978). *Nature* **276,** 634–636.

Johnston, K. H., Holmes, K. K., and Gotschlich, E. C. (1976). *J. Expt. Med.* **143,** 741–758.

Jones, D. (1975). *J. Gen. Microbiol.* **87,** 52–96.

Jones, G. W. and Isaacson, R. E. (1984). *CRC Crit. Rev. Microbiol.* **10,** 229–260.

Jones, R. B., Jemison, P. A., Newhall, W. J. and Haak, R. A. (1980). *Infect. Immun.* **30,** 773–780.

Jorgensen, R. A., Berg, D. E., Allet, B. and Reznikoff, W. S. (1979). *J. Bacteriol.* **137,** 681–685.

Juni, E. (1972). *J. Bacteriol.* **112,** 917–931.

Kadam, S. K., MacLachlan, P. R., Rehemtulla, A. and Sanderson, K. E. (1984). *Abstr. Ann. Meet. Am. Soc. Microbiol.*, p. 94, #H17.

Kado, C. I. and Liu, S.-T. (1981). *J. Bacteriol.* **145,** 1365–1373.

Kahl, G. and Schell, J. S. (1982). "Molecular Biology of Plant Tumors". Academic Press, London and Orlando.

Kahmann, R. and Kamp, D. (1979). *Nature* **280,** 247–250.

Kahn, M. L. and Timblin, C. R. (1984). *J. Bacteriol.* **128,** 1070–1077.

Kahn, M., Kolter, R., Thomas, C., Figurski, D., Meyer, R., Remaut, E. and Helinski, D. R. (1979). *Meth. Enzymol.* **68,** 268–280.

Kamberger, W. (1979). *Arch. Microbiol.* **121,** 83–90.

Kamimiya, S., Izaki, K. and Takahashi, H. (1972). *Agric. Biol. Chem.* **36,** 2367–2372.

Kamimiya, S., Nishiya, T., Izaki, K. and Takahashi, H. (1974). *Agric. Biol. Chem.* **38,** 1071–1078.

Kamio, Y. and Nikaido, H. (1976). *Biochemistry* **15,** 2561–2570.

Kamio, Y. and Nikaido, H. (1977). *Biochem. Biophys. Acta* **464,** 589–601.

Kamio, Y. and Takahashi, H. (1980). *J. Bacteriol.* **141,** 899–907.

Kamp, D. and Sandulache, R. (1983). *FEMS Microbiol. Lett.* **16,** 131–135.

Kao, J. C., Perry, K. L. and Kado, C. I. (1982). *Mol. Gen. Genet.* **188,** 425–432.

Katz, E. and Demain, A. L. (1977). *Bacteriol. Rev.* **41,** 449–474.

Kawaji, H., Mizuno, T. and Mizushima, S. (1979). *J. Bacteriol.* **140,** 843–847.

Kawaoka, Y., Otsuki, K. and Tsubokura, M. (1983). *J. Gen. Microbiol.* **129,** 2739–2747.

Keen, N. T. and Holliday, M. J. (1982). *In* "Phytopathogenic Prokaryotes" (M. S. Mount and G. H. Lacy, eds), Vol. II, pp. 179–217. Academic Press, London and Orlando.

Keen, N. T. and Williams, P. H. (1971). *Physiol. Plant Pathol.* **1,** 247–264.

Keen, N. T., Dahlbeck, D., Staskawicz, B. and Belser, W. (1984). *J. Bacteriol.* **159,** 825–831.

Keener, S. L., McNamee, K. P. and McEntee, K. (1984). *J. Bacteriol.* **160,** 153–160.

Kegoya, Y., Setoguchi, M., Yokohiki, K. and Hatanaka, C. (1984). *Agric. Biol. Chem.* **48,** 1055–1060.

Kelman, A. (1954). *Phytopathology* **44,** 693–695.

Kelman, A. and Hruschka, J. (1973). *J. Gen. Microbiol.* **76,** 177–188.

Kemper, J. (1974). *J. Bacteriol.* **117,** 94–99.

Kerr, A., Maningault, P. and Tempé, J. (1977). *Nature* **265,** 560–561.

Kersters, K. and DeLey, J. (1984). *In* "Bergey's Manual of Systematic Bacteriology" (N. R. Krieg and J. G. Holt, eds), Vol. 1, pp. 244–254. Williams & Wilkins Co., Baltimore.

Khan, S. and Novick, R. (1980). *Plasmid* **4,** 148–154.

Kirkpatrick, B. C., Stenger, D. C., Morris, T. J. and Purcell, A. H. (1985). *Phytopathology* **75**, 1351, #575 (Abstr.).

Klebe, R. J., Harriss, J. V., Sharp, Z. D. and Douglas, M. G. (1983). *Gene* **25,** 333–341.

Kleckner, N. (1981). *Ann. Rev. Genet.* **15,** 341–404.

Kleckner, N., Chan, R. K., Tye, B.-K. and Botstein, D. (1975). *J. Mol. Biol.* **97,** 561–575.

Kleckner, N., Roth, J. and Botstein, D. (1977). *J. Mol. Biol.* **116,** 125–159.

Klee, H. J., Gordon, M. P. and Nester, E. W. (1982). *J. Bacteriol.* **150,** 327–331.

Klein, D. T. and Klein, R. M. (1953). *J. Bacteriol.* **66,** 220–228.

Klein, D. T. and Klein, R. M. (1956). *J. Bacteriol.* **72,** 308–313.

Klement, Z. and Goodman, R. N. (1967). *Phytopathology* **57,** 322–323.

Knauf, V. C. and Nester, E. W. (1982). *Plasmid* **8,** 45–54.

Knox, K. W., Cullen, J. and Work, E. (1967). *Biochm. J.* **103,** 192–201.

Kobayashi, D. Y. and Keen, N. (1985). *Phytopathology* **75**, 1355 #607 (Abstr.).

Kolattukudy, P. E. (1984). *Can. J. Bot.* **62,** 2918–2933.

Kondorosi, A., Kondorosi, E., Pankhurst, C. E. and Broughton, W. J. (1982). *Mol. Gen. Genet.* **188,** 433–439.

Konisky, J. (1978). *In* "The Bacteria: A Treatise on Structure and Function" (L. N. Ornston and J. R. Sokatch, eds), Vol. VI, pp. 71–136. Academic Press, London and Orlando.

Kopecko, D. J. and Cohen, S. N. (1975). *Proc. Natl. Acad. Sci. USA* **72,** 1373–1377.

Kopecko, D. J., Brevet, J. and Cohen, S. N. (1976). *J. Mol. Biol.* **108,** 333–360.

Kopecko, D. J., Washington, O. and Formal, S. B. (1980). *Infect. Immun.* **29,** 207–214.

Koplow, J. and Goldfine, H. (1974). *J. Bacteriol.* **117,** 527–543.

Kosuge, T., Heskett, M. G. and Wilson, E. E. (1966). *J. Biol. Chem.* **241,** 3738–3744.

Kosuge, T., Comai, L. and Glass, N. L. (1983). *In* "Plant Molecular Biology" (R. D. Goldberg, ed.), pp. 167–177. Alan R. Liss, Inc., New York.

Kotoujansky, A., Lemattre, M. and Boistard, P. (1982). *J. Bacteriol.* **150,** 122–131.

Kotoujansky, A., Diolez, A., Boccara, M., Bertheau, Y., Andro, T. and Coleno, A. (1985). *EMBO J.* **4,** 781–785.

Kozloff, L. M., Lute, M. and Westaway, D. (1984). *Science* **226,** 845–846.
Kreft, J., Carrisius, J., Burger, K. J. and Goebel, D. (1982). *In* "Molecular Cloning and Gene Regulation in Bacilli" (A. T. Ganesar, S. Chang and J. A. Hoch, eds), pp. 145–157. Academic Press, Orlando.
Kreil, G. (1981). *Ann. Rev. Biochem.* **50,** 317–348.
Kretschmer, P. J. and Cohen, S. N. (1979). *J. Bacteriol.* **139,** 515–519.
Kretschmer, P. J., Chang, A. C. Y. and Cohen, S. N. (1975). *J. Bacteriol.* **124,** 225–231.
Kroll, J. (1973). *In* "A Manual of Quantitative Immunoelectrophoresis" (N. H. Axelsen, J. Kroll and B. Weeke, eds), pp. 61–68. Universitetsforlaget, Oslo.
Kropinski, A. M., Chan, L. C. and Milazzo, F. H. (1979). *Can. J. Microbiol.* **25,** 390–398.
Kuć, J. and Rush, J. S. (1985). *Arch. Biochem. Biophys.* **236,** 455–472.
Kumamoto, C. A. and Beckwith, J. (1983). *J. Bacteriol.* **154,** 253–260.
Kunkel, T. A. (1985). *Proc. Natl. Acad. Sci. USA* **82,** 488–492.
Kuo, T. T. and Kosuge, T. (1969). *J. Gen. Appl. Microbiol.* **15,** 51–63.
Kuykendall, L. E., Roy, M. A. and Baker, C. J. (1985). *Proc. 6th Intern. Conf. Plant Path. Bacteria*, Maryland (in press).
Lachica, R. V. and Zink, D. L. (1984). *Infect. Immun.* **44,** 540–543.
Lacy, G. H. (1978). *Phytopathology* **68,** 1323–1330.
Lacy, G. H. and Leary, J. V. (1976). *Genet. Res. Camb.* **27,** 363–368.
Lacy, G. H. and Leary, J. V. (1979). *Ann. Rev. Phytopathol.* **17,** 181–202.
Lacy, G. H. and Sparks, R. B., Jr. (1979). *Phytopathology* **69,** 1293–1297.
Laemmli, U. K. (1970). *Nature* **227,** 680–685.
Lambden, P. R., Heckels, J. E., James, L. T. and Watt, P. J. (1979). *J. Gen. Microbiol.* **114,** 305–312.
Lamed, R., Setter, E. and Bayer, E. A. (1983). *J. Bacteriol.* **156,** 828–836.
Langley, R. A. and Kado, C. I. (1972). *Mutat. Res.* **14,** 277–286.
Langsford, M. L., Gilkes, N. R., Wakarchuk, W. W., Kilburn, D. G., Miller, R. C. Jr. and Warren, A. J. (1984). *J. Gen Microbiol.* **130,** 1367–1376.
Larson, J. L. and Hershberger, C. L. (1984). *J. Bacteriol.* **157,** 314–317.
Lathe, R. F., Lecocq, J. P. and Everett, R. (1983). *In* "Genetic Engineering (R. Williamson, ed.), Vol. 4, pp. 1–56. Academic Press, London and Orlando.
Lawson, E. N., Gantotti, B. V. and Starr, M. P. (1982). *Curr. Microbiol.* **7,** 327–332.
Lawton, W. D., Morris, B. C. and Burrows, T. W. (1968). *J. Gen. Microbiol.* **52,** 25–34.
Lê, T. K. T., Leccas, D. and Boucher, C. (1978). *In* "Proc. 4th Int. Conf. Plant Path. Bact." (Station de Pathologie Végétale et Phytobactériologie, ed.), pp. 819–822. Institut National de la Recherche Agronomique, Angers.
Leach, J. E., Cantrell, M. A. and Sequeira, L. (1982a). *Plant Physiol.* **70,** 1353–1358.
Leach, J. E., Cantrell, M. A. and Sequeira, L. (1982b). *Physiol. Plant Pathol.* **21,** 319–325.
Leach, J. E., Sherwood, J., Fulton, R. W. and Sequeira, L. (1983). *Physiol. Plant Pathol.* **23,** 377–386.
Leary, J. V. (1979). *Can. J. Microbiol.* **25,** 637–640.
Leary, J. V. and Fulbright, D. W. (1982). *In* "Phytopathogenic Prokaryotes" (M. S. Mount and G. H. Lacy, eds), pp. 229–253. Academic Press, London and Orlando.
Leben, C. (1974). "Survival of Plant Pathogenic Bacteria". Ohio Agric. Res. and Dev. Ctr., Spec. Circ. 100.
Lee, D. R., Schnaitman, C. A. and Pugsley, A. P. (1979). *J. Bacteriol.* **138,** 861–870.
Lee, I. M. and Davis, R. E. (1980). *Can. J. Microbiol.* **26,** 1356–1363.
Lee, J. H., Heffernan, L. and Wilcox, G. (1980). *J. Bacteriol.* **143,** 1325–1331.

Lee, S.-C. and West, C. A. (1981). *Plant Physiol.* **67,** 633–639.
Leemans, J., Villarroel, R., Silva, B., Van Montagu, M. and Schell, J. (1980). *Gene* **10,** 319–328.
Leemans, J., Langenakens, J., DeGreve, H., Deblaere, R., Van Montagu, M. and Schell, J. (1982). *Gene* **19,** 361–364.
Lehmann, V. (1977). *Eur. J. Biochem.* **75,** 257–266.
Lehmann, V. and Rupprecht, E. (1977). *Eur. J. Biochem.* **81,** 443–452.
Lehmann, V., Rupprecht, E. and Osborn, M. J. (1977). *Eur. J. Biochem.* **76,** 41–49.
Lehmann, V., Redmond, J., Egan, A. and Minner, I. (1978). *Eur. J. Biochem.* **86,** 487–496.
Lehrbach, P. R., Kung, A. H. C. and Lee, B. T. O. (1978). *J. Gen. Microbiol.* **108,** 119–124.
Lei, S.-P., Lin, H.-C., Heffernan, L. and Wilcox, G. (1985). *Gene* **35,** 63–70.
Leisinger, T., Hutter, R., Cook, A. M. and Nuesch, J. (1981). "Microbial Degradation of Xenobiotics and Recalcitrant Molecules". Academic Press, London and Orlando.
Leive, L. (1965). *Proc. Natl. Acad. Sci. USA* **53,** 745–750.
Leive, L. (1974). *Ann. N. Y. Acad. Sci.* **235,** 109–129.
Leive, L., Shovlin, V. K. and Mergenhagen, S. E. (1968). *J. Biol. Chem.* **243,** 6384–6391.
Lejeune, P., Mergeay, M., Van Gijsegem, F., Faelen, M., Gerits, M. and Toussaint, A. (1983). *J. Bacteriol.* **155,** 1015–1026.
Lelliott, R. A. (1974). In "Bergey's Manual of Determinative Bacteriology", 8th Edn (R. E. Buchanan and N. E. Gibbon, eds), pp. 332–340. Williams & Wilkins Co., Baltimore.
Lelliott, R. A. and Dickey, R. S. (1984). *In* "Bergey's Manual of Systematic Bacteriology" (N. R. Krieg and J. G. Holt, eds), Vol. 1, pp. 469–476. Williams & Wilkins Co., Baltimore.
Leong, S. A. and Neilands, J. B. (1982). *Arch. Biochem. Biophys.* **218,** 351–359.
Levin, D. H. and Racker, E. (1959). *J. Biol. Chem.* **234,** 2532–2539.
Levy, S. B., Clowes, R. C. and Koenig, E. L. (1981). "Molecular Biology, Pathogenicity, and Ecology of Bacterial Plasmids". Plenum Press, New York.
Liao, C. H. and Heberlein, G. T. (1979). *Can. J. Microbiol.* **25,** 185–191.
Lichtenstein, C. and Brenner, S. (1981). *Mol. Gen. Genet.* **183,** 380–387.
Lim, R. and Cohen, S. S. (1966). *J. Biol. Chem.* **241,** 4304–4315.
Lindberg, A. A. (1973). *Ann. Rev. Microbiol.* **27,** 205–241.
Lindberg, A. A. (1977). *In* "Surface Carbohydrates of the Prokaryotic Cell" (I. Sutherland, ed.), pp. 289–356. Academic Press, London and Orlando.
Lindgren, P. B., Peet, R. C. and Panopoulos, N. J. (1985). *Proc. 6th Intern. Conf. Plant Path. Bacteria*, Maryland (in press).
Lindow, S. E., Arny, D. C. and Upper, C. D. (1978). *Phytopathology* **68,** 523–527.
Lipmann, F. (1980). *In* "Advances in Microbial Physiology" (A. H. Rose and J. G. Morris, eds), Vol. 21, pp. 227–266. Academic Press, London and Orlando.
Lippincott, B. B. and Lippincott, J. A. (1966). *J. Bacteriol.* **92,** 937–945.
Lippincott, B. B. and Lippincott, J. A. (1969). *J. Bacteriol.* **94,** 620–628.
Lippincott, J. A. and Lippincott, B. B. (1976). *In* "Physiological Plant Pathology" (R Heitefuss and P. H. Williams, eds), pp. 356–388. Springer-Verlag, New York.
Lippincott, J. A. and Lippincott, B. B. (1980). *In* "Bacterial Adherence, Ser. B, Receptors and Recognition" (E. H. Beachey, ed.), Vol. 6, pp. 439–451. Chapman and Hall, London.

Little, J. W. and Mount, D. W. (1982). *Cell* **29,** 11–22.
Liu, S. C. Y. (1973). *In* "Proc. 2nd Int. Cong. Plant Pathol.", #0654 (Abstr.).
Liu, S. T. and Kado, C. I. (1979). *Biochem. Biophys. Res. Commun.* **90,** 171–178.
Liu, S. T., Perry, K. L., Schardl, C. L. and Kado, C. I. (1982). *Proc. Natl. Acad. Sci. USA* **79,** 2812–2816.
Ljungdahl, L. G. and Eriksson, K.-E. (1985). *Adv. Microb. Ecol.* **8,** 237–299.
Long, M., Barton-Willis, P., Staskawicz, B. J., Dahlbeck, D. and Keen, N. T. (1985). *Phytopathology* **75,** 235–239.
Loper, J. E., Orser, C. S., Panopoulos, N. J., Schroth, M. N., and Lindow, S. E. (1983). *Phytopathology* **73,** 826, #A471 (Abstr.).
Loper, J. E., Orser, C. S., Panopoulos, N. J. and Schroth, M. N. (1984). *J. Gen. Microbiol.* **130,** 1507–1515.
Lory, S., Tai, P. C. and Davis, B. D. (1983). *J. Bacteriol.* **156,** 695–702.
Low, K. B. (1972). *Bacteriol. Rev.* **36,** 587–607.
Low, K. B. and Porter, D. D. (1978). *Ann. Rev. Genet.* **12,** 249–287.
Lozano, J. C. (1981). "Proceedings of the 5th International Conference on Plant Pathogenic Bacteria". Centro Internacional de Agricultura Tropical, Angers.
Lüderitz, O., Westphal, O., Staub, A. M. and Nikaido, H. (1971). In "Microbial Toxins" (G. Weinbaum, S. Kadis and S. J. Ajl, eds), Vol. IV, pp. 145–233. Academic Press, Orlando and London.
Ludwig, R. A., de Vries, G. E. and Raymond, C. K. (1984). "Proc. 2nd Int. Symp., The Molecular Genetics of the Bacteria–Plant Interactions", #129 (Abstr.), Ithaca, New York.
Lugtenberg, B. and van Alphen, L. (1983). *Biochim. Biophys. Acta.* **737,** 51–115.
Lugtenberg, B., Meijers, J., Peters, R., van der Hoek, P. and van Alphen, L. (1975). *FEBS Lett.* **58,** 254–258.
Lugtenberg, E. J. J. and Peters, R. (1976). *Biochem. Biophys. Acta* **441,** 38–47.
Lund, B. M. and Brocklehurst, T. F. (1978). *J. Gen. Microbiol.* **104,** 59–66.
Maas Geesteranus, H. P. (1972). "Proceedings of the 3rd International Conference on Plant Pathogenic Bacteria". Centre for Agricultural Publishing and Documentation, Wageningen.
Macrina, F. L., Tobian, J. A., Jones, K. R., Evans, R. P. and Clewell, D. B. (1982). *Gene* **19,** 345–353.
Macrina, F. L., Evans, R. P., Tobian, J. A., Hartley, D. L., Clewell, D. B. and Jones, K. R. (1983). *Gene* **25,** 145–150.
Mäkelä, P. H. and Stocker, B. A. D. (1969). *Ann. Rev. Genet.* **3,** 291–322.
Maki, L. R., Galyan, E. L., Chang-Chien, M. M. and Caldwell, D. R. (1974). *Appl. Microbiol.* **28,** 456–459.
Malik, A. N., Veal, D., Taylor, J. D. and Vivian, A. (1984). *In* "Proc. 2nd Int. Symp., The Molecular Genetics of the Bacteria–Plant Interaction", #130 (Abstr.), Ithaca, New York.
Manasse, R. J. and Corpe, W. A. (1967). *Can. J. Microbiol.* **13,** 1591–1603.
Mandel, M. and Higa, A. (1970). *J. Mol. Biol.* **53,** 159–162.
Maniatis, T., Fritsch, E. F. and Sambrook, J. (1982). "Molecular Cloning: A Laboratory Manual". Cold Spring Harbor Press, Cold Spring Harbor, New York.
Manniello, J. M., Heymann, H. and Adair, F. W. (1979). *J. Gen. Microbiol.* **112,** 397–400.
Manning, P. A. and Actman, M. (1979). *In* "Bacterial Outer Membranes: Biogenesis and Functions" (M. Inouye, ed.), pp. 409–447. John Wiley, New York.
Marcoli, R., Iida, S. and Bickle, T. (1980). *FEBS Lett.* **110,** 11–14.

Markovitz, A. (1977). *In* "Surface Carbohydrates of the Prokaryotic Cell" (I. Sutherland, ed.), pp. 415–462. Academic Press, London and Orlando.
Marsch, D. G. and Crutchley, M. J. (1967). *J. Gen. Microbiol.* **47,** 405–420.
Marshall, K. C. (1976). "Interfaces in Microbial Ecology". Harvard University Press, Cambridge, Mass.
Martin, J. F. and Demain, A. L. (1976). *Biochem. Biophys. Res. Commun.* **71,** 1103–1109.
Martin, J. F. and Demain, A. L. (1980). *Microbiol. Rev.* **44,** 230–251.
Matthysse, A. G., Holmes, K. V. and Gurlitz, R. H. G. (1981). *J. Bacteriol.* **145,** 583–595.
Maurelli, A. T., Blackman, B. and Curtiss, R., III. (1984). *Infect. Immun.* **43,** 195–201.
Mazzucchi, U. (1983). *In* "Biochemical Plant Pathology" (J. A. Callow, ed.), pp. 299–324. John Wiley, Chichester.
Mazzucchi, U. and Pupillo, P. (1976). *Physiol. Plant Pathol.* **9,** 101–112.
Mazzucchi, U., Alberghira, A. and Garibaldi, A. (1974). *Phytopathol. Mediterr.* **13,** 27–35.
McCammon, S. L. and Coplin, D. L. (1982). *Phytopathology* **72,** 1001, #570 (Abstr.).
McCammon, S. L. and Davis, R. E. (1985). *Proc. 6th Intern. Conf. Plant Pathogenic Bacteria,* Maryland (in press).
McCammon, S. L., Rowan, R. G. and Coplin, D. L. (1981). *Phytopathology* **71,** 240 (Abstr.).
McCoy, R. E. (1982). *In* "Phytopathogenic Prokaryotes" (M. S. Mount and G. H. Lacy, eds), pp. 475–489. Academic Press, London and Orlando.
McEvoy, J. L. and Chatterjee, A. K. (1983). *Phytopathology* **73,** 809, #A341 (Abstr.).
McEvoy, J. L., Murata, H., Engwall, J. K. and Chatterjee, A. K. (1985). *Proc. 6th Intern. Conf. Plant Path. Bacteria,* Maryland (in press).
McNeil, M., Darvill, A. G., Fry, S. C. and Albersheim, P. (1984). *Ann. Rev. Biochem.* **53,** 625–663.
Meade, H. M. and Signer, E. R. (1977). *Proc. Natl. Acad. Sci. USA* **74,** 2076–2078.
Meade, H. M., Long, S. R., Ruvkun, G. B., Brown, S. E. and Ausubel, F. M. (1982). *J. Bacteriol.* **149,** 114–122.
Mellano, M. A. and Cooksey, D. A. (1985). *Phytopathology* **75,** 1326, #382 (Abstr.).
Mergaert, J., Verdonck, L., Kersters, K., Swings, J., Boeufgras, J.-M. and De Ley, J. (1984). *J. Gen. Microbiol.* **130,** 1893–1910.
Merlo, D. J. (1982). *In* "Advances in Plant Pathology" (D. S. Ingram and P. H. Williams, eds), Vol. 1, pp. 139–178. Academic press, London and Orlando.
Message, B., Boistard, P., Pitrat, M., Schmit, J. and Boucher, C. (1978). *In* "Proc. 4th Int. Conf. Plant Path. Bact." (Station de Pathologie Végétale et Phytobactériologie, ed.), pp. 823–833. Institut National de la Recherche Agronomique, Angers.
Meyer, R. J. and Shapiro, J. A. (1980). *J. Bacteriol.* **143,** 1362–1373.
Michaelis, S. and Beckwith, J. (1982). *Ann. Rev. Microbiol.* **36,** 436–465.
Milani, V. and Heberlein, G. T. (1972). *J. Virol.* **10,** 1722.
Mildenhall, J. P., Prior, B. A. and Trollope, L. A. (1981). *J. Gen. Microbiol.* **127,** 27–34.
Miller, J. H. (1972). "Experiments in Molecular Genetics". Cold Spring Harbor Laboratory, Cold Spring Habor, New York.
Mills, D. (1985). *Ann. Rev. Phytopathol.* **23** (in press).
Mills, D. and Gonzalez, C. F. (1982). *In* "Phytopathogenic Prokaryotes" (M. S. Mount and G. H. Lacy, eds), Vol. 1, pp. 77–119. Academic Press, London and Orlando.
Minsavage, G. V. and Schaad, N. W. (1983). *Phytopathology* **73,** 747–755.
Minton, N. P. (1984). *Gene* **31,** 269–273.

Mitchell, R. E. (1976). *Phytochemistry* **15,** 1941–1947.
Mitchell, R. E. (1978). *Physiol. Plant Pathol.* **13,** 37–49.
Mitchell, R. E. (1981). *In* "Toxins in Plant Disease" (R. D. Durbin, ed.), pp. 259–293. Academic Press, London and Orlando.
Mitchell, R. E. (1984). *Ann. Rev. Phytopathol.* **22,** 215–245.
Mitchell, R. E. and Bieleski, R. L. (1977). *Plant Physiol.* **60,** 723–729.
Mitchell, R. E. and Hart, P. A. (1983). *Phytochemistry* **22,** 1425–1428.
Mitchell, R. E. and Parsons, E. A. (1977). *Phytochemistry* **16,** 280–281.
Mitchell, R. E., Hale, C. N. and Shanks, J. C. (1983). *Physiol. Plant Pathol.* **23,** 315–322.
Modrich, P., Anraku, Y. and Lehman, I. R. (1973). *J. Biol. Chem.* **248,** 7495–7501.
Mogen, B. D., Oleson, A. E. and Secor, G. A. (1985) *Phytopathology* **75**, 1378, #793 (Abstr.).
Moldow, C., Robertson, J. and Rothfield, L. (1972). *J. Membr. Biol.* **10,** 137–152.
Monner, D. A., Jonsson, S. and Boman, H. G. (1971). *J. Bacteriol.* **107,** 420–432.
Mooi, F. R., Harms, N., Bakker, D. and de Graff, F. K. (1981). *Infect. Immun.* **32,** 1155–1163.
Moore, L. W. and Cooksey, D. A. (1981). *In* "International Review Cytology Supplement 13" (K. L. Giles and A. G. Atherly, eds), pp. 15–46. Academic Press, London and Orlando.
Moore, L., Warren, G. and Strobel, G. (1979). *Plasmid* **2,** 617–626.
Moore, R. E., Niemczura, W. P., Kwok, O. C. H. and Patil, S. S. (1984). *Tetrahedron Lett.* **25,** 3931–3934.
Morales, V. M. and Sequeira, L. (1985a). *Phytopathology* **75,** 767–771.
Morales, V. M. and Sequeira, L. (1985b). *Proc. 6th Intern. Conf. Plant Path. Bacteria*, Maryland (in press).
Morales, V. M. and Sequeira, L. (1985c). *J. Bacteriol.* **163,** 1263–1264.
Moran, F. and Starr, M. P. (1969). *Eur. J. Biochem.* **11,** 291–295.
Moran, F., Nasuno, S. and Starr, M. P. (1968). *Arch. Biochem. Biophys.* **125,** 734–741.
Morell, S., Baur, L. and Link, K. P. (1934) *J. Biol. Chem.* **105,** 1–13.
Moreno, E., Pitt, M. W., Jones, L. M., Schurig, G. G. and Berman, D. T. (1979). *J. Bacteriol.* **138,** 361–369.
Morgan, M. K. and Chatterjee, A. K. (1984a). *In* "Proc. 2nd Int. Symp., The Molecular Genetics of the Bacteria—Plant Interaction", (Abstr.) (in press).
Morgan, M. K. and Chatterjee, A. K. (1984b). *Phytopathology* **74,** 879, #A718 (Abstr.).
Morgan, M. K. and Chatterjee, A. K. (1985). *J. Bacteriol.* **164,** 14–18.
Mori, M., Aoyagi, K., Tatibana, M., Ishikawa, T. and Ishii, H. (1977). *Biochem. Biophys. Res. Commun.* **76,** 900–904.
Mouches, C., Barroso, G. and Bore, J. M. (1983). *J. Bacteriol.* **156,** 952–955.
Mount, M. S. and Lacy, G. H. (1982). "Phytopathogenic Prokaryotes". Academic Press, London and Orlando.
Mount, M. S., Bateman, D. F. and Basham, H. G. (1970). *Phytopathology* **60,** 924–931.
Mount, M. S., Berman, P. M., Mortlock, R. P. and Hubbard, J. P. (1979). *Phytopathology* **69,** 117–120.
Mudd, J. B., Lee, I.-M., Liu, H.-Y. and Calavan, E. C. (1979). *J. Bacteriol.* **137,** 1056–1058.
Muhlradt, P., Wray, V. and Lehmann, V. (1977). *Eur. J. Biochem.* **81,** 193–203.
Mulligan, J. T., Margolin, W., Krueger, J. H. and Walker, G. C. (1982). *J. Bacteriol.* **151,** 609–619.

Munson, R. S., Jr., Rasmussen, N. S. and Osborn, M. J. (1978). *J. Biol. Chem.* **253,** 1503–1511.
Murai, N., Skoog, F., Doyle, M. E. and Hanson, R. S. (1980). *Proc. Natl. Acad. Sci. USA* **77,** 619–623.
Murata, N. and Starr, M. P. (1974). *Can. J. Microbiol.* **20,** 1545–1566.
Murooka, Y., Takizawa, N. and Harada, T. (1981). *J. Bacteriol.* **145,** 358–368.
Myers, R. M., Lerman, L. S. and Maniatis, T. (1985). *Science* **299,** 242–247.
Nagahari, K. and Sakaguchi, K. (1978). *J. Bacteriol.* **133,** 1527–1529.
Nakae, T. (1975). *Biochem. Biophys. Res. Commun.* **64,** 1224–1230.
Nakamura, K. and Inouye, M. (1979). *Cell* **18,** 1109–1117.
Nakamura, K. and Inouye, M. (1982). *EMBO J.* **1,** 771–775.
Nakamura, K. and Mizushima, S. (1976). *J. Biochem.* (Tokyo) **80,** 1411–1422.
Nano, F. E., Shepard, W. D., Watkins, M. M., Kuhl, S. A. and Kaplan, S. (1984). *Gene* **34,** 219–226.
Napoli, C., Swanson, J., Dahlbeck, D. and Staskawicz, B. (1985). *Phytopathology* **75,** 1354–1355, #606 (Abstr.).
Nasuno, S. and Starr, M. P. (1966). *J. Biol. Chem.* **241,** 5298–5306.
Neill, R. J., Ivias, B. E. and Holmes, R. K. (1983). *Science* **221,** 289–291.
Nester, E. W. and Kosuge, T. (1981). *Ann. Rev. Microbiol.* **35,** 531–565.
Nester, E. W., Gordon, M. P., Amasino, R. M. and Yanotsky, M. F. (1984). *Ann. Rev. Plant Physiol.* **35,** 387–413.
Neukom, H. and Deuel, H. (1958). *Chem. Industry* (London), p. 683.
Ng, T. K. and Zeikus, J. G. (1981). *Appl. Environ. Microbiol.* **42,** 231–240.
Nieland, J. B. (1982). *Ann. Rev. Microbiol.* **36,** 285–309.
Niepold, F., Anderson, D. M. and Mills, D. (1985a). *In* "Advances in Molecular Genetics of the Bacteria–Plant Interaction" (A. A. Szalay and R. P. Legocki, eds), pp. 173–175. Media Services, Cornell University, Ithaca, New York.
Niepold, F., Anderson, D. and Mills, D. (1985b). *Proc. Natl. Acad. Sci. USA* **82,** 406–410.
Nikaido, H. (1973). *In* "Membranes and Walls" (L. Lieve, ed.), pp. 131–208. Dekker, New York.
Nikaido, H. (1979). *In* "Bacterial Outer Membranes: Biogenesis and Functions" (M. Inouye, ed.), pp. 361–407. John Wiley, New York.
Nilsson, B., Uhlen, M., Josephson, S., Gatenbeck, S. and Philipson, L. (1983). *Nucl. Acids Res.* **11,** 8019–8030.
Nishimura, A., Nishimura, Y. and Caro, L. (1973). *J. Bacteriol.* **116,** 1107–1112.
Nishimura, Y., Caro, L., Berg, C. M. and Hirota, Y. (1971). *J. Mol. Biol.* **55,** 441–456.
Nixdorff, K., Fitzer, H., Gmeiner, J. and Martin, H. H. (1977). *Eur. J. Biochem.* **81,** 63–69.
Noda, T., Sato, Z., Kobayashi, H., Iwasaki, S. and Okuda, S. (1980). *Ann. Phytopath. Soc. Japan* **46,** 663–666.
Nordeen, R. O. and Currier, T. C. (1983). *App. Env. Microbiol.* **45,** 1884–1889.
Norris, K. E., Iserentant, D., Couteras, R. and Fiers, W. (1979). *Gene* **7,** 355–362.
Notani, N. K. and Setlow, J. K. (1974). *Prog. Nucl. Acids Res. Mol. Biol.* **14,** 39–100.
Novick, R. P., Clowes, R. C., Cohen, S. N., Curtiss, R., III, Datta, N. and Falkow, S. (1976). *Bacteriol. Rev.* **40,** 168–189.
Novick, R., Edelman, I., Schwesinger, M., Gruss, A., Swanson, E. and Pattee, P. (1979). *Proc. Natl. Acad. Sci. USA* **76,** 400–404.
Nugent, M., Bone, D. and Datta, N. (1979). *Nature* **282,** 422–423.
Obukowicz, M. and Shaw, P. D. (1983). *J. Bacteriol.* **155,** 438–442.

Obukowicz, M. and Shaw, P. D. (1985). *Appl. Environ. Microbiol.* **49,** 468–473.
O'Farrell, P. H. (1975). *J. Biol. Chem.* **250,** 4007–4021.
O'Farrell, P. Z., Goodman, H. M. and O'Farrell, P. H. (1977). *Cell* **12,** 1133–1142.
Ofuya, C. O. (1984). *Curr. Microbiol.* **10,** 141–146.
Ohuchi, A. and Tominaga, T. (1973). *Ann. Phytopathol. Soc. Japan* **39,** 417–424.
Oka, A., Nomura, N., Sugimoto, K., Sugisaki, H. and Takanami, M. (1978). *Nature* **276,** 845–847.
Okabe, N. and Goto, M. (1963). *Ann. Rev. Phytopathol.* **1,** 397–418.
Okamoto, K., Hatanaka, C. and Ozawa, J. (1964). *Agr. Biol. Chem.* **28,** 331–336.
Oliver, D. B. (1985). *J. Bacteriol.* **161,** 285–291.
Oliver, D. B. and Beckwith, J. (1981). *Cell* **25,** 765–772.
Olsen, R. H., DeBusscher, G. and McCombie, W. R. (1982). *J. Bacteriol.* **150,** 60–69.
Orser, C., Staskawicz, B. J., Loper, J., Panopoulos, N. J., Dahlbeck, D., Lindow, S. E. and Schroth, M. N. (1983). *In* "Molecular Genetics of the Bacteria–Plant Interaction" (A. Pühler, ed.), pp. 353–361. Springer-Verlag, Berlin.
Orser, C., Staskawicz, B. J., Panopoulos, N. J., Dahlbeck, D. and Lindow, S. E. (1985). *J. Bacteriol.* **164,** 359–366.
Ørskov, I., Ørskov, F., Jann, B. and Jann, K. (1977). *Bacteriol. Rev.* **41,** 667–710.
Osborn, M. J. (1969). *Ann. Rev. Biochem.* **38,** 501–538.
Osborn, M. J. (1971). *In* "Structure and Function of Biological Membranes" (L. I. Rothfield, ed.), pp. 343–400. Academic Press, Orlando and London.
Osborn, M. J. (1979). *In* "Bacterial Outer Membranes: Biogenesis and Functions" (M. Inouye, ed.), pp. 15–34. John Wiley, New York.
Osborn, M. J. and Wu, H. C. P. (1980). *Ann. Rev. Microbiol.* **34,** 369–422.
Osborn, M. J. and Rothfield, L. I. (1971). *In* "Microbial Toxins" (G. Weinbaum, S. Kadis and S. I. Ajl, eds), Vol. IV, pp. 331–350. Academic Press, Orlando and London.
Osborn, M. J., Gander, J. E. and Parisi, E. (1972). *J. Biol. Chem.* **247,** 3973–3986.
Osman, S. F. and Fett, W. F. (1985). *Proc. 6th Intern. Conf. Plant Path. Bacteria,* Maryland (in press).
Otta, J. D. (1976). *Phytopathology* **66,** 249–252.
Ottow, J. C. G. (1975). *Ann. Rev. Microbiol.* **29,** 79–108.
Overbeeke, N. and Lugtenberg, B. (1980). *FEBS Lett.* **112,** 229–232.
Owen, P. and Salton, M. J. R. (1975). *Proc. Natl. Acad. Sci. USA* **72,** 3711–3715.
Ozaki, A., Katsumata, R., Oka, T. and Furuya, A. (1984). *Mol. Gen. Genet.* **196,** 175–178.
Palleroni, N. J. (1984). *In* "Bergey's Manual of Systematic Bacteriology" (N. R. Krieg and J. G. Holt, eds), Vol. 1, pp. 141–199. Williams & Wilkins Co., Baltimore.
Palva, E. T., Liljestrom, P. and Harayama, S. (1981). *Mol. Gen. Genet.* **181,** 153–157.
Panopoulos, N. J. and Schroth, M. N. (1974). *Phytopathology* **64,** 1389–1397.
Panopoulos, N. J. and Staskawicz, B. J. (1981). *In* "Toxins in Plant Disease" (R. D. Durbin, ed.), pp. 79–107. Academic Press, London and Orlando.
Panopoulos, N. J. and Peet, R. C. (1985) *Ann. Rev. Phytopathol.* **23,** 381–419.
Panopoulos, N. J., Guimares, W. V., Hua, S.-S., Sabersky-Lehman, C., Resnik, S., Lai, M. and Shaffer, S. (1978). *In* "Microbiology 1978" (D. Schlessinger, ed.), pp. 238–241. American Society for Microbiology, Washington, D. C.
Panopoulos, N. J., Lindgren, P. B., Peet, R. C., Thom, R. F., Hickman, M. J., Gies, D. R. and Willis, D. K. (1985). *In* "Advances in Molecular Genetics of the Bacteria–Plant Interaction" (A. A. Szalay and R. P. Legocki, eds), pp. 188–192. Media Services, Cornell University, Ithaca, New York.

Pappenheimer, A. M., Jr. (1977). *Ann. Rev. Biochem.* **46,** 69–94.

Pappenheimer, A. M., Jr. and Gill, D. M. (1973). *Science* **182,** 353–358.

Paranchych, W., Sastry, P. A., Frost, L. S., Carpenter, M., Armstrong, G. D. and Watts, T. H. (1979). *Can. J. Microbiol.* **25,** 1175–1181.

Passmoor, M. and Epton, H. A. S. (1978). *In* "Proc. 4th Int. Conf. Plant Path. Bact." (Station de Pathologie Végétale et Phytobactériologie, ed.), pp. 675–685. Institut National de la Recherche Agronomique, Angers.

Pastan, I. and Adhya, S. (1976). *Bacteriol. Rev.* **40,** 527–551.

Patil, S. S. (1974). *Ann. Rev. Phytopathol.* **12,** 259–279.

Patil, S. S., Kolattukudy, P. E. and Dimond, A. E. (1970). *Plant Physiol.* **46,** 752–753.

Patil, S. S., Hayward, A. C. and Emmons, R. (1974). *Phytopathology* **64,** 590–595.

Pattee, P. A., Thompson, N. E., Haubrich, D. and Novick, R. P. (1977). *Plasmid* **1,** 38–51.

Paulin, J. P. and Luisetti, J. (1978). *In* "Proc. 4th Int. Conf. Plant Path. Bact." (Station de Pathologie Végétale et Phytobactériologie, ed.), pp. 725–731. Institut National de la Recherche Agronomique, Angers.

Paulin, J. P. and Nassan, N. A. (1978). *In* "Proc. 4th Int. Conf. Plant Path. Bact." (Station de Pathologie Végétale et Phytobactériologie, ed.), pp. 539–545. Institut National de la Recherche Agronomique, Angers.

Pearce, W. A. and Buchann, T. M. (1980). *In* "Bacterial Adherence, Ser. B, Receptors and Recognition" (E. H. Beachey, ed.), Vol. 6, pp. 289–344. Chapman and Hall, London.

Pecknold, P. C. and Grogan, R. G. (1973). *Int. J. Syst. Bacteriol.* **23,** 111–121.

Peet, R. C., Lingren, P. B. and Panopoulos, N. J. (1985). *Proc. 6th Intern. Conf. Plant Path. Bacteria*, Maryland (in press).

Pemberton, J. M. and Fisher, P. R. (1977). *Nature* **268,** 732–733.

Penner, D., DeVay, J. E. and Backman, P. (1969). *Plant Physiol.* **44,** 806–808.

Pérombelon, M. C. M. and Boucher, C. (1978). *In* "Proc. 4th Int. Conf. Plant Path. Bact." (Station de Pathologie Végétale et Phytobactériologie, ed.), pp. 47–52. Institut National de la Recherche Agronomique, Angers.

Pérombelon, M. C. M. and Kelman, A. (1980). *Ann. Rev. Phytopathol.* **18,** 361–387.

Perreaux, D., Maraite, H. and Meyer, J. A. (1982). *Physiol. Plant Pathol.* **20,** 313–319.

Perry, K. L., Simonitch, T., Harrison-Lavoie, K. and Liu, S. T. (1985). *Proc. 6th Intern. Conf. Plant Path. Bacteria*, Maryland (in press).

Perry, M. B., Daoust, V., Diena, B. B., Ashton, F. E. and Wallace, R. (1975). *Can. J. Biochem.* **53,** 623–629.

Peterson, P. K. and Quie, P. G. (1981). *Ann. Rev. Med.* **32,** 29–43.

Piggott, N. H., Sutherland, I. W. and Jarman, T. R. (1981). *Eur. J. Appl. Microbiol. Biotechnol.* **13,** 179–183.

Pischl, A. L. and Farrand, S. K. (1982). *Abstr. Ann. Meet. Am. Soc. Microbiol.*, p. 113, #H2.

Pischl, D. L. and Farrand, S. K. (1983). *J. Bacteriol.* **153,** 1451–1460.

Pischl, D. L. and Farrand, S. K. (1984). *J. Bacteriol.* **159** (in press).

Piwowarski, J. M. and Shaw, P. D. (1982). *Plasmid* **7,** 85–94.

Planque, K. and Kijne, J. W. (1977). *FEBS Lett.* **73,** 64–66.

Plasterk, R. H. A. and Van De Putte, P. (1984). *Biochem. Biophys. Acta.* **782,** 111–119.

Politis, D. J. and Goodman, R. N. (1980). *Appl. Environ. Microbiol.* **40,** 596–607.

Pomponi, M., Spanò, L., Sabbadini, M. G. and Costantino, P. (1983). *Plasmid* **10,** 119–129.

Poplawsky, A. R. and Mills, D. (1983). *Phytopathology* **73,** 826, #A473 (Abstr.).

Portnoy, D. A., Blank, H. F., Kingsbury, D. T. and Falkow, S. (1983). *J. Infect. Dis.* **148,** 297–304.
Portnoy, D. A., Wolf-watz, H., Bolin, I., Beeder, A. B. and Falkow, S. (1984). *Infect. Immun.* **43,** 108–114.
Postma, P. W. and Roseman, S. (1976). *Biochim. Biophys. Acta.* **457,** 213–257.
Prakash, R. K. and Schilperoort, R. A. (1982). *J. Bacteriol.* **149,** 1129–1134.
Prehm, P., Stirm, S., Jann, B. and Jann, K. (1975). *Eur. J. Biochem.* **56,** 41–55.
Prentki, P. and Krisch, H. M. (1982). *Gene* **17,** 189–196.
Priefer, U. B., Simon, R. and Puhler, A. (1985). *J. Bacteriol.* **163,** 324–330.
Primrose, S. B. (1979). *J. Appl. Bacteriol.* **46,** 1–25.
Prokulevich, V. A. and Fomichev, Y. K. (1981). *Genetika* **17,** 2048–2051.
Prokulevich, V. A., Belyasova, N. A. and Fomichev, Y. K. (1983). *Genetika* **19,** 1624–1629.
Prokulevich, V. A., Belyasova, N. A. and Fomichev, Y. K. (1984). *Geneticka* **20,** 1961–1967.
Pugashetti, B. K. (1976). "Genetics and physiology of virulence in *Erwinia amylovora*". Ph.D. Thesis, University of California, Davis. 135 pp.
Pugashetti, B. K. and Starr, M. P. (1975). *J. Bacteriol.* **122,** 485–491.
Pugashetti, B. K., Chatterjee, A. K. and Starr, M. P. (1978). *Can. J. Microbiol.* **24,** 448–454.
Pugashetti, B. K., Vadas, L., Prihar, H. S. and Feingold, D. S. (1983). *J. Bacteriol.* **153,** 1107–1110.
Pugsley, A. P. and Schnaitman, C. A. (1978). *J. Bacteriol.* **133,** 1181–1189.
Pupillo, P., Mazzucchi, U. and Pierini, G. (1976). *Physiol. Plant Pathol.* **9,** 113–120.
Putman, E. W., Potter, A. L., Hodgson, R. and Hassid, W. Z. (1950). *J. Am. Chem. Soc.* **72,** 5024–5026.
Putnoky, P., Kiss, G. B., Ott, I. and Kondorosi, A. (1983). *Mol. Gen. Genet.* **191,** 288–294.
Quant, R. and Mills, D. (1981). *In* "Proc. 5th Int. Conf. on Plant Path. Bact." (J. C. Lozano, ed.), pp. 412–419. Centro Internacional de Agricultura Tropical, Cali.
Quant, R. L. and Mills, D. (1984). *Mol. Gen. Genet.* **193,** 459–466.
Quantick, P., Cervone, F. and Wood, R. K. S. (1983). *Physiol. Plant Pathol.* **22,** 77–86.
Quirk, A. V., Sletten, A. and Hignett, R. C. (1976). *J. Gen. Microbiol.* **96,** 375–381.
Ramaley, R. F. (1979). *Adv. Appl. Microbiol.* **25,** 37–55.
Ranhand, J. M., Mitchell, W. O., Popkin, T. J. and Cole, R. M. (1980). *J. Bacteriol.* **143,** 1194–1199.
Rao, C. S. and Devadath, S. (1977). *Indian Phytopathol.* **30,** 233–236.
Rao, R. N. (1976). *J. Bacteriol.* **128,** 356–362.
Ravetch, J., Ohsumi, M., Model, P., Vovis, G., Fischhoff, D. and Zinder, N. (1979). *Proc. Natl. Acad. Sci. USA* **76,** 2195–2198.
Raymundo, A. K. and Ries, S. M. (1980). *Phytopathology* **70,** 1066–1069.
Raymundo, A. K. and Ries, S. M. (1981a). *In* "Proc. 5th Int. Conf. Plant Path. Bact." (J. C. Lozano, ed.), pp. 308–322. Centro Internacional de Agricultura Tropical, Cali.
Raymundo, A. K. and Ries, S. M. (1981b). *Phytopathology* **71,** 45–49.
Ream, L. W. and Gordon, M. P. (1982). *Science* **218,** 854–859.
Reed, R. R., Young, R. A., Steitz, J. A., Grudley N. D. F. and Guyer, M. S. (1979). *Proc. Natl. Acad. Sci. USA* **76,** 4882–4886.
Reese, E. T., Siu, R. G. H. and Levinson, H. S. (1950). *J. Bacteriol.* **59,** 485–497.

Reeves, P. (1979). *In* "Bacterial Outer Membranes: Biogenesis and Functions" (M. Inouye, ed.), pp. 255–291. John Wiley, New York.

Reijnders, B., Hoekstra, W. P. M., Bergmans, J. E. N. and van Die, I. M. (1979). *In* "Transformation in *Escherichia coli*" (S. W. Glover and L. O. Butler, eds), pp. 113–121. Cotswold Press, Oxford.

Resibois, A., Colet, M., Faelen, M., Schoonejans, E. and Toussaint, A. (1984). *Virology* **137,** 102–112.

Reverchon, S., Hugouvieux-Cotte-Pattat, N. and Robert-Baudouy, J. (1985). *Gene* **35,** 121–130.

Rexova-Benkova, L. and Markovic, O. (1976). *Adv. Carbohydr. Chem. Biochem.* **33,** 323–385.

Rhodes-Roberts, M. E. and Skinner, F. A. (1982). "Bacteria and Plants". Academic Press, London and Orlando.

Richardson, M. A., Mabe, J. A., Beerman, N. E., Nakatsukasa, W. M. and Fayerman, J. T. (1982). *Gene* **20,** 451–457.

Rick, P. D. and Osborn, M. J. (1977). *J. Biol. Chem.* **252,** 4895–4903.

Rick, P. D., Fung, L. W.-M., Ho, C. and Osborn, M. J. (1977). *J. Biol. Chem.* **252,** 4904–4912.

Ried, J. L. and Collmer, A. (1985). *Proc. 6th Intern. Conf. Plant. Path. Bacteria*, Maryland (in press).

Riess, G., Holloway, B. W. and Pühler, A. (1980). *Genet. Res. Camb.* **36,** 99–109.

Rietschel, E. T., Hase, S., King, M. T., Redmond, J. and Lehmann, V. (1977). *In* "Microbiology – 1977" (D. Schlessinger, ed.), pp. 262–268. American Society for Microbiology, Washington, D.C.

Righetti, P. G. and Drysdale, J. W. (1979). "Laboratory Techniques in Biochemistry and Molecular Biology: Isoelectric Focusing". North-Holland Publishing Co., Amsterdam.

Riley, M., O'Reilly, C. and McConnell, D. (1984). *J. Bacteriol.* **157,** 655–657.

Ritchie, D. F. and Klos, E. J. (1977). *Phytopathology* **67,** 101–104.

Ritchie, D. F. and Klos, E. J. (1979). *Phytopathology* **69,** 1078–1083.

Roantree, R. J., Kuo, T. T. and MacPhee, D. G. (1977). *J. Gen. Microbiol.* **103,** 223–231.

Robbins, P. W. and Uchida, T. (1962). *Biochemistry* **1,** 323–335.

Robbins, P. W. and Wright, A. (1971). *In* "Microbial Toxins" (G. Weinbaum, S. Kadis, and S. J. Ajl, eds), Vol. IV, pp. 351–368. Academic Press, Orlando and London.

Roberts, D. P., Berman, P. M., Lacy, G. H., Mount, M. S. and Allen, C. (1984). *Phytopathology* **74,** 797, #A63 (Abstr.).

Roberts, D. P., Berman, P. M., Lacy, G. H. and Mount, M. S. (1985). *Proc. 6th Intern. Conf. Plant Path. Bacteria*, Maryland (in press.).

Roberts, R. J. (1978). *In* "Microbiology – 1978" (D. Schlessinger, ed.), pp. 5–9. American Society for Microbiology, Washington, D. C.

Roberts, R. J. (1980a). *Meth. Enzymol.* **65,** 1–15.

Roberts, R. J. (1980b). *Nucl. Acids Res.* **8,** r63–r80.

Roberts, W. P. (1982). *Int. Rev. Cytol.* **80,** 63–92.

Robeson, D. J. and Cook, D. R. (1985). *Physiol. Plant Pathol.* **26,** 219–230.

Robinson, J. B. and Tuovinen, O. H. (1984). *Microbiol. Rev.* **48,** 95–124.

Robinson, M. K., Bennett, P. M., Falkow, S. and Dodd, H. M. (1980). *Plasmid* **3,** 343–347.

Roby, D., Toppan, A. and Esquerré-Tugayé, M.-T. (1985). *Plant Physiol.* **77,** 700–704.
Roeder, D. L. and Collmer, A. (1985a). *J. Bacteriol.* **164,** 51–56.
Roeder, D. L. and Collmer, A. (1985b). *Proc. 6th Intern. Conf. Plant Path. Bacteria*, Maryland (in press).
Rogers, H. J. (1983). "Bacterial Cell Structure". American Society for Microbiology, Washington, D.C.
Rogers, S. W., Gilleland, H. E., Jr. and Eagon, R. G. (1969). *Can. J. Microbiol.* **15,** 743–748.
Rombouts, F. W. (1972). "Agricultural Research Reports", No. 779. Centre for Agricultural Publishing and Documentation, Wageningen.
Romeiro, R. D. S., Karr, A. L. and Goodman, R. (1981a). *Physiol. Plant Pathol.* **19,** 383–390.
Romeiro, R., Karr, A. L. and Goodman, R. (1981b). *Plant Physiol.* **68,** 772–777.
Romeo, C. J. and Patil, S. S. (1984). *Phytopathology* **74,** 837, #A384 (Abstr.).
Rosenberg, C., Casse-Delbart, F., Dusha, I., David, M. and Boucher, C. (1982). *J. Bacteriol.* **150,** 402–406.
Rosenberg, M. (1981). *Appl. Environ. Microbiol.* **42,** 375–377.
Rosenberg, M., Gutnick, D. and Rosenberg, E. (1980). *FEMS Microbiol. Lett.* **9,** 29–33.
Rosenberg, M., Perry, A., Bayer, E. A., Gutnick, D. L., Rosenberg, E. and Ofek, I. (1981). *Infect. Immun.* **33,** 29–33.
Rosenbusch, J. P. (1974). *J. Biol. Chem.* **249,** 8019–8029.
Rosenfeld, S. A. and Brenchly, J. E. (1980). *J. Bacteriol.* **144,** 848–851.
Rosner, M. R., Khorana, H. G. and Satterthwait, A. C. (1979a). *J. Biol. Chem.* **254,** 5906–5917.
Rosner, M. R., Tang, J., Barzilay, I. and Khorana, H. G. (1979b). *J. Biol. Chem.* **254,** 5918–5925.
Ross, L. M., McEvoy, J. L., Zink, R. T., Thurn, K. K. and Chatterjee, A. K. (1984). *Abstr. Ann. Meet. Am. Soc. for Microbiol.*, p. 113, #H132.
Rossetti, O. L., Altman, R. and Young, R. (1984). *Gene* **32,** 91–98.
Roth, I. L. (1977). *In* "Surface Carbohydrates of the Prokaryotic Cell" (I. Sutherland, ed.), pp. 5–26. Academic Press, London and Orlando.
Roth, J. R. (1970). *Meth. Enzymol.* **17a,** 3–35.
Rothfield, L. and Pearlman-Kothencz, M. (1969). *J. Mol. Biol.* **44,** 477–492.
Roychoudhury, R., Jay, E. and Wu, R. (1976). *Nucl. Acids Res.* **3,** 101–116.
Rudolph, K. (1976). *In* "Encyclopedia of Plant Physiology" (A. Pirson and M. H. Zimmermann, eds), Vol. 4, pp. 270–315. Springer-Verlag, Berlin.
Rudolph, K. and Rasche, E. (1981). *Phytopathol. Z.* **101,** 298–313.
Rudolph, K., Rasche, E. and Zellar, W. (1978). *Arch. Microbiol.* **119,** 219–226.
Rudolph, K., El-Banoby, F. and Gross, M. (1985). *Proc. 6th Intern. Conf. Plant Path. Bacteria*, Maryland (in press).
Russa, R., Urbanik, T., Kowalczuk, E. and Lorkiewicz, Z. (1982). *FEMS Microbiol. Lett.* **13,** 161–165.
Ruvken, G. B. and Ausubel, F. M. (1981). *Nature* **289,** 85–88.
Ryan, C. A., Bishop, P. D., Walker-Simmons, M., Brown, W. E. and Graham, J. S. (1985). *In* "Cellular and Molecular Biology of Plant Stress" (J. L. Key and T. Kosuge, ed), pp. 319–334. Alan R. Liss, Inc., New York.
Ryu, D. D. Y. and Mandels, M. (1980). *Enzyme Microb. Technol.* **2,** 91–102.
Saedler, H., Cornelis, G., Cullum, J., Schumacher, B. and Sommer, H. (1981). *Cold Spring Harbor Symp. Quant. Biol.* **45,** 93–98.
Saier, M. H. (1977). *Bacteriol. Rev.* **41,** 856–871.

Salkinoja-Salonen, M. and Boeck, R. (1978). *J. Gen. Microbiol.* **105,** 119–125.
Samson, R. (1973). *Ann. Phytopathol.* **5,** 377–388.
Samson, R. and Nassan-Agha, N. (1978). *In* "Proc. 4th Int. Conf. Plant Path. Bact." (Station de Pathologie Végétale et Phytobactériologie, ed.), pp. 547–553. Institut National de la Recherche Agronomique, Angers.
Sanderson, K. E. (1976). *Ann. Rev. Microbiol.* **30,** 327–349.
Sanderson, K. E. and Hartman, P. E. (1978). *Microbiol. Rev.* **42,** 471–519.
Sanderson, K. E. and Roth, J. R. (1983). *Microbiol. Rev.* **47,** 410–453.
Sanderson, K. E., Ross, H., Ziegler, L. and Makela, P. H. (1972). *Bacteriol. Rev.* **36,** 608–637.
Sanderson, K. E., Van Wyngaarden, J., Lüderitz, O. and Stocker, B. A. D. (1974). *Can. J. Microbiol.* **20,** 1127–1134.
Sandulache, R. and Prehm, P. (1985). *J. Bacteriol.* **161,** 1226–1227.
Santamaría, R., Gil, J. A., Mesas, J. M. and Martín, J. F. (1984). *J. Gen. Microbiol.* **130,** 2237–2246.
Santaolalla, M. (1971). *Microbiol. esp.* **24,** 243–256.
Santaolalla, M. (1974). *Revista de la Real Academia de Ciencias Exactas, Fisicas v Naturales de Madrid* **68,** 357–394.
Sasser, M. (1978). *Phytopathology* **68,** 361–364.
Sato, M. (1984). *Ann. Phytopath. Soc. Japan* **50,** 39–45.
Sato, M. and Kaji, A. (1975). *Agric. Biol. Chem.* **39,** 819–824.
Sato, M. and Kaji, A. (1977a). *Agric. Biol. Chem.* **41,** 2193–2197.
Sato, M. and Kaji, A. (1977b). *Agric. Biol. Chem.* **41,** 2199–2203.
Sato, M. and Kaji, A. (1979). *Agric. Biol. Chem.* **43,** 1547–1551.
Sato, M. and Kaji, A. (1980). *Agric. Biol. Chem.* **44,** 717–721.
Sato, M., Staskawicz, B. J., Panopoulos, N. J., Peters, S. and Honma, M. (1981). *Plasmid* **6,** 325–331.
Sato, M., Nishiyama, K. and Shirata, A. (1983). *Ann. Phytopath. Soc. Japan* **49,** 522–528.
Schaad, N. W. (1979). *Ann. Rev. Phytopathol.* **17,** 123–147.
Schaller, H. (1979). *Cold Spring Harbor Symp. Quant. Biol.* **43,** 401–408.
Schell, M. A. and Vinson, D. J. (1985). *In* "Advances in Molecular Genetics of the Bacteria–Plant Interaction" (A. A. Szalay and R. P. Legocki, eds), pp. 182–187. Media Services, Cornell University, Ithaca, New York.
Schellhorn, H. E. and Forsberg, C. W. (1984). *Can. J. Microbiol.* **30,** 930–937.
Schlabach, A. J. (1970). *Diss. Abstr. Int. B.* **31,** 1427–1428.
Schleif, R. F. and Wensink, P. C. (1981). "Practical Methods in Molecular Biology". Springer-Verlag, New York.
Schleifer, K. H. and Stackebrandt, E. (1983). *Ann. Rev. Microbiol.* **37,** 143–187.
Schlemmer, A., Mayama, M. and Keen, N. T. (1985). *Phytopathology* **75,** 1321, #341 (Abstr.).
Schmeiger, H. (1972). *Mol. Gen. Genet.* **119,** 75–88.
Schmitt, R., Bernhard, E. and Mattes, R. (1979). *Mol. Gen. Genet.* **172,** 53–65.
Schnaitman, C. A. (1971). *J. Bacteriol.* **108,** 545–552.
Schnaitman, C. A. (1974). *J. Bacteriol.* **118,** 454–464.
Schnaitman, C., Smith, D. and Forn de Salsas, M. (1975). *J. Virol.* **15,** 1121–1130.
Schoedel, C., Luttrell, K. and Collmer, A. (1985). *Proc. 6th Intern. Conf. Plant Path. Bacteria*, Maryland (in press).
Schoonejans, E. and Toussaint, A. (1983). *J. Bacteriol.* **154,** 1489–1492.
Schroth, M. N. and Hancock, J. G. (1981). *Ann. Rev. Microbiol.* **35,** 453–476.
Schroth, M. N. and Hildebrand, D. C. (1980). *In* "Laboratory Guide for Identifica-

tion of Plant Pathogenic Bacteria" (N. W. Schaad, ed.), pp. 26–30. American Phytopathological Society, St. Paul.
Schukin, N. N., Avdienko, I. D. and Goldfarb, D. M. (1980). *Plasmid* **4,** 228–230.
Seeberg, A. H. and Wiedemann, B. (1984). *J. Bacteriol.* **157,** 89–94.
Selvaraj, G. and Iyer, V. N. (1983). *J. Bacteriol.* **156,** 1292–1300.
Selvaraj, G. and Iyer, V. N. (1984). *J. Bacteriol.* **158,** 580–589.
Selvaraj, G., Fong, Y. C. and Iyer, V. N. (1984). *Gene* **32,** 235–241.
Sequeira, L. (1979). *In* "Host–Parasite Interfaces" (B. B. Nickol, ed.), pp. 71–84. Academic Press, Orlando and London.
Sequeira, L. (1983). *Ann. Rev. Microbiol.* **37,** 51–79.
Sequeira, L. (1984). *Biol. Cell* **51,** 281–286.
Sequeira, L. and Graham, T. L. (1977). *Physiol. Plant Pathol.* **11,** 43–54.
Sequeira, L., Gaard, G. and De Zoeten, G. A. (1977). *Physiol. Plant Pathol.* **10,** 43–50.
Setlow, J. K. and Hollaender, A. (1979). "Genetic Engineering, Principles and Methods", Vol. 1. Plenum Press, New York.
Setlow, J. K. and Hollaender, A. (1980). "Genetic Engineering, Principles and Methods", Vol. 2. Plenum Press, New York.
Sgaramella, V. (1972). *Proc. Natl. Acad. Sci. USA* **69,** 3389–3393.
Shapiro, J. A. (1983). "Mobile Genetic Elements". Academic Press, London and Orlando.
Shapiro, J. A. (1984). *In* "The Microbe 1984. Part II. Prokaryotes and Eukaryotes", 26th Symp. Soc. Gen. Microbiol. (D. P. Kelly and N. G. Carr, eds), pp. 169–193. Cambridge University Press, Cambridge.
Shapiro, J. A. and Sporn, P. (1977). *J. Bacteriol.* **129,** 1632–1635.
Shaw, J. J., Close, T. J., Engebrecht, J. and Kado, C. (1985). *Phytopathology* **75,** 1288, #88 (Abstr.).
Shaw, P. D. (1981). *In* "Toxins in Plant Disease" (R. D. Durbin, ed.), pp. 21–44. Academic Press, London and Orlando.
Sherratt, D. (1981). *In* "Genetics as a Tool in Microbiology" (S. W. Glover and D. A. Hopwood, eds), pp. 35–47. Cambridge University Press, Cambridge.
Shockman, G. D. and Barrett, J. F. (1983). *Ann. Rev. Microbiol.* **37,** 501–527.
Shoemaker, S., Schweickaert, V., Ladner, M., Geifand, D., Kwok, S., Myambo, K. and Innis, M. (1983). *Biotechnology* **1,** 691–696.
Silhavy, T. J., Bassford, P. J., Jr. and Beckwith, J. R. (1979). *In* "Bacterial Outer Membranes: Biogenesis and Functions" (M. Inouye, ed.), pp. 203–254. John Wiley, New York.
Silhavy, T., Benson, S. A. and Emr, S. D. (1983). *Microbiol. Rev.* **47,** 313–344.
Silverman, M. and Simon, M. (1983). *In* "Mobile Genetic Elements" (J. E. Shapiro, ed.), pp. 537–557. Academic Press, London and Orlando.
Simon, M., Zieg, J., Silverman, M., Mandel, G. and Doolittle, R. (1980). *Science* **209,** 1370–1374.
Simon, R., Priefer, U. and Pühler, A. (1983). *Biotechnology* **1,** 784–791.
Sinden, S. L. (1966). Ph.D. Thesis, University of California, Davis, California. *Diss. Abstr. Int. B.* **27,** 4048.
Sinden, S. L., DeVay, J. E. and Backmann, P. A. (1971). *Physiol. Plant Pathol.* **1,** 199–213.
Singer, B. and Kusmierek, J. T. (1982). *Ann. Rev. Biochem.* **51,** 655–693.
Singer, J. T. and Finnerty, W. R. (1984). *J. Bacteriol.* **157,** 607–711.
Sinsheimer, R. L. (1977). *Ann. Rev. Biochem.* **46,** 415–438.
Sjulin, T. M. and Beer, S. V. (1978). *Phytopathology* **68,** 89–94.

Skerman, V. B. D., McGowan, V. and Sneath, P. H. A. (1980). "Approved Lists of Bacterial Names". American Society for Microbiology, Washington, D.C.
Skurray, R. A., Nagaishi, H. and Clark, A. J. (1978). *Plasmid* **1,** 174–186.
Slota, J. E. and Farrand, S. K. (1982). *Plasmid* **8,** 175–186.
Smidt, M. and Kosuge, T. (1978). *Physiol. Plant Pathol.* **13,** 203–214.
Smidt, M. L. and Vidaver, A. K. (1982). *Can. J. Microbiol.* **28,** 600–604.
Smit, J. and Nikaido, H. (1978). *J. Bacteriol.* **135,** 687–702.
Smit, J., Kamio, Y. and Nikaido, H. (1975). *J. Bacteriol.* **124,** 942–958.
Smith, A. R. W., Zamze, S. E., Munro, S. M., Carter, K. J. and Hignett, R. C. (1984). *In* "Proc. 2nd Working Group on *Pseudomonas syringae* Pathovars" (C. G. Panagopoulos, P. G. Psallidas, and A. S. Ativizatos, eds), pp. 90–91. Hellenic Phytopathol. Soc., Athens.
Smith, E. F. (1911). *In* "Bacteria in Relation to Plant Diseases", Vol. 2. Carnegie Institution of Washington, Washington, D.C.
Smith, H. O. (1979). *Science* **205,** 455–462.
Smith, H. O., Danner, D. B. and Deich, R. A. (1981). *Ann. Rev. Biochem.* **50,** 41–68.
Smith, J. J. and Mansfield, J. W. (1982). *Physiol. Plant Pathol.* **21,** 259–266.
Smith, J. M. and Gots, J. S. (1980). *J. Bacteriol.* **132,** 870–875.
Smyth, C. J., Friedman-Kein, A. E. and Salton, M. R. J. (1976). *Infect. Immun.* **13,** 1273–1288.
Smyth, C. J., Jonsson, P., Olsson, E., Soderlind, O., Rosengren, J., Hjerten, S. and Wadstrom, T. (1978). *Infect. Immun.* **22,** 462–472.
So, M. and McCarthy, B. (1980). *Proc. Natl. Acad. Sci. USA* **77,** 4011–4015.
So, M., Heffron, F. and McCarthy, B. (1979). *Nature* **277,** 453–456.
Sokatch, J. R. (1979). *In* "The Bacteria: A Treatise on Structure and Function" (J. R. Sokatch and L. N. Ornston, eds), Vol. VII, pp. 229–289. Academic Press, London and Orlando.
Sommer, H., Cullum, J. and Saedler, H. (1979). *Mol. Gen. Genet.* **177,** 85–89.
Spanò, L., Pomponi, M., Costantino, P., Van Slogteren, G. M. S. and Tempé, J. (1982). *Plant Mol. Biol.* **1,** 291–300.
Sparks, R. B., Jr. and Lacy, G. H. (1980). *Phytopathology* **70,** 369–372.
Sprang, M. L. and Lindow, S. E. (1981). *Phytopathology* **71,** 256 (Abstr.).
Sprenger, G. A. and Lengeler, J. W. (1984). *J. Bacteriol.* **157,** 39–35.
Springer, W. and Goebel, W. (1980). *J. Bacteriol.* **144,** 53–59.
Stack, J. P., Mount, M. S., Berman, P. M. and Hubbard, J. P. (1980). *Phytopathology* **70,** 267–272.
Stall, R. E. (1985). *Proc. 6th Intern. Conf. Plant Path. Bacteria*, Maryland (in press).
Stall, R. E., Loschke, D. C. and Rice, R. W. (1984). *Phytopathology* **74,** 797, #A64 (Abstr.).
Stapleton, M. J., Jagger, K. S. and Warren, R. L. (1984). *J. Bacteriol.* **157,** 7–12.
Stapp, C. (1958). "Pflanzenpathogene Bakterien". Verlag Paul Parey, Berlin and Hamburg.
Starlinger, P. (1980). *Plasmid* **3,** 241–59.
Starlinger, P. and Saedler, H. (1976). *Curr. Top. Microbiol. Immunol.* **75,** 111–152.
Starr, M. P. (1981a). *In* "The Prokaryotes" (M. P. Starr, H. Stolp, H. G. Truper, A. Balows and H. G. Schlegel, eds), Vol. I, pp. 742–763. Springer-Verlag, New York.
Starr, M. P. (1981b). *In* "The Prokaryotes" (M. P. Starr, H. Stolp, H. G. Truper, A. Balows and H. G. Schlegel, eds), Vol. II, pp. 1260–1271. Springer-Verlag, New York.
Starr, M. P. and Chatterjee, A. K. (1972). *Ann. Rev. Microbiol.* **26,** 389–426.

Starr, M. P., Chatterjee, A. K., Starr, P. B. and Buchanan, G. E. (1977). *J. Clin. Microbiol.* **6,** 379–386.

Starr, M. P., Stolp, H., Truper, H. G., Balows, A. and Schlegel, H. G. (1981). "The Prokaryotes". Springer-Verlag, New York.

Staskawicz, B. J. (1983). *In* "Biochemical Plant Pathology" (J. A. Callow, ed.), pp. 199–213. John Wiley, Chichester.

Staskawicz, B. J. and Panopoulos, N. J. (1979). *Phytopathology* **69,** 663–666.

Staskawicz, B. J., Panopoulos, N. J., and Hoogenraad, N. J. (1980). *J. Bacteriol.* **142,** 720–723.

Staskawicz, B., Dahlbeck, D., Miller, J. and Damm, D. (1982). *Phytopathology* **72,** 1000, #565 (Abstr.).

Staskawicz, B. J., Dahlbeck, D. and Keen, N. T. (1984). *Proc. Natl. Acad. Sci. USA* **81,** 6024–6028.

Staskawicz, B., Dahlbeck, D., Napoli, C. and Keen, N. (1985). *In* "Advances in Molecular Genetics of the Bacteria–Plant Interaction" (A. A. Szalay and R. P. Legocki, eds), pp. 195–198. Media Services, Cornell University, Ithaca, New York.

Steinberger, E. M. and Beer, S. V. (1983). *Phytopathology* **73,** 1347 (Abstr.).

Steinberger, E. M. and Beer, S. V. (1985). *Proc. 6th Intern. Conf. Plant Path. Bacteria*, Maryland (in press).

Stemmer, P. and Sequeira, L. (1981). *Phytopathology* **71,** 906 (Abstr.).

Stemmer, W. P. C. and Sequeira, L. (1985). *In* "Advances in Molecular Genetics of the Bacteria–Plant Interaction" (A. A. Szalay and R. P. Legocki, eds), pp. 199–201. Media Services, Cornell University, Ithaca, New York.

Stephens, D. S. and McGee, Z. A. (1983). *J. Infect. Dis.* **147,** 282–292.

Stephens, G. J. and Wood, R. K. S. (1975). *Physiol. Plant Pathol.* **5,** 165–181.

Sternberg, N., Tiemeier, D. and Enquist, L. (1977). *Gene* **1,** 255–280.

Stock, J. B., Rauch, B. and Roseman, S. (1977). *J. Biol. Chem.* **252,** 7850–7861.

Stocker, B. A. D. and Makela, P. H. (1971). *In* "Genetic Aspects of Biosynthesis and Structure of Salmonella Lipopolysaccharide" (George Weinbaum, S. Kadis and S. J. Ajl, eds), pp. 369–438. Academic Press, Orlando and London.

Strobel, G. A. (1974). *Ann. Rev. Plant Physiol.* **25,** 541–566.

Strobel, G. A. (1977). *Ann. Rev. Microbiol.* **31,** 205–224.

Strobel, G. A. (1982). *Ann. Rev. Biochem.* **51,** 309–333.

Sullivan, M. A., Yasbin, R. E. and Young, F. E. (1984). *Gene* **29,** 21–26.

Summers, A. O. and Silver, S. (1978). *Ann. Rev. Microbiol.* **32,** 637–672.

Surico, G. and DeVay, J. E. (1982). *Physiol. Plant Pathol.* **21,** 39–53.

Surico, G., Comai, L. and Kosuge, T. (1984). *Phytopathology* **74,** 490–493.

Surico, G., Varvaro, L. and Frisullo, S. (1985). *Proc. 6th Intern. Conf. Plant Path. Bacteria*, Maryland (in press).

Susskind, M. M. and Botstein, D. (1978). *Microbiol. Rev.* **42,** 385–413.

Sutherland, I. W. (1977a). *In* "Surface Carbohydrates of the Prokaryotic Cell" (I. Sutherland, ed.), pp. 27–96. Academic Press, London and Orlando.

Sutherland, I. W. (1977b). *In* "Extracellular Microbial Polysaccharides" (P. Sandford and A. Laskin, eds), pp. 40–57. American Chemical Society, Washington, D.C.

Sutherland, I. W. (1982). *Adv. Microbiol. Physiol.* **23,** 79–150.

Sutton, J. C. and Williams, P. H. (1970). *Can. J. Bot.* **48,** 645–651.

Suzuki, H., Nishimura, Y., Iketani, H., Campisi, J., Hirashima, A., Inouye, M. and Hirota, Y. (1976). *J. Bacteriol.* **127,** 1494–1501.

Szabo, L. J. and Mills, D. (1984). *J. Bacteriol.* **157,** 821–827.

Szabo, L. J., Volpe, J. and Mills, D. (1981). *In* "Proc. 5th Int. Conf. Plant Path. Bact."

(J. C. Lozano, ed.), pp. 403–411. Centro Internacional de Agricultura Tropical, Cali.
Tait, R. C., Close, T. J., Lundquist, R. C., Hagiya, M., Rodriguez, R. L. and Kado, C. I. (1983). *Biotechnology* **1,** 269–275.
Taylor, J. D., Vivian, A. and Malik, A. (1984). *In* "Proc. 2nd Working Group on *Pseudomonas syringae* Pathovars" (C. G. Panagopoulos, P. G. Psallidas and A. S. Alivizatos, eds), pp. 90–91. Hellenic Phytopath. Soc., Athens.
Tejerina, G., Serra, M. T., Castresana, M. C. and Rodriguez, J. F. (1982). *Phytopathol. Z.* **105,** 97–108.
Terawaki, Y., Kishi, H. and Nakaya, R. (1975). *J. Bacteriol.* **121,** 857–862.
Thiry, G. and Faelen, M. (1984). *J. Gen. Microbiol.* **130,** 1633–1640.
Thomas, M. D. and Durbin, R. D. (1985). *J. Gen. Microbiol.* **131,** 1061–1067.
Thomas, M. D. and Leary, J. V. (1980). *J. Gen. Microbiol.* **121,** 349–355.
Thomas, M. D., Langston-Unkefer, P. J., Uchytie, T. F. and Durbin, R. D. (1983). *Plant Physiol.* **71,** 912–915.
Thurn, K. K. and Chatterjee, A. K. (1982a). *Abstr. Ann. Meet. Am. Soc. Microbiol.*, p. 126. #H79.
Thurn, K. K. and Chatterjee, A. K. (1982b). *Curr. Microbiol.* **7,** 87–92.
Thurn, K. K. and Chatterjee, A. K. (1984). *Abstr. Ann. Meet. Am. Soc. Microbiol.*, p. 167, #K167.
Thurn, K. K. and Chatterjee, A. K. (1985a). *Appl. Environ. Microbiol.* **50,** 894–898.
Thurn, K. K. and Chatterjee, A. K. (1985b). *Proc. 6th Intern. Conf. Plant Path. Bacteria*, Maryland (in press).
Thurn, K. K., Trainer, A. J. and Chatterjee, A. K. (1981). *Phytopathology* **71,** 261 (Abstr.).
Thurn, K. K., Tyrell, D. J. and Chatterjee, A. K. (1982). *Phytopathology* **72,** 934, #41 (Abstr.).
Thurn, K. K., Tyrell, D. J. and Chatterjee, A. K. (1983). *Phytopathology* **73,** 827, #A477 (Abstr.).
Thurn, K. K., Kegoya-Yoshino, Y. and Chatterjee, A. K. (1985). *Proc. 6th Intern. Conf. Plant Path. Bacteria*, Maryland (in press).
Timmis, K. N. (1981). *In* "Genetics as a Tool in Microbiology", (S. W. Glover and D. A. Hopwood, eds), pp. 49–109. Cambridge University Press, Cambridge.
Timmis, K. N. and Puhler, A. (1979). "Plasmids of Medical, Environmental and Commercial Importance". Elsevier/North-Holland, Amsterdam.
Tipper, D. J. and Wright, A. (1979). In "The Bacteria: A Treatise on Structure and Function" (J. R. Sokatch, and L. N. Ornston, eds), vol. VII, pp. 291–426. Academic Press, London and Orlando.
Tomasz, A. (1969). *Ann. Rev. Genet.* **3,** 217–232.
Tomich, P., An, F. Y. and Clewell, D. B. (1979). *Cold Spring Harbor Symp. Quant. Biol.* **43,** 1217–1221.
Tomizawa, H. and Takahashi, H. (1971). *Agric. Biol. Chem.* **35,** 191–200.
Tomoeda, M., Inuzuka, M. and Date, T. (1975). *Prog. Biophys. Molec. Biol.* **30,** 23–56.
Toppan, A., Roby, D. and Esquerré-Tugayé, M.-T. (1982). *Plant Physiol.* **70,** 82–86.
Torti, S. V. and Park, J. T. (1976). *Nature* **263,** 323–326.
Toussaint, A., Faelen, M. and Bukhari, A. I. (1977). *In* "DNA Insertion Elements, Plasmids and Episomes" (A. I. Bukhari, J. A. Shapiro and S. L. Adhya, eds), pp. 275–285. Cold Spring Harbor Laboratory, Cold Spring Harbor, New York.
Townsend, R., Markham, P. G., Plaskitt, K. A. and Daniels, M. J. (1977). *J. Gen. Microbiol.* **100,** 15–21.

Townsend, R., Archer, D. B. and Plaskitt, K. A. (1980). *J. Bacteriol.* **142,** 694–700.
Troy, F. A., II. (1979). *Ann. Rev. Microbiol.* **33,** 519–560.
Truchet, G., Rosenberg, C., Vasse, J., Julliot, J. S., Camut, S. and Denarie, J. (1984). *J. Bacteriol.* **157,** 134–142.
Tsuchiya, K., Mew, T. W. and Wakimoto, S. (1982). *Phytopathology* **72,** 43–46.
Tsuyumu, S. (1977). *Nature* **269**, 237–238.
Tsuyumu, S. (1979). *J. Bacteriol.* **137,** 1035–1036.
Tsuyumu, S. and Chatterjee, A. K. (1983a). *Abstr. Ann. Meet. Am. Soc. Microbiol.*, p. 215, #K230.
Tsuyumu, S. and Chatterjee, A. K. (1983b). *In* "Proc. 4th Int. Cong. Plant Pathol.", p. 195, #778 (Abstr.).
Tsuyumu, S. and Chatterjee, A. K. (1984). *Physiol. Plant Pathol.* **24,** 291–302.
Tsuyumu, S. and Goto, M. (1978). *Ann. Phytopathol. Soc. Japan.* **44,** 380 (Abstr.).
Tsuyumu, S., Funakubo, T., Takikawa, Y. and Goto, M. (1984). *Ann. Phytopath. Soc. Japan* **50,** 410.
Tsuyumu, S., Funakubo, T. and Chatterjee, A. K. (1985). *Physiol. Plant Pathol.* **27,** 119–130.
Tu, C. P. D. and Cohen, S. N. (1980). *Cell* **19,** 151–160.
Tully, J. G. and Whitcomb, R. F. (1981). *In* "The Prokaryotes" (M. P. Starr, H. Stolp, H. G. Truper, A. Balows and H. G. Schlegel, eds), Vol. II, pp. 2271–2278. Springer-Verlag, New York.
Turner, J. G. and Novacky, A. (1974). *Phytopathology* **64,** 885–890.
Turner, P., Barber, C. and Daniels, M. (1984). *Mol. Gen. Genet.* **195,** 101–107.
Turner, P., Barber, C. and Daniels, M. (1985). *Mol. Gen. Genet.* (in press).
Twiddy, W. and Liu, S. C. Y. (1972). *Phytopathology* **62,** 794 (Abstr.).
Tyrell, D. J., Thurn, K. K. and Chatterjee, A. K. (1983). *Phytopathology* **73,** 827, #A479 (Abstr.).
Uchida, T., Gill, D. M. and Pappenheimer, A. M., Jr. (1971). *Nature New Biol.* **233,** 8–11.
Uchida, Y., Tsukada, Y. and Sugimori, T. (1973). *Agric. Biol. Chem.* **37,** 2105–2110.
Uemura, J. and Mizushima, S. (1975). *Biochem. Biophys. Acta* **413,** 163–176.
Ulrich, J. M. (1975). *Physiol. Plant Pathol.* **5,** 37–44.
Van Alphen, W. and Lugtenberg, B. (1977). *J. Bacteriol.* **131,** 623–630.
Vance, C. P. (1983). *Ann. Rev. Microbiol.* **37,** 399–424.
Van den Mooter, M., Swings, J., Gossele, F., Kersters, K. and De Ley, J. (1985). *Proc. 6th Intern. Conf. Plant Path. Bacteria*, Maryland (in press).
Van der Zwet, T. and Keil, H. L. (1979). "Agriculture Handbook", No. 510. United States Department of Agriculture, Washington, D.C.
van Embden, J. D. A., de Graff, F. K., Schouls, L. M. and Teppema, J. S. (1980). *Infect. Immun.* **29,** 1125–1133.
Van Gijsegem, F. and Toussaint, A. (1982). *Plasmid* **7,** 30–44.
Van Gijsegem, F. and Toussaint, A. (1983). *J. Bacteriol.* **154,** 1227–1235.
Van Gijsegem, F., Toussaint, A. and Schoonejans, E. (1984). *In* "Proc. 2nd Int. Symp., The Molecular Genetics of the Bacteria–Plant Interaction", #124 (Abstr.).
Van Gijsegem, F., Toussaint, A. and Schoonejans, E. (1985a). *EMBO J.* **4,** 787–792.
Van Gijsegem, F., Hugouvieux-Cotte-Pattat, N. and Robert-Baudouy, J. (1985b). *J. Bacteriol.* **161,** 702–708.
Van Larebeke, N., Genetello, C., Hernalsteens, J. P., De Picker, A., Zaenen, I., Messens, E., Van Montagu, M. and Schell, J. (1977). *Mol. Gen. Genet.* **152,** 119–124.
Van Montagu, M. and Schell, J. (1982). *Curr. Top. Microbiol. Immunol.* **96,** 237–254.

Varvaro, L. and Surico, G. (1985). *Proc. 6th Intern. Conf. Plant Path. Bacteria*, Maryland (in press).

Vasil'eva, S. V. and D'jakav, J. T. (1970). *Biol. Nauki.* **13,** 103–106.

Vidaver, A. K. (1976). *Ann. Rev. Phytopathol* **14.** 451–465

Vidaver, A. K. (1982). *Ann. Rev. Microbiol.* **36,** 495–517.

Vidaver, A. K. and Starr, M. P. (1981). *In* "The Prokaryotes" (M. P. Starr, H. Stolp, H. G. Truper, A. Balows and H. G. Schlegel, eds), Vol. II, pp. 1879–1887. Springer-Verlag, New York.

Vidaver, A. K., Lambrecht, P. A., Muenchau, D. and Carson, R. R. (1981). *In* "Proc. of the 5th Int. Conf. Plant Path. Bact." (J. C. Lozano, ed.), pp. 129–134. Centro Internacional de Agricultura Tropical, Cali.

Vidaver, A. K., Haskins, W. P. and Hendrick, C. A. (1985). *Proc. 6th Intern. Conf. Plant Path. Bacteria*, Maryland (in press).

Vieira, J. and Messing, J. (1982). *Gene* **19,** 259–268.

Vincent, J. R. and Fulbright, D. W. (1983). *J. Bacteriol.* **156,** 1349–1356.

Volk, W. A. (1968). *J. Bacteriol.* **95,** 980–982.

Volk, W. A., Salomonsky, N. L. and Hunt, D. (1972). *J. Biol. Chem.* **247,** 3881–3887.

von Graevenitz, A. (1977). *Ann. Rev. Microbiol.* **31,** 447–471.

von Riesen, V. L. (1975). *J. Clin. Microbiol.* **2,** 552–553.

von Riesen, V. L. (1976). *Int. J. Syst. Bacteriol.* **26,** 143–145.

Wacek, T. J., and Sequeira, L. (1973). *Physiol. Plant Pathol.* **3,** 363–369.

Wadstrom, T., Smyth, C. J., Faris, A., Jonsson, P., and Freer, J. H. (1978). *In* "Proc. Int. Symp. On Neonatal Diarrhea" (S. D. Acres, ed.), pp. 29–58. Stuart Brandle Publishing Service, Saskatoon, Canada.

Wagner, W., Vogel, M. and Goebel, W. (1983). *J. Bacteriol.* **154,** 200–210.

Wakarchuk, W. W., Kilburn, D. G., Miller, R. C. Jr. and Warren, R. A. J. (1984). *J. Gen. Microbiol.* **130,** 1385–1389.

Walker, G. C. (1984). *Microbiol. Rev.* **48,** 60–93.

Walker, G. C. (1985). *Ann. Rev. Biochem.* **54,** 425–457.

Walker, J. B. (1978). *In* "Cyclitols and Phosphoinositides" (W. Wells and F. Eisenberg, eds), pp. 423–438. Academic Press, Orlando and London.

Walsh, P. M., Haas, M. J. and Somkuti, G. A. (1984). *App. Environ. Microbiol.* **47,** 253–257.

Wang, J. S. and Kelman, A. (1982). *Phytopathology* **72,** 936, #59 (Abstr.).

Ward, L. J. and Copeman, R. J. (1985). *Proc. 6th Intern. Conf. Plant Path. Bacteria*, Maryland (in press).

Way, J. C., Davis, M. A., Morisato, D., Roberts, D. E. and Kleckner, N. (1984). *Gene* **32,** 369–379.

Weckesser, J., Drews, G. and Mayer, H. (1979). *Ann. Rev. Microbiol.* **33,** 215–239.

Weiss, A. A. and Falkow, S. (1984). *Infect. Immun.* **43,** 263–269.

Weiss, B., Jacquemin-Sablon, A., Live, T. R., Fareed, G. C. and Richardson, C. C. (1968). *J. Biol. Chem.* **243,** 4543–4555.

West, C. A., Moesta, P., Jin, D. F., Lois, A. F. and Wickham, K. A. (1985). *In* "Cellular and Molecular Biology of Plant Stress" (J. L. Key and T. Kosuge, eds), pp. 335–349. Alan R. Liss, Inc., New York.

Westphal, O. and Jann, K. (1965). *Meth. Carbohydr. Chem.* **5,** 83–91.

Whatley, M. H., Bodwin, J. S., Lippincott, B. B. and Lippincott, J. A. (1976). *Infect. Immun.* **13,** 1080–1083.

Whatley, M. H., Hunter, N., Cantrell, M. A., Hendrick, C., Keegstra, K. and Sequeira, L. (1980). *Plant Physiol.* **65,** 557–559.

Whitaker, J. R. (1984). *Enzyme Microb. Technol.* **6,** 341–349.
White, F. F. and Nester, E. W. (1980). *J. Bacteriol.* **144,** 710–720.
White, F. F., Ghidossi, G., Gordon, M. P. and Nester, E. W. (1982). *Proc. Natl. Acad. Sci. USA* **79,** 3193–3197.
White, F. F., Klee, H. J. and Nester, E. W. (1983). *J. Bacteriol.* **153,** 1075–1078.
Whitefield, C., Sutherland, I. W. and Cripps, R. E. (1981). *J. Gen. Microbiol.* **124,** 385–392.
Wickner, W. (1979). *Ann. Rev. Biochem.* **48,** 23–45.
Wilkinson, S. G. (1977). *In* "Surface Carbohydrates of the Prokaryotic Cell" (I. Sutherland, ed.), pp. 97–177. Academic Press, London and Orlando.
Wilkinson, S. G. and Taylor, D. P. (1978). *J. Gen. Microbiol.* **109,** 367–370.
Willetts, N. S., Crowther, C. and Holloway, B. W. (1981). *Plasmid* **6,** 30–52.
Williams, J. W. and Silver, S. (1984). *Enz. Microbiol. Tech.* **6,** 530–537.
Willis, J. W. and Chatterjee, A. K. (1985). *Proc. 6th Intern. Conf. Plant Path. Bacteria,* Maryland (in press).
Willis, J. W., Thurn, K. K. and Chatterjee, A. K. (1985). *Proc. 6th Intern. Conf. Plant Path. Bacteria,* Maryland (in press).
Willis, J. W., Leary, J. V. and Tarollinger, D. B. (1986). *J. Bacteriol.* (submitted).
Willmitzer, L., Sanchez-Serrano, J., Buschfeld, E. and Schell, J. (1982). *Mol. Gen. Genet.* **183,** 16–22.
Witkin, E. M. (1976). *Bacteriol. Rev.* **40,** 869–907.
Wojciechowicz, M., Heinrichova, K. and Ziotecki, A. (1980). *J. Gen. Microbiol.* **117,** 193–199.
Wood, R. K. S., Ballio, A. and Graniti, A. (1972). "Phytotoxins in Plant Diseases". Academic Press, London and Orlando.
Woodward, J. and Wiseman, A. (1982). *Enz. Microbiol. Tech.* **4,** 73–79.
Wretlind, B., Sjoberg, L. and Wadstrom, T. (1977). *J. Gen. Microbiol.* **103,** 329–336.
Wroblewski, H., Johansson, K.-E. and Hjerten, S. (1977). *Biochem. Biophys. Acta* **465,** 275–289.
Wu, R. (1979). "Methods in Enzymology", Vol. 68. Academic Press, London and Orlando.
Wu, T. T. (1966). *Genetics* **54,** 405–410.
Wu, W. C. (1972a). *Ann. Phytopathol. Soc. Japan* **38,** 146–155.
Wu, W. C. (1972b). *Ann. Phytopathol. Soc. Japan* **38,** 333–341.
Wu, W. C., Middleton, R. B. and Hsu, W. H. (1977). *Chin. J. Microbiol.* **10,** 37–47.
Xu, G. W. and Gross, D. C. (1985). *Phytopathology* **75,** 1325, #374 (Abstr.).
Yakrus, M. and Schaad, N. W. (1979). *Phytopathology* **69,** 517–522.
Yamada, K. and Komagata, K. (1970). *J. Gen. Appl. Microbiol.* **16,** 103–113.
Yamada, T., Palm, C. J., Brooks, B. and Kosuge, T. (1985). *Proc. Natl. Acad. Sci. USA* **82,** 6522–6526.
Yamada, Y., Calame, K. L., Grindley, J. N. and Nakada, D. (1979). *J. Bacteriol.* **137,** 990–999.
Yamagata, H., Nakamura, K. and Inouye, M. (1981). *J. Biol. Chem.* **256,** 2194–2198.
Yamasaki, Y., Murata, N. and Suwa, T. (1966). *Proc. Jap. Acad.* **42,** 946–949.
Yano, H. and Fujii, H. (1983). *Ann. Phytopathol. Soc. Japan* **49,** 266–269.
Yano, H., Fujii, H., Mukoo, H., Fukuyasu, T., Terakado, N. and Isayama, Y. (1979). *Ann. Phytopathol. Soc. Japan* **45,** 201–206.
Yanofsky, C. (1981). *Nature* **289,** 751–758.
Yem, D. W. and Wu, H. C. (1977). *J. Bacteriol.* **131,** 759–764.
Yem, D. W. and Wu, H. C. (1978). *J. Bacteriol.* **133,** 1419–1426.

Yoder, O. C. (1980). *Ann. Rev. Phytopathol.* **18,** 103–129.
Yoder, O. C. (1981). *In* "Toxins in Plant Disease" (R. D. Durbin, ed.), pp. 45–78. Academic Press, London and Orlando.
Yoshimoto, H. and Yoshikawa, M. (1975). *J. Bacteriol.* **124,** 661–667.
Young, J. M., Dye, D. W., Bradbury, J. F., Panagopoulos, C. G. and Robbs, C. F. (1978). *N.Z.J. Agr. Res.* **21,** 153–177.
Young, R., Way, J., Way, S., Yin, J. and Syvanen, M. (1979). *J. Mol. Biol.* **132,** 307–322.
Zaenen, I., van Larebeke, N., Teuchy, H., Van Montagu, M. and Schell, J. (1974). *J. Mol. Biol.* **86,** 109–127.
Zagursky, R. T. and Berman, M. L. (1984). *Gene* **27,** 183–191.
Zakour, R. A. and Loeb, L. A. (1982). *Nature (London)* **295,** 708–710.
Zambryski, P., Holsters, M., Kruger, K., Depicker, A., Schell, J., Van Montagu, M. and Goodman, H. (1980). *Science* **209,** 1385–1391.
Zambryski, P., Goodman, H. M., Van Montagu, M. and Schell, J. (1983). *In* "Mobile Genetic Elements" (J. A. Shapiro, ed.), pp. 505–535. Academic Press, London and Orlando.
Zevenhuizen, L. P. T. M. (1973). *Carbohydr. Res.* **26,** 409–419.
Zevenhuizen, L. P. T. M. and Scholten-Koerselman, H. J. (1979). *Antonie van Leeuwenhoek* **45,** 165–175.
Zink, R. T. and Chatterjee, A. K. (1984a). *In* "Proc. 2nd Int. Symp., The Molecular Genetics of the Bacteria-Plant Interaction", #139 (Abstr.).
Zink, R. T. and Chatterjee, A. K. (1984b). *Phytopathology* **74,** 876, #A692 (Abstr.).
Zink, R. T. and Chatterjee, A. K. (1985). *Appl. Environ. Microbiol.* **49,** 714–717.
Zink, R. T., Feese, D. A. and Chatterjee, A. K. (1982). *Phytopathology* **72,** 933–934.
Zink, R. T., Kemble, R. J. and Chatterjee, A. K. (1984). *J. Bacteriol.* **157,** 809–814.
Zink, R. T., Engwall, J. K., McEvoy, J. L. and Chatterjee, A. K. (1985a). *Proc. 6th Intern. Conf. Plant Path. Bacteria,* Maryland (in press).
Zink, R. T., Engwall, J. K., McEvoy, J. L. and Chatterjee, A. K. (1985b). *J. Bacteriol.* **164,** 390–396.
Zischek, C., Tuyen, L. T. K., Boistard, P. and Boucher, C. (1981). *In* "Proc. 5th Int. Conf. Plant Path. Bact." (J. C. Lozano, ed.), pp. 427–437. Centro Internacional de Agricultura Tropical, Cali.
Zucker, M. and Hankin, L. (1970). *J. Bacteriol.* **104,** 13–18.
Zucker, M. and Hankin, L. (1971). *Can. J. Microbiol.* **17,** 1313–1318.

ADDENDUM

Since this manuscript went to press, a number of relevant publications have appeared. We provide below a synopsis of the findings that extend earlier observations and confirm or strengthen hypotheses presented in this article.

A. Genetics

By extending the previous findings of Cma with R68::Mu c^+ (see pp 44–45 and Pérombelon and Boucher, 1978), Forbes and Pérombelon (1985) ordered 17 chromosomal loci in *E. carotovora* subsp. *carotovora* strain SCRI 193. In addition to generalized Cma, the Mu-directed chromosomal transfer occurred at a 10-fold higher frequency. A comparison of the linkage maps of this strain and other *Erwinia* spp. revealed both conserved and dispersed gene sequences (Forbes and Pérombelon, 1985). The availability of amber-suppressing mutants (Hinton *et al.*, 1985) and a partial linkage map in the strain SCRI 193 should facilitate genetic analysis of its pathogenicity factors.

Additional technological developments potentially significant in the genetic and biochemical analyses of soft-rot *Erwinia* spp. include: (i) the construction of a RecA$^-$ strain in *E. chrysanthemi* B374 (Faelen *et al.*, 1985); (ii) transformation of pBR322 DNA into *E. chrysanthemi* strains 3937j and B374 (Reverchon and Robert-Baudouy, 1985); (iii) *in vivo* cloning of the *cya* (adenylate cyclase) gene of *E. chrysanthemi* B374 and the characterization of the gene product (Hedegaard and Danchin, 1985); and (iv) sensitization of *E. carotovora* to the coliphage lambda [see pp. 54–55 and Salmond *et al.* (1986)].

The findings that the broad host range conjugative plasmid $pBPW_1$ of *P. syringae* pv. *tabaci* (see p. 27) was unable to replicate in various strains of other *P. syringae* pathovars suggested its potential use as a transpositional vehicle (Sato *et al.*, 1985). In *P. syringae* pv. *atropurpurea* (NIAES 1309), $pBPW_1$ caused alterations in cell surface properties resulting in cell aggregation and antigen alteration (Sato *et al.*, 1985). While the underlying mechanism is not yet known, plasmid-mediated cell surface alterations occur in other bacterial systems (see p. 33). The plasmid $pBPW_1$ has also been genetically characterized (Quinnett and Shaw, 1985). They localized a DNA segment that specified receptivity of the bacteriophage PRDl. In addition, the *tra*$^+$ and mob$^+$ functions were genetically separated.

In *P. solanacearum*, as in several other bacteria (see p. 26), Tn*7* transposition was highly site-specific (Boucher *et al.*, 1985). Physical analysis of avirulent TN*5* mutants of *P. solanacearum* revealed insertions in at least 10 different *Eco*RI fragments (Boucher *et al.*, 1985). This indicated that a considerable amount of *P. solanacearum* DNA is involved in pathogenicity. Similar observations have been made in several other phytopathogenic bacteria (see pp. 67–74).

Phenotypic changes accompanying lysogenization of *Xanthomonas campestris* pv. *citri* with the filamentous phage Cf16 strongly suggested genomic rearrangements (Wu *et al.*, 1985). The molecular basis for this and other similar observations (see p. 55) awaits elucidation.

B. Cell Surface Components

1. Extracellular polysaccharide (EPS)

The EPS of *X. campestris* pv. *citri*, like that of many other phytopathogenic bacteria (see Section IV.A.1.a.), appears crucial for plant pathogenicity (Takahashi and Doke, 1984; Goto and Hyodo, 1985). Although the precise role of EPS is not fully understood, recent findings suggest the following functions: (i) EPS interacts with a citrus leaf glycoprotein to determine attachment of virulent bacteria to plant cells (Takahashi and Doke, 1984); and (ii) EPS causes water-soaking, as previously noted in other host–pathogen systems (see pp. 80–81), thereby facilitating pathogen establishment (Goto and Hyodo, 1985). Additional studies are needed to assess if attachment of *X. campestris* pv. *citri* is a prerequisite to infection. If confirmed, this would be unusual since in most plant–bacterial interactions, with the exception of agrobacteria and rhizobia, attachment is associated with an incompatible interaction (see Section IV.A.2.a.vi). The finding of Goto and Hyodo (1985) of EPS-stimulated ethylene production may indicate a generalized host resistance in response to pathogenicity factors (see p. 112). We should, however, note that the stimulation of ethylene production in bean leaves by the phytotoxin coronatine led Ferguson and Mitchell (1985) to suggest a specific role of ethylene in the coronatine toxic syndrome.

2. Lipopolysaccharide (LPS)

An analysis of sugar composition of *E. carotovora* LPS (DeBoer *et al.*, 1985) revealed the presence of 2-keto-3-deoxyoctonic acid, glucosamine, heptose, and glucose (also galactose in most LPS) as the probable constituents of the core-oligosaccharide (see Fig. 12 and p. 98). In addition, either fucose or rhamnose was present along with one or more unidentified sugars in various LPS preparations. The immunological data indicated that the determinants of *E. carotovora* serogroups probably resided in the O-antigen, as in other enterobacteria (see p. 99) and Wilkinson, 1977). The occurrence of the O-antigen specific sugars (viz. rhamnose, mannose, fucose) in *E. carotovora* LPS (DeBoer *et al.*, 1985) contrasts with the findings of Sanulache and Prehm (1985) (see p. 100).

The findings of Jayaswal *et al.* (1985) may be relevant, when considering the role of LPS in the host association. They isolated a Mu-insertion *E. carotovora* subsp. *carotovora* mutant which was deficient in UDP glucose-pyrophosphorylase and consequently was galactose sensitive, altered in LPS and nonpathogenic. Although Jayaswal *et al.* (1985) considered galactose-sensitivity a probable reason for nonpathogenicity, the effect of altered LPS was not rigorously tested.

As a prelude to understanding the significance of the LPS O-side chain in the host–pathogen interaction, Smith *et al.* (1985) determined the structure of this region of the *P. syringae* pv. *morsprunorum* (C28) LPS. The repeat unit had the following structure:

$$\rightarrow 3)\text{-D-Rha}p\text{-}(\alpha\ 1 \rightarrow 3)\text{-D-Rha}p\text{-}(\alpha\ 1 \rightarrow 2)\text{-D-Rha}p\text{-}\alpha\ 1 \rightarrow$$

with a degree of polymerization of 70.

The hypothesis that the LPS of *A. tumefaciens* functions in its attachment to plant cells (see p. 107) was not supported in subsequent studies. For example, Pueppke and Benny (1983) did not detect inhibition of tumorigenesis in potato disks by the *A. tumefaciens* LPS. Moreover, the recent findings of Puvanesarajah *et al.* (1985) strongly implicated 2-linked-β-D-glucan of *A. tumefaciens* in attachment to plant cells.

C. Pectolytic Enzymes and Polygalacturonate Catabolism

Major developments in this area include (i) nucleotide sequencing of *pel*E and *pel*B of *E. chrysanthemi* (EC16) (Keen *et al.*, in prep.), (ii) Mu-*lac* insertion mutagenesis of *pel*C (Diolez and Coleno, 1985) and the ketodeoxygluconate catabolic genes, *kdgK* and *kdgA* (see Fig. 17, p. 131) and *E. chrysanthemi* strain 3937 (Hugouvieux-Cotte-Pattat and Robert-Baudouy, 1985), and (iii) cloning of a polygalacturonase (*peh*) gene of *E. carotovora* strain EC and the demonstration that the enzyme has tissue-macerating ability (Lei *et al.*, 1985).

The nucleotide sequence data of Keen *et al.* (1984) revealed th presence of RNA polymerase and CRP binding sites as well as an *N*-terminal leader signal sequence which is probably involved in enzyme export (see Section C.2.b.ii, pp. 132–136). The sequence data indicated little homology between *pel*B and *pel*E; this conclusion is also supported in immunological studies (Thurn and Chatterjee, 1985b).

The utility of *lac* fusions (see pp. 28–31 and also Silhavy and Beckwith, 1985) in studies on pectolytic enzymes and polygalacturonate catabolism has been demonstrated (Diolez and Coleno, 1985; Hugouvieux-Cotte-Pattat and Robert-Baudouy, 1985). Starting with a cloned *pel*C gene in pBR322, Diolez and Coleno (1985) constructed, by insertion of Mu dII1734 (see p. 29), a plasmid pAD1 which was Pel$^-$ and Lac$^+$. By marker exchange, an *E. chrysanthemi* strain was constructed which was PelC$^-$ but Lac$^+$. As expected, the Lac$^+$ phenotype was induced by polygalacturonate. A similar approach is being used in several laboratories with other cloned *pel* genes to elucidate the regulation and function of these enzymes.

By using the Mu d(Ap, *lac*) phage (see p. 29), Hugouvieux-Cotte-Pattat and Robert-Baudouy (1985) isolated mutants of *E. chrysanthemi* containing genetic fusions of the *kdgA* and *kdgK* genes to the *lacZ*. The synthesis of β-galactosidase in these strains was induced by polygalacturonate and its catabolic products. The insensitivity of β-galactosidase expression to glucose repression in these mutants contrasts with glucose sensitivity of several polygalacturonate catabolic genes including *pel*.

In addition to the Mu-*lac* gene fusion elements (see pp. 28–31), two Tn*3*-*lacZ* transposons (Tn*3*-HoHo1 and TN*3*-HoHo2) (Stachel *et al.*, 1985) should prove useful in constructing gene fusions in various Gram-negative bacteria. While Tn*3*-HoHo1 generates both transcriptional and translational fusions, Tn*3*-HoHo2 mediates only transcriptional fusions.

Lei *et al.* (1985) cloned an 8-kb DNA segment of *E. carotovora* (strain EC) that specified PEH but not PEL. Although the enzyme has not yet been fully characterized, it caused maceration of plant tissue. Willis *et al.* (1985) have also cloned a *peh* gene of *E. carotovora* subsp. *carotovora* strain Ecc71 which is tightly linked with a *pel*$^+$ gene. The Ecc71 PEH, like the enzyme from the strain EC, macerated plant tissue; however, the Ecc71 enzyme did not cause electrolyte loss or cell death. In addition to the endo-PEH, the strain Ecc71 appears to produce an exo-PEH (Willis *et al.*, in prep.).

The detection of cutinolytic and pectolytic enzymes within 48 hr of infection of tomato leaves with *P. syringae* pv. *tomato* suggests a role of these activities in the primary stages of disease development (Bashan *et al.*, 1985). This hypothesis awaits genetic verification. Of particular significance is the occurrence in *P. syringae* pv. *tomato* of cutinolytic activity which has rarely been detected in bacteria but is common in many phytopathogenic fungi (see Kolattukudy, 1984).

D. Genetics of Phaseolotoxin Production

Earlier observations of Jamieson *et al.* (1981) (see p. 169) had suggested that the phaseolotoxin gene(s) of *P. syringae* pv. *phaseolicola* resided on the chromosome and not on a plasmid(s). A study by Quigley *et al.* (1985) with Tn*5* insertion Tox$^-$ mutants substantiates this conclusion. The mutants carried Tn*5* insertions in different chromosomal fragments. Linkage between TN*5* (Km^r) and the Tox$^-$ phenotype was established by marker-exchange mutagenesis. Physical evidence suggested the involvement of at least five genes in phaseolotoxin synthesis and export; these did not appear to be closely clustered. The Tox$^-$::Tn*5* mutants, like the parent strains, were pathogenic on bean plants and resistant to phaseolotoxin. The isolation of the wild-type *tox*$^+$ sequences by Quigley *et al.* (1985) should now facilitate critical analysis of genes and enzymes of phaseolotoxin biosynthesis.

Quigley *et al.* (1985) also investigated the genetic basis of toxin resistance in phaseolotoxin-producing bacteria. Physiological evidence of Staskawicz *et al.* (1980) (see p. 167) indicated that this resistance was due to toxin-insensitive ornithine carbamoyltransferase. Further, Quigley *et al.* (1985) suggested that the Tox^r bacteria also possessed two genes for ornithine carbamoyltransferase (*oct*). It would be of considerable interest to determine the relationship between the expression of *pst* and the two *oct* genes.

References

Bashan, Y., Okon, Y. and Henis, Y. (1985). *Phytopathology* **75,** 940.

Boucher, C. A., Barberis, P. A., Trigalet, A. Ph. and Demery, D. A. (1985). *J. Gen Microbiol.* **131,** 2449–2457.

De Boer, S. H., Bradshaw-Rouse, J. J., Sequeira, L. and McNaughton, M. E. (1985). *Can. J. Microbiol.* **31,** 583.

Diolez, A. and Coleno, A. (1985). *J. Bacteriol.* **163,** 913–917.

Faelen, M., Schoonejans, E. and Toussaint, A. (1985). *Mol. Gen. Genet.* **201,** 126–128.

Ferguson, I. B. and Mitchell, R. E. (1985). *Plant Physiol.* **77,** 969–973.

Forbes, K. J. and Pérombelon, M. C. M. (1985). *J. Bacteriol.* **164,** 1110–1116.

Goto, M. and Hyodo, H. (1985). *Ann. Phytopath. Soc. Japan* **51,** 22–31.

Hedegaard, L. and Danchin, A. (1985). *Mol. Gen. Genet.* **201,** 38–42.

Hinton, J. C. D., Pérombelon, M. C. M. and Salmond, G. P. C. (1985). *FEMS Microbiol. Lett.* **28,** 103–106.

Hugouvieux-Cotte-Pattat, N. and Robert-Baudouy, L. (1985). *J. Gen Microbiol.* **131,** 1205–1211.

Jayaswal, R. K., Bressan, R. A. and Handa, A. K. (1985). *J. Bacteriol.* **164,** 473–476.

Lei, S.-P., Lin, H.-C., Heffernan, L. and Wilcox, G. (1985). *J. Bacteriol.* **164,** 831–835.

Pueppke, S. G. and Benny, U. K. (1983). *Physiol. Plant Pathol.* **23,** 439–446.

Puvanesarajah, V., Schell, F. M., Stacey, G., Douglas, C. J. and Nester, E. W. (1985). *J. Bacteriol.* **164,** 102–106.

Quigley, N. B., Lane, D. and Bergquist, P. L. (1985). *Curr. Microbiol.* **12,** 295–300.

Quinnett, M. P. and Shaw, P. D. (1985). *Phytopathology* **75,** 1321, #342 (Abstr.).

Reverchon, S. and Robert-Baudouy, J. (1985). *Biochemie* **67,** 253–257.

Sato, M., Nishiyama, K. and Panopoulos, N. J. (1985). *Ann. Phytopath. Soc. Japan* **51,** 426–434.

Silhavy, T. J. and Beckwith, J. R. (1985). *Microbiol. Rev.* **49,** 398–418.
Smith, A. R. W., Zamze, S. E., Munro, S. M., Carter, K. J. and Hignett, R. C. (1985). *Euro. J. Biochem.* **149,** 73–78.
Stachel, S. E., An, G., Flores, C. and Nester, E. W. (1985). *EMBO J.* **4,** 891–898.
Takahashi, T. and Doke, N. (1984). *Ann. Phytopath. Soc. Japan* **50,** 565–573.

Index

D

E

Q

R

S

T

V

W

X

Y